Rings of Continuous Functions

Leonard Gillman & Meyer Jerison

Dover Publications, Inc.
Mineola, New York

Bibliographical Note

This Dover edition, first published in 2018, is an unabridged republication of the work originally published in "The University Series in Higher Mathematics" in 1960 by the D. Van Nostrand Company, Inc., Princeton, New Jersey.

Library of Congress Cataloging-in-Publication Data

Names: Gillman, Leonard, author. | Jerison, Meyer, 1922- author.
Title: Rings of continuous functions / Leonard Gillman and Meyer Jerison.
Description: Mineola, New York : Dover Publications, Inc., 2018. | Series: Dover books on mathematics | Reprint of the 1960 edition published by Van Nostrand, Princeton, N.J. | Includes bibliographical references and index.
Identifiers: LCCN 2017031467| ISBN 9780486816883 (paperback) | ISBN 0486816885 (paperback)
Subjects: LCSH: Function spaces. | Functions, Continuous. | Rings (Algebra) | Ideals (Algebra) | BISAC: MATHEMATICS / Algebra / Abstract.
Classification: LCC QA323 .G5 2018 | DDC 515/.73—dc23
LC record available at https://lccn.loc.gov/2017031467

Manufactured in the United States by LSC Communications
81688501 2017
www.doverpublications.com

PREFACE

This book is addressed to those who know the meaning of each word in the title: none is defined in the text. The reader can estimate the knowledge required by looking at Chapter 0; he should not be discouraged, however, if he finds some of its material unfamiliar or the presentation rather hurried.

Our objective is a systematic study of the ring $C(X)$ of all real-valued continuous functions on an arbitrary topological space X. We are concerned with algebraic properties of $C(X)$ and its subring $C^*(X)$ of bounded functions and with the interplay between these properties and the topology of the space X on which the functions are defined. Major emphasis is placed on the study of ideals, especially maximal ideals, and on their associated residue class rings. Problems of extending continuous functions from a subspace to the entire space arise as a necessary adjunct to this study and are dealt with in considerable detail.

The contents of the book fall naturally into three parts. The first, comprising Chapters 1 through 5 and the beginning of Chapter 10, presents the fundamental aspects of the subject insofar as they can be discussed without introducing the Stone-Čech compactification. In Chapter 3, the study is reduced to the case of completely regular spaces.

In the second part, Chapters 6 through 11, the Stone-Čech compactification βX is constructed, investigated in great detail, and applied to the study of $C(X)$. The fundamental theorem of Gelfand and Kolmogoroff characterizing the maximal ideals is presented in Chapter 7; it falls out as a natural consequence of the particular way in which βX was constructed in Chapter 6. Čech's more familiar construction of βX is presented in Chapter 11. Chapter 8 deals with the "real" maximal ideals in $C(X)$ and the associated realcompact spaces (Q-spaces), introduced by Hewitt.

The third part is devoted to various additional topics. The different chapters in this section of the book are almost entirely independent of one another, the main exception being Chapter 14 (*Prime ideals*), the middle third of which leans heavily on portions of Chapter 13. Moreover, results depending upon βX are invoked at only a few places. Chapters 12, 15, and 16 include self-contained treatments of special subjects. Chapter 12 deals with Ulam's measure problem and the equivalent question of whether a discrete space can fail to be realcompact. Chapter 15 presents the elements of the theory of uniform spaces (exclusively in terms of pseudometrics); this chapter culminates in Shirota's theorem that the spaces that can admit a complete uniform structure are, barring Ulam measures, precisely the realcompact spaces. Finally, Chapter 16 contains a small but significant portion of dimension theory — as much as is needed in order to derive Katětov's algebraic characterization of dimension.

At the end of each chapter, there is a collection of problems, which provide much additional detail about the material covered in the text. The reader who wishes to master the subject should solve a good many of them. The problems vary widely in purpose and importance, ranging from mere exercises to detailed descriptions of pathological spaces that serve as warning posts and clarify the boundaries of the theory. The problems also vary widely in difficulty. The simpler ones can be solved directly by applying the results developed in the text in the same chapter (or principal theorems from earlier chapters). Those that depend upon secondary results from earlier chapters or upon other problems are supplied with references to those results. Problems whose solutions are not straightforward are accompanied by hints; these are often given in the form of assertions that themselves require proof. In cases of more than ordinary complexity, the earlier parts of a problem have been designed as preparation for the later parts.

The family $C(X)$ can be regarded as an ordered space or as a topological space, and there is a substantial literature dealing with these topics. Our own treatment of them is only incidental. The order structure of $C(X)$ is determined by the algebraic structure, and we study the former as a tool for deriving information about the latter. Topologies on $C(X)$ are considered only in scattered problems and, in a rather elementary way, in the last chapter. Algebras of *complex*-valued functions, about which there is a very extensive literature, are not dealt with here at all.

We have planned our book to serve both as a text for a graduate course and as a treatise for the active mathematician. A number of results that are presented as soon as they are provable can be given interesting alternative proofs by means of machinery developed later on; the student can profit from seeking such possibilities, which are not always pointed out. At the end of the book, there are comments on the historical development of the subject and references to original sources, listed in the bibliography. The index has been prepared in an attempt to render the book useful as a reference work.

Sections are numbered and problems are lettered consecutively within each chapter, with the chapter number included. In Chapter 1, for example, the sections are 1.1, 1.2, · · · , and the problems are 1A, 1B, · · · .

Leonard Gillman
University of Rochester
and
Meyer Jerison
Purdue University

ACKNOWLEDGMENTS

Our book had its origin in a seminar held at Purdue University during the year 1954-1955. The seminar was organized by our colleague Melvin Henriksen and was conducted by him jointly with us. The task of presenting and writing up the material was shared with four students: J. E. Kist, C. W. Kohls, M. J. Mansfield, and R. H. McDowell. Eventually, it was suggested — again by Henriksen — that the seminar notes be used as the starting point for a book, and this enterprise was begun in June, 1956. At the end of the summer, Henriksen left to accept a research fellowship, and shortly thereafter withdrew from the formal collaboration; however, he has continued to aid us with valuable, stimulating advice.

We are grateful to P. Civin for experimenting with an early draft in a seminar at the University of Oregon. A number of helpful comments were supplied by H. Banilower, D. G. Johnson, J. E. Mack, and E. C. Weinberg, as students of ours or Henriksen's at Purdue. Several valuable contributions were received from P. Dwinger, J. R. Isbell, W. Rudin, S. Warner, and R. E. Zink. We are deeply indebted to Edwin Hewitt and Robert H. McDowell, both of whom read the entire manuscript and contributed a large number of important criticisms and suggestions.

What polish our book may possess could never have been achieved without the inspiring help of Carl W. Kohls, who, after each new draft, harassed us with biting criticism, humored us with patient advice, and bombarded us with detailed suggestions for text and problems. The vastness of our debt to Kohls and Henriksen can perhaps be appreciated in full only by those who have enjoyed the pleasure of working with these men.

Most of our work was made possible by the generous support of the National Science Foundation. During the final stages, one of us was supported by a fellowship grant from the John Simon Guggenheim Memorial Foundation, and both of us were afforded the privilege of membership at the Institute for Advanced Study while on sabbatical leave from Purdue University. We wish to express here our gratitude to all these institutions.

CONTENTS

FOREWORD

0.1. The reader is presumed to have some background in general topology and abstract algebra, to the extent, at least, of feeling at home with the basic concepts. Here we set forth some conventions in notation and terminology, and record some preliminary results.

The set (space, field) of real numbers is denoted by $\mathbf{R}$. The reader is expected to be familiar with the elementary set-theoretic and topological properties of $\mathbf{R}$.

The subset of rational numbers is denoted by $\mathbf{Q}$, and the subset of positive integers, $\{1, 2, \cdots\}$, by $\mathbf{N}$.

The constant function, on any set, whose constant value is the real number r, is denoted by $\boldsymbol{r}$. The symbols $\mathbf{i}$ and $\mathbf{j}$, however, are reserved for special meanings: $\mathbf{i}$ is used for the identity function on $\mathbf{R}$ or its subsets, and $\mathbf{j}$ denotes the sequence $(1/n)_{n \in \mathbf{N}}$.

When dealing with rings of functions, one encounters each of the concepts *identity* and *inverse* in two different senses. The use of distinguishing terms is desirable in the interest of clarity. The choice of the word *identity* to denote the mapping $x \to x$ on any set (e.g., $\mathbf{i}$ above) seems indicated overwhelmingly. For the multiplicative identity in a ring, we shall use the term *unity*. The symbol a^{-1} is the obvious choice for the multiplicative inverse of a in a ring. For the inverse of a mapping φ, we introduce the symbol $\varphi^{\leftarrow}$.

THEORY OF SETS

0.2. *Mappings.* Square brackets are used to indicate the image of a set under a mapping:

$$\varphi[A] = \{\varphi x : x \in A\}.$$

For the inverse, we write

$$\varphi^{\leftarrow}(y) = \{x : \varphi x = y\},$$

and

$$\varphi^{\leftarrow}[B] = \{x : \varphi x \in B\},$$

i.e., $\varphi^{\leftarrow}[B] = \bigcup_{y \in B} \varphi^{\leftarrow}(y)$. When φ is known to be one-one, we write $x = \varphi^{\leftarrow}(y)$ instead of $\{x\} = \varphi^{\leftarrow}(y)$.

When φ is known to be real-valued, it is referred to as a function, and we write $\varphi(x)$ in place of φx. This is for emphasis, not logical distinction.

The restriction of a mapping φ to a set S is denoted, as usual, by $\varphi|S$.

Let φ be a mapping from A into B and ψ a mapping from B into E. The composite mapping from A into E is denoted by $\psi \circ \varphi$:

$$(\psi \circ \varphi)(x) = \psi(\varphi x) \qquad (x \in A).$$

The following abbreviations are useful for indicating unions or intersections of families of sets:

$$\bigcup \mathscr{S} = \bigcup_{S \in \mathscr{S}} S, \qquad \bigcap \mathscr{S} = \bigcap_{S \in \mathscr{S}} S.$$

The cardinal of S is denoted by $|S|$. *Countable* means finite or denumerably infinite. The cardinal $2^{\aleph_0}$ is denoted by $\mathfrak{c}$. Some use will be made of the elementary properties of cardinals and ordinals.

0.3. *Finite intersection property.* Let $\mathscr{S}$ be a nonempty family of sets. $\mathscr{S}$ is said to have the *finite* [resp. *countable*] *intersection property* provided that the intersection of any finite [resp. countable] number of members of $\mathscr{S}$ is nonempty.

In order that $\mathscr{S}$ have the finite intersection property, it is *not* enough that any *two* members of $\mathscr{S}$ have nonempty intersection. (E.g., let $\mathscr{S}$ consist of the three sets $\{0, 1\}$, $\{0, 2\}$, and $\{1, 2\}$.)

As usual, a class is said to be *closed* under an operation when the performance of the operation upon members of the class always yields a member of the class. For example, $\mathscr{S}$ is *closed* under finite intersection provided that the intersection of any finite number (>0) of members of $\mathscr{S}$ is a member of $\mathscr{S}$. Here it *is* enough that the intersection of any two members of $\mathscr{S}$ be in $\mathscr{S}$, for the stated property then follows by induction.

Thus, if $\emptyset \notin \mathscr{S}$, and if the intersection of any two members of $\mathscr{S}$ belongs to $\mathscr{S}$, then, certainly, $\mathscr{S}$ has the finite intersection property.

In the text, obvious inductions that lead, as above, from *two* to *finite*, will be taken for granted.

0.4. *Partially ordered sets.* For a partial order, we include the axiom that $a \leqq b$ and $b \leqq a$ implies $a = b$.

A mapping φ from a partially ordered set A into a partially ordered set E is said to *preserve order* if $a \leqq b$ in A implies $\varphi a \leqq \varphi b$ in E.

A *maximal* element of A is an element a such that $x \geqq a$ implies $x = a$. In contrast, the *largest* element of A—necessarily unique, if it exists—is the element c such that $c \geqq x$ for all $x \in A$. *Minimal* and *smallest* are defined similarly.

In easily recognizable situations, this terminology is applied to a *class* of sets, with the understanding that the partial order is that of set inclusion. Examples are maximal *chain* (0.7), and maximal *ideal* (0.15).

0.5. *Lattices.* In a partially ordered set, the symbol $a \vee b$ denotes $\sup \{a, b\}$, i.e., the smallest element c—if one exists—such that $c \geqq a$ and $c \geqq b$. Likewise, $a \wedge b$ stands for $\inf \{a, b\}$.

When both $a \vee b$ and $a \wedge b$ exist, for all $a, b \in A$, then A is called a *lattice.* A subset S is a *sublattice* of A provided that, for all $x, y \in S$, the elements $x \vee y$ and $x \wedge y$ *of* A belong to S. (Thus, it is not enough that x and y have a supremum and infimum *in* S.)

A mapping φ from a lattice A into a lattice E is a *lattice homomorphism into* E provided that

$$\varphi(a \vee b) = \varphi a \vee \varphi b \quad \text{and} \quad \varphi(a \wedge b) = \varphi a \wedge \varphi b.$$

It follows that $\varphi[A]$ is a sublattice of E.

A partially ordered set in which *every* nonempty subset has both a supremum and an infimum is said to be *lattice-complete.*

0.6. *Totally ordered sets.* A subset S of a totally ordered set A is said to be *cofinal* [resp. *coinitial*] if, for every $x \in A$, there exists $s \in S$ such that $s \geqq x$ [resp. $s \leqq x$].

A totally ordered set is said to be *Dedekind-complete* provided that every nonempty subset with an upper bound has a supremum—or, equivalently, every nonempty subset with a lower bound has an infimum. (For example, **R** is Dedekind-complete, but not lattice-complete.)

Every totally ordered set A has an essentially unique *Dedekind completion* B, characterized by the following properties: B is totally ordered and Dedekind-complete; A is a subset of B; and no proper subset of B that contains A is Dedekind-complete. Every element $\notin A$ is determined by a Dedekind cut of A. (For example, **R** is the Dedekind completion of **Q**.)

0.7. A totally ordered set is often referred to as a *chain.*

HAUSDORFF'S MAXIMAL PRINCIPLE. *Every partially ordered set contains a maximal chain* (i.e., maximal in the class of all chains as partially ordered by set inclusion). This proposition is equivalent to the axiom of choice and to the well-ordering theorem. All three forms will be used.

References. [B_2, Chapters 1–3], [B_3, Chapters 2–3], [H_1, pp. 45–83, 97–141], [K_9, pp. 31–36], and [S_5].

TOPOLOGY

0.8. Convergence will be described in terms of certain filter bases; the details are all in the text, and no prior knowledge about filters is required. The theory of nets is not used.

The closure of a set S in X is denoted by cl S or $\text{cl}_X S$, the interior by int S or $\text{int}_X S$.

The main classes of spaces to be considered are the completely regular spaces and their subclass, the compact spaces. The terms *completely regular* and *compact*, and also *normal*, will be applied to *Hausdorff* spaces only. All three are defined in Chapter 3.

From Chapter 4 on, all given spaces are assumed to be completely regular.

We state here for emphasis that a Hausdorff space is said to be *compact* provided that every family of closed sets with the finite intersection property has nonempty intersection—i.e., every open cover has a finite subcover.

A mapping φ from X into Y is said to be *closed* if for every closed subset A of X, $\varphi[A]$ is a closed set *in* Y. (It is not enough that $\varphi[A]$ be closed in $\varphi[X]$.) *Open* mapping is defined similarly.

0.9. A *neighborhood* of E is any set whose interior contains E.

LEMMA. *If E and p have disjoint neighborhoods, for each $p \in F$, and if F is compact, then E and F have disjoint neighborhoods.*

PROOF. Let U_p and V_p be disjoint neighborhoods of E and p, respectively. A finite collection

$$\{V_{p_1}, \cdots, V_{p_n}\}$$

covers F. Then $\bigcap_k U_{p_k}$ and $\bigcup_k V_{p_k}$ are disjoint neighborhoods of E and F.

0.10. COROLLARY. *In a Hausdorff space, a compact set and a point in its complement have disjoint neighborhoods. Hence every compact set in a Hausdorff space is closed.*

PROOF. The first assertion is immediate from the lemma (with E

the one-point set), and implies that the complement of a compact set is open.

More generally, we have:

0.11. Corollary. *Any two disjoint compact sets in a Hausdorff space have disjoint neighborhoods.*

PROOF. By the corollary, one set and each point of the other have disjoint neighborhoods, and the lemma now yields the result.

0.12. Constant use will be made of the following elementary results.

If X is dense in T, and V is open in T, then

$$\mathrm{cl}_T (V \cap X) = \mathrm{cl}_T V.$$

A continuous mapping from a space X into a Hausdorff space is determined by its values on any dense subset of X.

(a) *Let X be dense in each of the Hausdorff spaces S and T. If the identity mapping on X has continuous extensions σ from S into T, and τ from T into S, then σ is a homeomorphism onto, and $\sigma^{\leftarrow} = \tau$.*

By way of proof, observe that the mapping $\tau \circ \sigma$ must be the identity on S, because its restriction to X is the identity on X. Similarly, $\sigma \circ \tau$ is the identity on T. If $\sigma s_1 = \sigma s_2$, then $s_1 = \tau(\sigma s_1) = \tau(\sigma s_2) = s_2$; therefore σ is one-one. For $t \in T$ we have $\sigma(\tau t) = t$; hence σ is onto and $\sigma^{\leftarrow} = \tau$ (whence $\sigma^{\leftarrow}$ is continuous).

A continuous image of a compact space in a Hausdorff space is compact. A closed set in a compact space is compact. A continuous mapping of a compact space into a Hausdorff space is a closed mapping. A one-one, continuous mapping of a compact space onto a Hausdorff space is a homeomorphism.

0.13. A *discrete* subspace means a subspace that is discrete in its relative topology—but not necessarily closed in the space. (For example, $\{1/n\}_{n \in \mathbf{N}}$ is a discrete subspace of $\mathbf{R}$.) The following result will be needed in a number of proofs.

Theorem. *Every infinite Hausdorff space contains a copy of* $\mathbf{N}$ *(i.e., a countably infinite, discrete subset).*

PROOF. Given two distinct points, there is a neighborhood U of one whose closure does not contain the other. Either U or $X - \mathrm{cl}\, U$ is infinite. Hence there exists a point x_1, and an infinite, open set V_1, such that $x_1 \notin \mathrm{cl}\, V_1$. Similarly, there exists $x_2 \in V_1$, and an infinite, open set V_2 in V_1, such that $x_2 \notin \mathrm{cl}\, V_2$. The set $\{x_n\}_{n \in \mathbf{N}}$, constructed inductively in this way, is discrete.

References. [B_5, Chapter 1] and [K_9, Chapters 1 and 3].

ALGEBRA

0.14. *Ideals and homomorphisms.* In what follows, A will denote a *commutative* ring having a *unity* element, i.e., an element 1, necessarily unique, such that $1 \cdot a = a$ for all a. (However, much of the discussion is applicable to more general rings.)

A *unit* of A is an element a that has a multiplicative inverse a^{-1}, i.e., an element such that $aa^{-1} = 1$.

Ideal, unmodified, will always mean *proper* ideal, i.e., a subring $I \neq A$ such that $a \in I$ implies $xa \in I$ for all $x \in A$. Thus, an ideal cannot contain a unit.

Homomorphism, unmodified, will always mean *ring* homomorphism. The kernel of any nonzero homomorphism (i.e., the set of all elements that map to 0) is an ideal. Conversely, every ideal I is the kernel of some homomorphism. In particular, I is the kernel of the *canonical homomorphism* of A onto the residue class ring A/I, i.e., the homomorphism under which the image of a is the residue class $I + a$. If I is the kernel of a homomorphism of A onto B, then A/I is isomorphic with B.

The intersection of any nonempty family of ideals is an ideal. The smallest ideal—perhaps improper—containing an ideal I and an element a is denoted by (I, a); it consists of all elements of the form $i + xa$, where $i \in I$ and $x \in A$.

0.15. *Prime ideals and maximal ideals.* An ideal P in A is *prime* if $ab \in P$ implies $a \in P$ or $b \in P$, i.e., if A/P is an integral domain.

If M is a *maximal* ideal (with respect to set inclusion), then $a \notin M$ implies $1 \in (M, a)$, so that $1 \equiv xa \pmod{M}$ for some $x \in A$; conversely, $1 \equiv xa \pmod{M}$ implies $1 \in (M, a)$. Thus, an ideal M is maximal if and only if A/M is a field. In particular, every maximal ideal is prime.

The union of any nonempty chain of ideals is an ideal. (That the union is a proper subset of A follows from the presence of a unity element in A.) The maximal principle (0.7) now implies that *every ideal is contained in a maximal ideal,* and hence that every non-unit of A belongs to some maximal ideal.

0.16 The following results about prime ideals will not be needed, except incidentally, until Chapter 14.

THEOREM. *Let I be an ideal in A, and S a set that is closed under multiplication and disjoint from I. There exists an ideal P containing I, disjoint from S, and maximal with respect to this property. Such an ideal is necessarily prime.*

PROOF. By the maximal principle, there exists a maximal chain $\mathfrak{L}$ of ideals containing I and disjoint from S. Define $P = \bigcup \mathfrak{L}$; then P is an ideal containing I, disjoint from S, and maximal with respect to this property. Let $a \notin P$ and $b \notin P$. Because of the maximality of P, there exist $s, t \in S$ such that $s \in (P, a)$ and $t \in (P, b)$. Then $s \equiv xa \pmod P$ and $t \equiv yb \pmod P$, for suitable $x, y \in A$. Since S is closed under multiplication, we have $xyab \equiv st \not\equiv 0 \pmod P$. Therefore $ab \notin P$. This shows that P is prime.

0.17. COROLLARY. *Let I be an ideal. If no power of a belongs to I, then there exists a prime ideal containing I but not a.*

0.18. COROLLARY. *The intersection of all the prime ideals containing a given ideal I is precisely the set of all elements of which some power belongs to I.*

PROOF. If there exists a prime ideal P containing I but not a, then no power of a can belong to I, since no power of a belongs to P. Conversely, if no power of a belongs to I, then, by the preceding corollary, some prime ideal contains I but not a.

0.19. *Partially ordered rings.* Let a partial ordering relation be defined on the ring A. Then A is called a *partially ordered ring* provided that

$a \geqq b$ implies $a + x \geqq b + x$ for all x, and
$a \geqq 0$ and $b \geqq 0$ implies $ab \geqq 0$.

The following facts are evident: $a \geqq b$ if and only if $a - b \geqq 0$; $a \geqq 0$ if and only if $-a \leqq 0$; if $a \leqq r$ and $b \leqq s$, then $a + b \leqq r + s$.

To define such a partial ordering relation, it is enough to specify the elements $\geqq 0$, subject to:

$a \geqq 0$ and $-a \geqq 0$ if and only if $a = 0$, and
$a \geqq 0$ and $b \geqq 0$ implies $a + b \geqq 0$ and $ab \geqq 0$,

and then to define $a \geqq b$ to mean $a - b \geqq 0$.

To establish that a homomorphism φ from A into a partially ordered ring is order-preserving, it suffices to show that $a \geqq 0$ implies $\varphi a \geqq 0$.

If $a \vee b$ exists, for all a and b, then $a \wedge b$ exists, and

$$a \wedge b = -(-a \vee -b).$$

Therefore, to establish that A is a lattice—in which case it is called a *lattice-ordered* ring—it suffices to show that $a \vee b$ exists for each a and b.

In a lattice-ordered ring, $|a|$ denotes the element $a \vee -a$; it satisfies $|a| \geqq 0$ (see 5A).

To establish that A is *totally* ordered, it is enough to show that every element is comparable with 0.

0.20. *Totally ordered integral domains.* Let A be a totally ordered integral domain. Squares of nonzero elements are positive. In particular, $1 > 0$, so that $-1 < 0$; therefore -1 has no square root.

If $0 < a < b$, then $a^n < b^n$ (where $n \in \mathbf{N}$). Hence a positive element has at most one positive n^{th} root.

A contains a natural copy of the set of integers, in the form of the elements $m \cdot 1$. When A is a totally ordered *field*, the elements m/n (i.e., $(m \cdot 1)/(n \cdot 1)$), where m is an integer and $n \in \mathbf{N}$, constitute a copy of the rational field $\mathbf{Q}$.

0.21. *Ordered fields.* (In referring to totally ordered fields, one customarily drops the adverb.) An ordered field is said to be *archimedean* if the subset of integers is cofinal.

THEOREM. *An ordered field is archimedean if and only if it is isomorphic to a subfield of the ordered field* $\mathbf{R}$.

PROOF. Obviously, every subfield of $\mathbf{R}$ is archimedean. Conversely, let F be any archimedean field. Given $x < y$ in F, choose $n \in \mathbf{N}$ such that $n > 1/(y - x)$, and let m be the smallest integer $> nx$. Then $x < m/n < y$. This shows that $\mathbf{Q}$ is dense in F, so that every element of F is uniquely determined by a Dedekind cut of $\mathbf{Q}$. Consequently, F is embeddable in $\mathbf{R}$ in a unique way as an ordered *set*. Now, if r and s belong to the *ordered field* F, and if a, b, c, and d are rationals satisfying $a \leqq r < b$ and $c \leqq s < d$, then $a + c \leqq r + s < b + d$. It follows that sums in F—like sums in $\mathbf{R}$—are uniquely determined by Dedekind cuts of $\mathbf{Q}$. Products, likewise, are so determined. This shows that the embedding of F is an isomorphism.

0.22. Any nonzero homomorphism of a field is an isomorphism. For $\mathbf{R}$, we can say more.

THEOREM. *The only nonzero homomorphism of* $\mathbf{R}$ *into itself is the identity.*

PROOF. A real number is nonnegative if and only if it is a square. Since any homomorphism takes squares to squares, it takes nonnegative numbers to nonnegative numbers, and therefore is order-preserving. Now, if $\mathfrak{s}$ is a nonzero homomorphism, then, because $\mathfrak{s}r = (\mathfrak{s}r)(\mathfrak{s}1)$ for every r, we must have $\mathfrak{s}1 = 1$. It follows that $\mathfrak{s}$ is the identity on $\mathbf{Q}$. As $\mathbf{Q}$ is dense in $\mathbf{R}$, and $\mathfrak{s}$ preserves order, $\mathfrak{s}$ is the identity on $\mathbf{R}$ as well.

0.23. Corollary. *There is at most one isomorphism from a ring onto* **R**. *Any homomorphism onto* **R** *is uniquely determined by its kernel.*

PROOF. If $\mathfrak{u}$ and $\mathfrak{v}$ are isomorphisms from the same ring onto **R**, then $\mathfrak{u} \circ \mathfrak{v}^{\leftarrow}$ is an automorphism of **R**, and hence is the identity. Therefore $\mathfrak{u} = \mathfrak{v}$.

Given homomorphisms $\mathfrak{s}$ and $\mathfrak{t}$ from a ring A onto **R**, with common kernel I, we consider the associated isomorphisms $\bar{\mathfrak{s}}$ and $\bar{\mathfrak{t}}$ from A/I onto **R** (i.e., such that $\mathfrak{s} = \bar{\mathfrak{s}} \circ \mathfrak{h}$ and $\mathfrak{t} = \bar{\mathfrak{t}} \circ \mathfrak{h}$, where $\mathfrak{h}$ is the canonical homomorphism of A onto A/I). Since $\bar{\mathfrak{s}} = \bar{\mathfrak{t}}$, we have $\mathfrak{s} = \mathfrak{t}$.

References. [B_2, Chapters 2 and 14], [B_4, Chapter 1, and Chapter 6 pp. 1–34], [M_4, Chapters 1 and 3], and [W_1, §§ 14, 15, 19, and 20].

Chapter 1

FUNCTIONS ON A TOPOLOGICAL SPACE

1.1. The set $C(X)$ of all continuous, real-valued functions on a topological space X will be provided with an algebraic structure and an order structure.

Since their definitions do not involve continuity, we begin by imposing these structures on the collection $\mathbf{R}^X$ of *all* functions from X into the set $\mathbf{R}$ of real numbers. Addition and multiplication are defined by the formulas

$$(f + g)(x) = f(x) + g(x), \quad \text{and} \quad (fg)(x) = f(x)g(x).$$

It is obvious that both of the operations thus defined are associative and commutative, and that the distributive law holds: these conclusions are immediate consequences of the corresponding statements about the field $\mathbf{R}$.

In fact, it is clear that $\mathbf{R}^X$ is a commutative ring with unity element (provided that X is not empty). The zero element is the constant function $\mathbf{0}$, and the unity element is the constant function $\mathbf{1}$. The additive inverse $-f$ of f is characterized by the formula

$$(-f)(x) = -f(x).$$

The multiplicative inverse f^{-1}—in case it exists—is characterized by the formula

$$f^{-1}(x) = \frac{1}{f(x)}.$$

1.2. The partial ordering on $\mathbf{R}^X$ is defined by:

$$f \geqq g \text{ if and only if } f(x) \geqq g(x) \text{ for all } x \in X.$$

That this is a partial ordering relation follows from the fact that $\mathbf{R}$ is ordered. It is clear that for every h, $f + h \geqq g + h$ if and only if $f \geqq g$. Hence the ordering relation is invariant under translation. In addition, $f \geqq \mathbf{0}$ and $g \geqq \mathbf{0}$ implies $fg \geqq \mathbf{0}$. Therefore $\mathbf{R}^X$ is a partially ordered ring (0.19).

Next, for any f and g, the function k defined by the formula

$$k(x) = f(x) \vee g(x)$$

satisfies: $k \geqq f$ and $k \geqq g$; furthermore, for all h such that $h \geqq f$ and $h \geqq g$, we have $h \geqq k$. Therefore $f \vee g$ exists: it is k. Dually, $(f \wedge g)(x) = f(x) \wedge g(x)$. Thus, $\mathbf{R}^X$ is a lattice-ordered ring (0.19). The function $|f|$, defined as $f \vee -f$, satisfies

$$|f|(x) = |f(x)|.$$

Of course, $\mathbf{R}$ is also *totally* ordered (so that $r \vee s$ is simply $\max\{r, s\}$), but $\mathbf{R}^X$ is not if X contains at least two points.

The ambiguous use of the various symbols, which refer sometimes to $\mathbf{R}$ and sometimes to $\mathbf{R}^X$, should cause no difficulty.

1.3. The set of all *continuous* functions from the topological space X into the topological space $\mathbf{R}$ is denoted by $C(X)$—or, for short, by C.

The sum of two continuous functions is, of course, continuous; so is the product. And if f belongs to C, then so does $-f$. Therefore $C(X)$ is a commutative ring, a subring of $\mathbf{R}^X$. The constant function $\mathbf{1}$ belongs to C and is its unity element.

It is easy to see that if f is continuous, then the function $|f|$ is also continuous. Since

$$f \vee g = \mathbf{2}^{-1}(f + g + |f - g|),$$

$f, g \in C$ implies $f \vee g \in C$. Therefore C is a sublattice of $\mathbf{R}^X$ (0.5 and 0.19).

The symbol f^n ($n \in \mathbf{N}$) is used as in any ring. (Recall that $\mathbf{N}$ denotes the set of positive integers.) If $f \geqq \mathbf{0}$, then, more generally, f has a unique, nonnegative r^{th} power ($r \in \mathbf{R}$, $r > 0$), denoted by f^r and defined by

$$f^r(x) = f(x)^r \qquad (x \in X);$$

and if f is continuous, then f^r, as a composition of two continuous functions, is also continuous. In like manner, if n is odd ($n \in \mathbf{N}$), then $f^{1/n}$ may be defined as a function in C, for any $f \in C$.

If the space X is discrete, then every function on X is continuous, so that $\mathbf{R}^X$ is the same as $C(X)$. Conversely, if $\mathbf{R}^X = C(X)$, then the characteristic function of every set in X is continuous, which shows that the space is discrete.

1.4. The subset $C^* = C^*(X)$ of $C(X)$, consisting of all *bounded* functions in $C(X)$, is also closed under the algebraic and order operations discussed in 1.3. Therefore C^* is a subring and sublattice of C.

It can happen that the subring $C^*(X)$ is all of $C(X)$—i.e., every function in $C(X)$ is bounded. When this is the case, X is said to be *pseudocompact*. Every *compact* space is pseudocompact, as is well known.

More generally, as we shall now prove, every *countably compact* space is pseudocompact. By definition, X is countably compact provided that every family of closed sets with the finite intersection property has the countable intersection property—i.e., every countable open cover has a finite subcover. Suppose, now, that X is countably compact, and consider any function f in $C(X)$. The sets

$$\{x: |f(x)| < n\},$$

for $n \in \mathbf{N}$, constitute a countable open cover of X. Hence a finite subfamily covers X, i.e., f is bounded.

A pseudocompact space need not be countably compact; see 5I.

PROSPECTUS

1.5. A major objective of this book is to study relations between topological properties of a space X and algebraic properties of $C(X)$ and $C^*(X)$. It is obvious that each of these function rings is completely determined by the space X. One of the main problems will be to specify conditions under which, conversely, X is determined as a topological space by the algebraic structure of $C(X)$ or of $C^*(X)$. In other words, what restrictions on X and Y, if any are needed at all, will allow us to conclude that X is homeomorphic with Y, when we are given that $C(Y)$ is isomorphic with $C(X)$, or, perhaps, that $C^*(Y)$ is isomorphic with $C^*(X)$?

Another type of problem is that of determining the class of topological spaces whose function rings satisfy some natural algebraic conditions, or, conversely, of determining the effects on the function ring of imposing some natural topological condition on the space. An example that might fit into either classification is given in 1B: X is connected if and only if $C(X)$ is not a direct sum of proper subrings. Other classes of problems are to discover algebraic properties common to all function rings and to find relations between $C(X)$ and $C^*(X)$ for a given X.

INVARIANTS OF HOMOMORPHISMS

1.6. Even before embarking upon a detailed study of function rings, we can observe quickly that several important properties of the family of functions that may not seem to be determined by the ring structure are, in fact, so determined (see *Notes*). The most significant of these properties is the order structure. To describe the order, it is enough to specify the nonnegative functions; but the condition $f \geqq \mathbf{0}$ is simply the algebraic requirement that f be a square, i.e., $f = k^2$ for some k. It follows, moreover, that $|f|$ is determined algebraically: it is the unique nonnegative square root of f^2.

We have just proved that every isomorphism from $C(Y)$ into $C(X)$ preserves order. Moreover, if f is bounded and $f = k^2$, then k is bounded; hence an isomorphism from $C^*(Y)$ into $C(X)$ also preserves order. Here is a more conclusive result:

THEOREM. *Every (ring) homomorphism* $\mathfrak{t}$ *from* $C(Y)$ *or* $C^*(Y)$ *into* $C(X)$ *is a lattice homomorphism.*

PROOF. Since $g = l^2$ implies $\mathfrak{t}g = (\mathfrak{t}l)^2$, $\mathfrak{t}$ sends nonnegative functions into nonnegative functions, i.e., $\mathfrak{t}$ is order-preserving. Next,

$$(\mathfrak{t}|g|)^2 = \mathfrak{t}(|g|^2) = \mathfrak{t}(g^2) = (\mathfrak{t}g)^2,$$

and since $\mathfrak{t}|g| \geqq \mathbf{0}$, we have $\mathfrak{t}|g| = |\mathfrak{t}g|$. Combining this with the formula

$$(g \vee h) + (g \vee h) = g + h + |g - h|,$$

we get

$$\mathfrak{t}(g \vee h) + \mathfrak{t}(g \vee h) = \mathfrak{t}g + \mathfrak{t}h + |\mathfrak{t}g - \mathfrak{t}h| = (\mathfrak{t}g \vee \mathfrak{t}h) + (\mathfrak{t}g \vee \mathfrak{t}h).$$

But $\mathfrak{t}(g \vee h)$ and $\mathfrak{t}g \vee \mathfrak{t}h$ are real-valued functions (defined on X), and therefore $\mathfrak{t}(g \vee h) = \mathfrak{t}g \vee \mathfrak{t}h$.

1.7. Boundedness of functions is another property determined by the algebraic structure of C. More generally, we have the following result.

THEOREM. *Every (ring) homomorphism* $\mathfrak{t}$ *from* $C(Y)$ *or* $C^*(Y)$ *into* $C(X)$ *takes bounded functions to bounded functions.*

PROOF. As with any homomorphism, $\mathfrak{t}\mathbf{1} = \mathfrak{t}(\mathbf{1}\cdot\mathbf{1}) = (\mathfrak{t}\mathbf{1})(\mathfrak{t}\mathbf{1})$, so that the function $\mathfrak{t}\mathbf{1}$ in $C(X)$ is an idempotent. Therefore it can assume no values on X other than 0 or 1. Hence for each $n \in \mathbf{N}$, the function

$$\mathfrak{t}\boldsymbol{n} = \mathfrak{t}\mathbf{1} + \cdots + \mathfrak{t}\mathbf{1}$$

assumes no values other than 0 or n. Consider, now, any function g in $C^*(Y)$. Since $|g| \leqq \boldsymbol{n}$, for suitable $n \in \mathbf{N}$, we have $|\mathfrak{t}g| \leqq \mathfrak{t}\boldsymbol{n} \leqq \boldsymbol{n}$.

1.8. COROLLARY. *If* X *is not pseudocompact, then* $C(X)$ *is not a homomorphic image of* $C^*(Y)$*, for any* Y.

In particular, $C(X)$ and $C^*(X)$ are isomorphic only if they are identical.

1.9. Another consequence of Theorem 1.7 is that an *isomorphism* from $C(Y)$ onto $C(X)$ carries $C^*(Y)$ onto $C^*(X)$. This is also a corollary of the next theorem.

THEOREM. *Let* $\mathfrak{t}$ *be a homomorphism from* $C(Y)$ *into* $C(X)$ *whose image contains* $C^*(X)$. *Then* $\mathfrak{t}$ *carries* $C^*(Y)$ *onto* $C^*(X)$.

PROOF. We prove first that $\mathfrak{t}\mathbf{1} = \mathbf{1}$. Let $k \in C(Y)$ satisfy $\mathfrak{t}k = \mathbf{1}$; then $\mathfrak{t}\mathbf{1} = (\mathfrak{t}k)(\mathfrak{t}\mathbf{1}) = \mathfrak{t}(k \cdot \mathbf{1}) = \mathfrak{t}k = \mathbf{1}$. It follows that $\mathfrak{t}\boldsymbol{n} = \boldsymbol{n}$ for each $n \in \mathbf{N}$.

Now, given $f \in C^*(X)$, we are to find $g \in C^*(Y)$ such that $\mathfrak{t}g = f$. Choose $h \in C(Y)$ for which $\mathfrak{t}h = f$, and choose $n \in \mathbf{N}$ satisfying $|f| \leqq \boldsymbol{n}$. Now define $g = (-\boldsymbol{n} \vee h) \wedge \boldsymbol{n}$. Then $g \in C^*(Y)$, and, by Theorem 1.6, $\mathfrak{t}g = (-\boldsymbol{n} \vee f) \wedge \boldsymbol{n} = f$.

ZERO-SETS

1.10. In studying relations between topological properties of a space X and algebraic properties of $C(X)$, it is natural to look at the subsets of X of the form

$$\text{(a)} \qquad f^{\leftarrow}(r) = \{x \in X : f(x) = r\} \qquad (f \in C,\ r \in \mathbf{R}).$$

Clearly, these sets are closed.

We notice that if s is any real number, then

$$\{x : f(x) = r\} = \{x : (f - s)(x) = r - s\}.$$

Consequently, the family of sets of the form (a) obtained by allowing f to run through all of C, and r through all of $\mathbf{R}$, can also be obtained by holding r fixed. The algebraic aspect of the situation points to the choice of the number 0 as the fixed value of r to be considered.

The set $f^{\leftarrow}(0)$ will be called the *zero-set* of f. We shall find it convenient to denote this set by $\boldsymbol{Z}(f)$, or, for clarity, by $\boldsymbol{Z}_X(f)$:

$$\boldsymbol{Z}(f) = \boldsymbol{Z}_X(f) = \{x \in X : f(x) = 0\} \qquad (f \in C(X)).$$

Any set that is a zero-set of some function in $C(X)$ is called a zero-set in X. Thus, $\boldsymbol{Z}$ is a mapping from the ring C onto the set of all zero-sets in X.

Evidently, $\boldsymbol{Z}(f) = \boldsymbol{Z}(|f|) = \boldsymbol{Z}(f^n)$ (for all $n \in \mathbf{N}$), $\boldsymbol{Z}(\mathbf{0}) = X$, and $\boldsymbol{Z}(\mathbf{1}) = \emptyset$. Furthermore,

$$\boldsymbol{Z}(fg) = \boldsymbol{Z}(f) \cup \boldsymbol{Z}(g),$$

and

$$\boldsymbol{Z}(f^2 + g^2) = \boldsymbol{Z}(|f| + |g|) = \boldsymbol{Z}(f) \cap \boldsymbol{Z}(g).$$

If $f \in C$ and $g = |f| \wedge \mathbf{1}$, then $g \in C^*$ and $\mathbf{Z}(g) = \mathbf{Z}(f)$. Hence C and C^* yield the same zero-sets.

The formula

$$\mathbf{Z}(f) = \bigcap_{n \in \mathbf{N}} \{x \in X : |f(x)| < 1/n\}$$

shows that *every zero-set is a* G_δ, i.e., a countable intersection of open sets. Conversely, in a *normal* space, every closed G_δ is a zero-set (3D.*3*). This need not be so if the space is not normal, however (3K.*6*). On the other hand, in a *metric* space, every closed set is a zero-set, as it consists precisely of all points whose distance from it is zero.

1.11. *Cozero-sets.* Every set of the form $\{x : f(x) \geqq 0\}$ is a zero-set:

$$\{x : f(x) \geqq 0\} = \mathbf{Z}(f \wedge \mathbf{0}) = \mathbf{Z}(f - |f|).$$

Likewise,

$$\{x : f(x) \leqq 0\} = \mathbf{Z}(f \vee \mathbf{0}) = \mathbf{Z}(f + |f|).$$

Thus, the open sets

$$\operatorname{pos} f = \{x : f(x) > 0\}$$

and

$$\operatorname{neg} f = \{x : f(x) < 0\} = \operatorname{pos}(-f)$$

are *cozero-sets*, i.e., complements of zero-sets. Conversely, every cozero-set is of this form:

$$X - \mathbf{Z}(f) = \operatorname{pos} |f|.$$

1.12. *Units.* For a function f in $C(X)$, f^{-1} exists if and only if f vanishes nowhere on X; in other words,

f is a unit of C if and only if $\mathbf{Z}(f) = \emptyset$.

Likewise, *if f is a unit of C^*, then $\mathbf{Z}(f) = \emptyset$.* The converse need not hold, however, as the multiplicative inverse f^{-1} of f in C may not be a bounded function. In fact, the condition for C^* is clearly the following: a function f in C^* is a unit of C^* if and only if it is bounded away from zero, i.e., $|f| \geqq \mathbf{r}$ for some $r > 0$.

1.13. EXAMPLES. It is convenient to have examples of some specific topological spaces to illustrate the notions that are being discussed. A familiar, important example of a *compact* topological space is the closed interval $[0, 1]$ of $\mathbf{R}$. As we know, $C([0, 1]) = C^*([0, 1])$. Familiar examples of *noncompact* spaces are $\mathbf{R}$ itself, the subspace $\mathbf{Q}$ of rational numbers, and the subspace $\mathbf{N}$ of positive integers. Since $\mathbf{N}$ is discrete, every real-valued function on $\mathbf{N}$ is continuous, so that $C(\mathbf{N})$ [resp.

$C^*(\mathbf{N})$] is actually the ring of all [resp. all bounded] sequences of real numbers.

The identity function $\mathbf{i}$ on $\mathbf{N}$, defined by $\mathbf{i}(n) = n$, belongs to $C(\mathbf{N})$, and it is unbounded, so that $C(\mathbf{N}) \neq C^*(\mathbf{N})$. The zero-set $Z(\mathbf{i})$ is empty, of course, so that $\mathbf{i}^{-1}$ exists; indeed,

$$\mathbf{i}^{-1} = \mathbf{j}, \quad \text{the sequence} \quad (1/n)_{n \in \mathbf{N}}.$$

Evidently, $\mathbf{j} \in C^*(\mathbf{N})$, that is to say, $\mathbf{j}$ is a bounded function. Finally, $Z(\mathbf{j}) = \emptyset$, but $\mathbf{i} = \mathbf{j}^{-1} \notin C^*(\mathbf{N})$, and we have here perhaps the simplest example of a function in C^* whose zero-set is empty, but that is not a unit of C^*.

1.14. For $C' \subset C(X)$, we write $\mathbf{Z}[C']$ to designate the family of zero-sets $\{Z(f): f \in C'\}$. This is consistent with our notational convention for the image of a set under a mapping. On the other hand, the family $\mathbf{Z}[C(X)]$ of *all* zero-sets in X will also be denoted, for simplicity, by $\mathbf{Z}(X)$.

We have observed that $\mathbf{Z}[C^*(X)]$ is the same as $\mathbf{Z}(X)$, and that $\mathbf{Z}(X)$ is closed under the formation of finite unions and finite intersections.

(a) *$\mathbf{Z}(X)$ is closed under countable intersection.*

For, given $f_n \in C$, define $g_n = |f_n| \wedge \mathbf{2}^{-n}$, and let

$$g(x) = \sum_{n \in \mathbf{N}} g_n(x) \qquad (x \in X).$$

Since $|g_n| \leqq \mathbf{2}^{-n}$, the series converges uniformly, and therefore g is a continuous function. Clearly,

$$Z(g) = \bigcap_{n \in \mathbf{N}} Z(g_n) = \bigcap_{n \in \mathbf{N}} Z(f_n).$$

However, $\mathbf{Z}(X)$ need not be closed under infinite union. For example, every one-element set in $\mathbf{R}$ is a zero-set in $\mathbf{R}$, so that an infinite union of zero-sets need not even be closed. Moreover, in a general space, even a *closed*, countable union of zero-sets need not be a zero-set; see 6P.5. Nor need $\mathbf{Z}(X)$ be closed under arbitrary intersection; see 4N.

1.15. *Completely separated sets.* Two subsets A and B of X are said to be *completely separated* (from one another) *in* X if there exists a function f in $C^*(X)$ such that $\mathbf{0} \leqq f \leqq \mathbf{1}$,

$$f(x) = 0 \text{ for all } x \in A, \text{ and } f(x) = 1 \text{ for all } x \in B.$$

Clearly, it is enough to find a function g in $C(X)$ satisfying $g(x) \leqq 0$ for all $x \in A$ and $g(x) \geqq 1$ for $x \in B$: for then $(\mathbf{0} \vee g) \wedge \mathbf{1}$ has the required properties. And, of course, the numbers 0 and 1 may be replaced in the definition by any real numbers r and s (with $r < s$).

It is plain that two sets contained (respectively) in completely separated sets are completely separated, and that two sets are completely separated if and only if their closures are.

When a zero-set Z is a neighborhood of a set A, we refer to Z as a zero-set-neighborhood of A.

THEOREM. *Two sets are completely separated if and only if they are contained in disjoint zero-sets. Moreover, completely separated sets have disjoint zero-set-neighborhoods.*

PROOF. We begin with the sufficiency. If $\mathbf{Z}(f) \cap \mathbf{Z}(g) = \emptyset$, then $|f| + |g|$ has no zeros, and we may define

$$h(x) = \frac{|f(x)|}{|f(x)| + |g(x)|} \qquad (x \in X),$$

—in brief, $h = |f| \cdot (|f| + |g|)^{-1}$. Then $h \in C(X)$, and h is equal to 0 on $\mathbf{Z}(f)$ and to 1 on $\mathbf{Z}(g)$.

Conversely, if A and A' are completely separated, there exists $f \in C(X)$ equal to 0 on A and to 1 on A'. The disjoint sets

$$F = \{x : f(x) \leqq 1/3\}, \qquad F' = \{x : f(x) \geqq 2/3\}$$

are zero-set-neighborhoods of A and A', respectively.

The following result will also be useful.

(a) *If A and A' are completely separated, then there exist zero-sets F and Z such that*

$$A \subset X - Z \subset F \subset X - A'.$$

For, with A, A', f, and F as above, we simply take

$$Z = \{x : f(x) \geqq 1/3\}.$$

C-EMBEDDING AND C-EMBEDDING*

1.16. A major portion of our work will deal with the problem of extending continuous functions. We shall say that a subspace S of X is *C-embedded* in X if every function in $C(S)$ can be extended to a function in $C(X)$. Likewise, we say that S is *C*-embedded* in X if every function in $C^*(S)$ can be extended to a function in $C^*(X)$.

If a function f in $C^*(S)$ has an extension g in $C(X)$, then f also has a bounded extension: if n is a bound for $|f|$, then $(-\mathbf{n} \vee g) \wedge \mathbf{n}$ belongs to $C^*(X)$, and agrees with f on S. Thus, S is C^*-embedded in X if and only if every function in $C^*(S)$ can be extended to a function in $C(X)$.

It is obvious that if $S \subset X \subset Y$, and X is C-embedded in Y, then S is C-embedded in Y if and only if it is C-embedded in X. The corresponding transitivity is valid, of course, for C^*.

It is unusual for a subspace to be C^*-embedded. For instance, $\mathbf{R} - \{0\}$ is not C^*-embedded in $\mathbf{R}$: the function with value 1 for all positive r, and -1 for negative r, has no continuous extension. (Every *uniformly* continuous function on $\mathbf{R} - \{0\}$, however, does have a continuous extension to $\mathbf{R}$, as can be verified without difficulty. The result also follows from the general theory of uniform spaces (Chapter 15).) On the other hand, it is manifest that $\mathbf{N}$, for example, is not only C^*-embedded in $\mathbf{R}$, but even C-embedded.

1.17. The basic result about C^*-embedding is Urysohn's theorem, which we state in the following general form.

URYSOHN'S EXTENSION THEOREM. *A subspace S of X is C^*-embedded in X if and only if any two completely separated sets in S are completely separated in X.*

PROOF. *Necessity.* If A and B are completely separated sets in S, there exists a function f in $C^*(S)$ that is equal to 0 an A and 1 on B. By hypothesis, f has an extension to a function g in $C^*(X)$. Since g is 0 on A and 1 on B, these sets are completely separated in X.

Sufficiency. Let f_1 be a given function in $C^*(S)$. Then $|f_1| \leqq m$ for some $m \in \mathbf{N}$. For convenience of notation, define

$$r_n = \frac{m}{2}\left(\frac{2}{3}\right)^n \qquad (n \in \mathbf{N}).$$

Then $|f_1| \leqq m = 3r_1$. Inductively, given $f_n \in C^*(S)$, with $|f_n| \leqq 3r_n$, define

$$A_n = \{s \in S : f_n(s) \leqq -r_n\}, \quad \text{and} \quad B_n = \{s \in S : f_n(s) \geqq r_n\}.$$

Then A_n and B_n are completely separated in S, and so, by hypothesis, they are completely separated in X. Accordingly, there exists a function g_n in $C^*(X)$, equal to $-r_n$ on A_n, and to r_n on B_n, and with $|g_n| \leqq r_n$.

The values of f_n and g_n on A_n lie between $-3r_n$ and $-r_n$; on B_n, they lie between r_n and $3r_n$; and, elsewhere on S, they are between $-r_n$ and r_n. We now define

$$f_{n+1} = f_n - g_n|S,$$

and we have $|f_{n+1}| \leqq 2r_n$, i.e.,

$$|f_{n+1}| \leqq 3r_{n+1}.$$

This completes the induction step. Now put

$$g(x) = \sum_{n \in \mathbf{N}} g_n(x) \qquad (x \in X).$$

Because the series converges uniformly, this defines g as a continuous function on X. Next, we observe that

$$\begin{aligned}(g_1 + \cdots + g_n)|S &= (f_1 - f_2) + \cdots + (f_n - f_{n+1}) \\ &= f_1 - f_{n+1}.\end{aligned}$$

Since the sequence $(f_{n+1}(s))$ approaches 0 at every point s of S, this shows that $g(s) = f_1(s)$. Thus, g is an extension of f_1. This completes the proof.

In case X is a metric space, every closed set is a zero-set, so that any two disjoint closed sets are completely separated. Now, if S is closed, then closed sets in S are also closed in X; therefore completely separated sets in S have disjoint closures in X. It follows from the theorem that every closed set in a metric space is C^*-embedded. This result is Tietze's extension theorem. Urysohn's generalization to normal spaces will be discussed in Chapter 3.

In particular, every closed set in $\mathbf{R}$ is C^*-embedded in $\mathbf{R}$.

1.18. A C^*-embedded subspace need not be C-embedded. Later we shall see innumerable examples of this phenomenon. The simplest is given by the space Σ of 4M, which contains $\mathbf{N}$ as a dense, C^*-embedded subset that is not C-embedded. A more striking example is the pseudocompact space Λ of 6P, which contains $\mathbf{N}$ as a *closed*, C^*-embedded subset; since Λ is pseudocompact, *no* unbounded function on $\mathbf{N}$ can be extended continuously to Λ.

The relation between C^*-embedding and C-embedding is clarified by the next theorem.

THEOREM. *A C^*-embedded subset is C-embedded if and only if it is completely separated from every zero-set disjoint from it.*

PROOF. Let S be C^*-embedded in X.

Necessity. Given a zero-set $\mathbf{Z}(h)$ in X, disjoint from S, put $f(s) = 1/h(s)$ for $s \in S$. This defines f as a continuous function on S. Let g be a continuous extension of f to all of X. Then gh belongs to $C(X)$, and is equal to 1 on S and to 0 on $\mathbf{Z}(h)$.

Sufficiency. Consider any function f in $C(S)$. Then $\arctan \circ f$ belongs to $C^*(S)$, and so has an extension to a function g in $C(X)$. The set

$$Z = \{x \in X : |g(x)| \geqq \pi/2\}$$

belongs to $\mathbf{Z}(X)$, and is disjoint from S. By hypothesis, there is a function h in $C(X)$ equal to 1 on S and to 0 on Z, and with $|h| \leqq 1$. The function gh then agrees with arctan $\circ f$ on S, and satisfies $|(gh)(x)| < \pi/2$ for every x. Hence $\tan \circ (gh)$ is a real-valued, continuous extension of f to all of X.

1.19. In particular, every closed set in $\mathbf{R}$ is C-embedded. This leads to the following special sufficient condition for a set S in X to be C-embedded.

THEOREM. *If there exists a function in $C(X)$ that carries S homeomorphically onto a closed set in* $\mathbf{R}$, *then S is C-embedded in X.*

PROOF. Let h denote the postulated function in $C(X)$. Then $\theta = (h|S)^{\leftarrow}$ is a continuous mapping from $H = h[S]$ onto S, with $\theta(h(s)) = s$ (for $s \in S$). Consider, now, an arbitrary function f in $C(S)$. The composition $f \circ \theta$ belongs to $C(H)$. Since H is closed in $\mathbf{R}$, by hypothesis, it is C-embedded, and so there is a function g in $C(\mathbf{R})$ that agrees with $f \circ \theta$ on H. Then $g \circ h$ is in $C(X)$, and for all $s \in S$, we have

$$(g \circ h)(s) = f(\theta(h(s))) = f(s),$$

i.e., $g \circ h$ is an extension of f.

1.20. COROLLARY. *Let $E \subset X$, and suppose that some function h in $C(X)$ is unbounded on E. Then E contains a copy of* $\mathbf{N}$, *C-embedded in X, on which h approaches infinity.*

1.21. COROLLARY. *X is pseudocompact if and only if it contains no C-embedded copy of* $\mathbf{N}$.

PROBLEMS

1A. CONTINUITY ON SUBSETS.

Let $f \in \mathbf{R}^X$.

1. If the restriction of f to each of a finite number of closed sets, whose union is X, is continuous, then f is continuous.

2. If the restriction of f to each of an arbitrary number of open sets, whose union is X, is continuous, then f is continuous.

3. Let $\mathscr{S}$ be a family of closed sets whose union is X and such that every point of X has a neighborhood that meets only finitely many members of $\mathscr{S}$. ($\mathscr{S}$ is then said to be *locally finite.*) If the restriction of f to each member of $\mathscr{S}$ is continuous, then f is continuous.

1B. COMPONENTS OF X.

1. In $C(X)$ (or $C^*(X)$), all positive units have the same number of square roots.

2. X is connected if and only if $\mathbf{1}$ has exactly two square roots.

3. For finite $\mathfrak{m}$, X has $\mathfrak{m}$ components if and only if $\mathbf{1}$ has $2^{\mathfrak{m}}$ square roots. For infinite $\mathfrak{m}$, the statement is false. [Consider the subspace $\{1, 1/2, \cdots, 1/n, \cdots, 0\}$ of $\mathbf{R}$.]

4. X is connected if and only if $\mathbf{0}$ and $\mathbf{1}$ are the only idempotents in $C(X)$.

5. If X is connected, then $C(X)$ is not a direct sum of any two rings (except trivially).

6. If X is the union of disjoint nonempty open sets A and B, then $C(X)$ is isomorphic to the direct sum of $C(A)$ and $C(B)$.

1C. C AND C^* FOR VARIOUS SUBSPACES OF $\mathbf{R}$.

Consider the subspaces $\mathbf{R}$, $\mathbf{Q}$, $\mathbf{N}$, and $N^* = \{1, 1/2, \cdots, 1/n, \cdots, 0\}$ of $\mathbf{R}$, and the rings C and C^* for each of these spaces. Each of these rings is of cardinal $\mathfrak{c}$.

1. For each $\mathfrak{m} \leqq \aleph_0$, each ring on $\mathbf{R}$, $\mathbf{N}$, or N^* contains a function having exactly $2^{\mathfrak{m}}$ square roots. If a member of $C(\mathbf{Q})$ has more than one square root, it has $\mathfrak{c}$ of them.

2. $C(\mathbf{R})$ has just two idempotents, $C(N^*)$ has exactly $\aleph_0$, and $C(\mathbf{Q})$ and $C(\mathbf{N})$ have $\mathfrak{c}$.

3. Every nonzero idempotent in $C(\mathbf{Q})$ is a sum of two nonzero idempotents. In $C(\mathbf{N})$, and in $C(N^*)$, some, but not all idempotents have this property.

4. Except for the obvious identity $C(N^*) = C^*(N^*)$, *no two of the rings in question are isomorphic.*

5. Each of $C(\mathbf{Q})$ and $C(\mathbf{N})$ is isomorphic with a direct sum of two copies of itself. $C(N^*)$ is isomorphic with a direct sum of two subrings, just one of which is isomorphic with $C(N^*)$.

6. The ring $C(\mathbf{R})$ is isomorphic with a proper subring. [Consider the functions that are constant on $[0,1]$.] But $C(\mathbf{R})$ has no proper summand.

1D. DIVISORS OF FUNCTIONS.

1. If $\mathbf{Z}(f)$ is a neighborhood of $\mathbf{Z}(g)$, then f is a multiple of g—that is, $f = hg$ for some $h \in C$. Furthermore, if $X - \operatorname{int} \mathbf{Z}(f)$ is compact, then h can be chosen to be bounded. [Define $h(x) = f(x)/g(x)$ for $x \notin \operatorname{int} \mathbf{Z}(f)$, and $h(x) = 0$ for $x \in \mathbf{Z}(f)$, and apply 1A.*1*.]

2. Construct an example in which $\mathbf{Z}(f) \supset \mathbf{Z}(g)$, but f is not a multiple of g.

3. If $|f| \leqq |g|^r$ for some real $r > 1$, then f is a multiple of g. [Define $h(x) = f(x)/g(x)$ for $x \notin \mathbf{Z}(g)$, and $h(x) = 0$ otherwise.] Hence if $|f| \leqq |g|$, then f^r is a multiple of g for every $r > 1$ for which f^r is defined.

1E. UNITS.

1. Let $f \in C$. There exists a positive unit u of C such that

$$(-\mathbf{1} \vee f) \wedge \mathbf{1} = uf.$$

2. The following are equivalent.

(1) For every $f \in C$, there exists a unit u of C such that $f = u|f|$.

(2) For every $g \in C^*$, there exists a unit v of C^* such that $g = v|g|$.

[(1) *implies* (2). $g = u|g|$ for some unit u *of* C; consider pos u and neg u.]

3. Describe the functions f in $C(\mathbf{N})$ for which there exists a unit u of $C(\mathbf{N})$ satisfying $f = u|f|$. Do the same for $C(\mathbf{Q})$ and $C(\mathbf{R})$.

4. Do the same for the equation $f = k|f|$, where k belongs to C but is not necessarily a unit.

1F. C-EMBEDDING.

1. Every C^*-embedded zero-set is C-embedded.

2. Let $S \subset X$; if every zero-set in S is a zero-set in X, then S is C^*-embedded in X.

3. A discrete zero-set is C^*-embedded if and only if all of its subsets are zero-sets.

4. A subset S of $\mathbf{R}$ is C-embedded [resp. C^*-embedded] if and only if it is closed. [A point of cl $S - S$ is the limit of a sequence in S.]

5. If a (nonempty) subset S of X is C-embedded in X, then $C(S)$ is a homomorphic image of $C(X)$. The corresponding result holds for C^*.

1G. PSEUDOCOMPACT SPACES.

1. Any continuous image of a pseudocompact space is pseudocompact.

2. X is pseudocompact if and only if $f[X]$ is compact for every f in $C^*(X)$.

3. Let X be a Hausdorff space. If, of any two disjoint closed sets, at least one is compact, or even countably compact, then X is countably compact. [A Hausdorff space is countably compact if and only if every infinite set has a limit point.]

4. If, of any two disjoint *zero-sets* in X, at least one is compact, or even pseudocompact, then X is pseudocompact. [If $C \neq C^*$, then some function f in C assumes the values 0 and 1 infinitely often on a C-embedded copy of $\mathbf{N}$.] (But X need not be countably compact; see 8.20.)

1H. BASICALLY AND EXTREMALLY DISCONNECTED SPACES.

A space X is said to be ***extremally disconnected*** if every open set has an open closure; X is ***basically disconnected*** if every cozero-set has an open closure. Hence any extremally disconnected space is basically disconnected. (The converse fails; see 4N.)

1. X is extremally disconnected if and only if every pair of disjoint open sets have disjoint closures. What is the analogous condition for basically disconnected spaces?

2. In an extremally disconnected space, any two disjoint open sets are completely separated. In a basically disconnected space, any two disjoint cozero-sets are completely separated; equivalently, for every $f \in C$, pos f and neg f are completely separated.

3. If X is basically disconnected, then for every $f \in C$, there exists a unit $\boldsymbol{u}$ of C such that $f = u|f|$.

4. Every dense subspace X of an extremally disconnected space T is extremally disconnected. In fact, disjoint open sets in X have disjoint open closures in T.

5. Every open subspace of an extremally disconnected space is extremally disconnected. (A *closed* subspace, however, need not even be basically disconnected; see 6W.)

6. X is extremally disconnected if and only if every open subspace is C^*-embedded. [*Necessity.* Apply Urysohn's extension theorem, invoking *2.* *Sufficiency.* Apply *1.*]

1I. ALGEBRA HOMOMORPHISMS.

Let $\mathfrak{t}$ be a (ring) homomorphism from $C(Y)$ or $C^*(Y)$ into $C(X)$.

1. $\mathfrak{t}\boldsymbol{r} = \boldsymbol{r}\cdot\mathfrak{t}\mathbf{1}$ for each $r \in \mathbf{R}$. [For each $x \in X$, the mapping $r \to (\mathfrak{t}\boldsymbol{r})(x)$ is a homomorphism from $\mathbf{R}$ into $\mathbf{R}$, and hence is either the zero homomorphism or the identity (0.22). Also, $(\mathfrak{t}\mathbf{1})(x) = 0$ or 1.]

2. $\mathfrak{t}$ is an algebra homomorphism, i.e., $\mathfrak{t}(\boldsymbol{r}g) = \boldsymbol{r}\cdot\mathfrak{t}g$ for all $r \in \mathbf{R}$ and $g \in C(Y)$.

1J. PRESERVATION OR REDUCTION OF NORM.

For $f \in C^*(X)$, define $\|f\| = \sup_{x\in X} |f(x)|$.

1. $\|f\| = \inf\{r \in \mathbf{R}: |f| \leqq \boldsymbol{r}\}$.

2. If $\mathfrak{t}$ is a nonzero homomorphism of $C^*(Y)$ into $C^*(X)$, then $\|\mathfrak{t}\boldsymbol{r}\| = |r|$. [1I.*1.*]

3. $\mathfrak{t}\boldsymbol{r} \leqq \boldsymbol{r}$ for $r \geqq 0$ in $\mathbf{R}$.

4. $\|\mathfrak{t}g\| \leqq \|g\|$ for every $g \in C^*(Y)$. [Theorem 1.6.]

5. If $\mathfrak{t}g \leqq \boldsymbol{r}$, then $\mathfrak{t}g \leqq \mathfrak{t}\boldsymbol{r}$. [1I.*1.*]

6. If $\mathfrak{t}$ is an *isomorphism* into $C^*(X)$, then $\|\mathfrak{t}g\| = \|g\|$ for all $g \in C^*(Y)$. [If $\|g\| > r$, then $|g| \neq |g| \wedge \boldsymbol{r}$.]

Chapter 2

IDEALS AND z-FILTERS

2.1. Continuing our study of the relations between algebraic properties of $C(X)$ and topological properties of X, we now examine the special features of the family of zero-sets of an *ideal* of functions. Such a family turns out to possess properties analogous to those of a filter; this fact will play a central role in the development.

We recall that a proper subset I of C is an ideal in C provided that I is a subring such that $gf \in I$ whenever $f \in I$, for arbitrary $g \in C$. A subset having these algebraic properties is a proper subset if and only if it contains no unit. We shall occasionally refer to the ring C itself as an improper ideal. Thus, the word *ideal*, unmodified, will always mean *proper* ideal.

The intersection of any nonempty family of ideals is an ideal. *Every ideal is embeddable in a maximal ideal.* Every maximal ideal M is prime, that is, if $fg \in M$, then $f \in M$ or $g \in M$.

The smallest ideal (perhaps improper) containing a given collection of ideals $I, \cdots$, and elements $f, \cdots$, is denoted by

$$(I, \cdots, f, \cdots).$$

It consists of all elements of C expressible as (finite) sums $i + \cdots + sf + \cdots$, where $i \in I, \cdots$, and where $s, \cdots$ are arbitrary functions in C.

Corresponding remarks apply to C^* (in fact, to any commutative ring with unity element). Evidently, if I is an ideal in C, then $I \cap C^*$ is an ideal in C^*.

2.2. A nonempty subfamily $\mathscr{F}$ of $\mathbf{Z}(X)$ is called a *z-filter* on X provided that

(i) $\emptyset \notin \mathscr{F}$;
(ii) if $Z_1, Z_2 \in \mathscr{F}$, then $Z_1 \cap Z_2 \in \mathscr{F}$; and
(iii) if $Z \in \mathscr{F}$, $Z' \in \mathbf{Z}(X)$, and $Z' \supset Z$, then $Z' \in \mathscr{F}$.

By (iii), X belongs to every z-filter. Because of (iii), (ii) may be replaced in the above list by

(ii′) if $Z_1, Z_2 \in \mathscr{F}$, then $Z_1 \cap Z_2$ contains a member of $\mathscr{F}$.

Every family $\mathscr{B}$ of zero-sets that has the finite intersection property is contained in a z-filter: the smallest such is the family $\mathscr{F}$ of all zero-sets containing finite intersections of members of $\mathscr{B}$. We say that $\mathscr{B}$ generates the z-filter $\mathscr{F}$. When $\mathscr{B}$ itself is closed under finite intersection, it is called a base for $\mathscr{F}$.

The above definition is, of course, an analogue of the familiar definition of *filter*: a nonempty family of subsets of X, closed under the formation of finite intersections and of supersets, and that does not contain the empty set. A z-filter is a topological object, while a filter is a purely set-theoretic one. In a *discrete* space, every set is a zero-set, so that filters and z-filters are the same in discrete spaces.

In any space X, the intersection with $\mathbf{Z}(X)$ of any filter is a z-filter. Conversely, if $\mathscr{F}'$ is the smallest filter containing a given z-filter $\mathscr{F}$ (i.e., $\mathscr{F}$ is a base for $\mathscr{F}'$), then $\mathscr{F}' \cap \mathbf{Z}(X) = \mathscr{F}$.

2.3. THEOREM.

(a) *If I is an ideal in $C(X)$, then the family*

$$\mathbf{Z}[I] = \{\mathbf{Z}(f) : f \in I\}$$

is a z-filter on X.

(b) *If $\mathscr{F}$ is a z-filter on X, then the family*

$$\mathbf{Z}^{\leftarrow}[\mathscr{F}] = \{f : \mathbf{Z}(f) \in \mathscr{F}\}$$

is an ideal in C.

PROOF. (a). (i). Since I contains no unit, $\emptyset \notin \mathbf{Z}[I]$.

(ii). Let $Z_1, Z_2 \in \mathbf{Z}[I]$. Let $f_1, f_2 \in I$ satisfy $Z_1 = \mathbf{Z}(f_1)$, $Z_2 = \mathbf{Z}(f_2)$. Since I is an ideal, $f_1^2 + f_2^2 \in I$. Hence

$$Z_1 \cap Z_2 = \mathbf{Z}(f_1^2 + f_2^2) \in \mathbf{Z}[I].$$

(iii). Let $Z \in \mathbf{Z}[I]$, and $Z' \in \mathbf{Z}(X)$. Let $f \in I$, $f' \in C$ satisfy $Z = \mathbf{Z}(f)$, $Z' = \mathbf{Z}(f')$. Since I is an ideal, we have $ff' \in I$. Hence if $Z' \supset Z$, then

$$Z' = Z \cup Z' = \mathbf{Z}(ff') \in \mathbf{Z}[I].$$

(b). Let $J = \mathbf{Z}^{\leftarrow}[\mathscr{F}]$. By 2.2(i), J contains no unit. Let $f, g \in J$, and let $h \in C$. Then

$$\mathbf{Z}(f - g) \supset \mathbf{Z}(f) \cap \mathbf{Z}(g) \in \mathscr{F},$$

by 2.2(ii), and $\mathbf{Z}(hf) \supset \mathbf{Z}(f) \in \mathscr{F}$. Hence $\mathbf{Z}(f - g) \in \mathscr{F}$ and $\mathbf{Z}(hf) \in \mathscr{F}$, by 2.2(iii). Therefore $f - g \in J$ and $hf \in J$. Thus, J is an ideal in C.

REMARK. In particular, $(f, g) \neq C$ if and only if $\mathbf{Z}(f)$ meets $\mathbf{Z}(g)$, hence if and only if $f^2 + g^2$—or $|f| + |g|$—is not a unit of C.

2.4. Like any mapping, $\mathbf{Z}$ satisfies (for $\mathscr{F} \subset \mathbf{Z}(X)$)

$$\mathbf{Z}[\mathbf{Z}^{\leftarrow}[\mathscr{F}]] = \mathscr{F} \quad \text{and} \quad \mathbf{Z}^{\leftarrow}[\mathbf{Z}[I]] \supset I.$$

The first relation implies that every z-filter is of the form $\mathbf{Z}[J]$ for some ideal J in C. In the second relation, the inclusion may be proper, as the following example shows.

EXAMPLES. Consider the principal ideal $I = (\mathbf{i})$ in $C(\mathbf{R})$ ($\mathbf{i}$ denoting the identity function on $\mathbf{R}$). This consists of all functions f in $C(\mathbf{R})$ such that $f(x) = xg(x)$ for some $g \in C(\mathbf{R})$. In particular, every function in I vanishes at 0. Hence every zero-set in $\mathbf{Z}[I]$ contains the point 0. As a matter of fact, since $\mathbf{Z}[I]$ is a z-filter that includes the set $\{0\}$, it must be the family of *all* zero-sets containing 0. Additional properties of the ideal $(\mathbf{i})$ are given in 2H.

The ideal $M_0 = \mathbf{Z}^{\leftarrow}[\mathbf{Z}[I]]$ evidently consists of all functions in $C(\mathbf{R})$ that vanish at 0. Hence M_0 certainly contains I. However, $M_0 \neq I$. For instance, $\mathbf{i}^{1/3} \in M_0 - I$. That $\mathbf{i}^{1/3} \in M_0$ is obvious. And if $\mathbf{i}^{1/3} \in I$, then $\mathbf{i}^{1/3} = g\mathbf{i}$ for some $g \in C(\mathbf{R})$; but then $g(x) = x^{-2/3}$ for $x \neq 0$, so that g cannot be continuous at 0.

Note that $\mathbf{Z}[M_0] = \mathbf{Z}[I]$, in spite of the fact that $M_0 \neq I$.

Finally, we observe that M_0 is a *maximal* ideal. For, if $f \notin M_0$, then $\mathbf{Z}(f)$ is disjoint from $\mathbf{Z}(\mathbf{i})$, whence we have $(M_0, f) \supset (\mathbf{i}, f) = C$ (see 2.3, REMARK).

The analogue of Theorem 2.3(a), with C^* in place of C, is false, in general. If J is an ideal in C^*, then $\mathbf{Z}[J]$ does satisfy the properties (ii) and (iii) of a z-filter (as the proof of (a) shows); however, (i) need not hold. For example, the set J of all sequences that converge to zero is obviously an ideal in $C^*(\mathbf{N})$; but since $\mathbf{j} \in J$, and $\mathbf{Z}(\mathbf{j}) = \emptyset$, it follows that $\emptyset \in \mathbf{Z}[J]$, and hence that $\mathbf{Z}[J]$ is the family $\mathbf{Z}(\mathbf{N})$ of all subsets of $\mathbf{N}$. Observe that J is not an ideal in C; in fact, $\mathbf{j}$ is a unit of C.

2.5. By a *z-ultrafilter* on X is meant a maximal z-filter, i.e., one not contained in any other z-filter. Thus, a z-ultrafilter is a maximal subfamily of $\mathbf{Z}(X)$ with the finite intersection property. It follows from the maximal principle (0.7) that *every subfamily of* $\mathbf{Z}(X)$ *with the finite intersection property is contained in some z-ultrafilter.*

In a discrete space, z-ultrafilters are the same as ultrafilters, i.e., maximal filters.

THEOREM.

(a) *If M is a maximal ideal in $C(X)$, then* $\mathbf{Z}[M]$ *is a z-ultrafilter on X.*

(b) *If $\mathscr{A}$ is a z-ultrafilter on X, then $\mathbf{Z}^{\leftarrow}[\mathscr{A}]$ is a maximal ideal in C.*

The mapping $\mathbf{Z}$ is one-one from the set of all maximal ideals in C onto the set of all z-ultrafilters.

PROOF. Since $\mathbf{Z}$ and $\mathbf{Z}^{\leftarrow}$ preserve inclusion, the result follows at once from Theorem 2.3.

As we saw in 2.4, we cannot conclude that an ideal is maximal from the fact that its z-filter is maximal.

2.6. THEOREM.

(a) *Let M be a maximal ideal in $C(X)$; if $\mathbf{Z}(f)$ meets every member of $\mathbf{Z}[M]$, then $f \in M$.*

(b) *Let $\mathscr{A}$ be a z-ultrafilter on X; if a zero-set Z meets every member of $\mathscr{A}$, then $Z \in \mathscr{A}$.*

PROOF. By Theorem 2.5, the two statements are equivalent. In (b), $\mathscr{A} \cup \{Z\}$ generates a z-filter. As this contains the maximal z-filter $\mathscr{A}$, it must be $\mathscr{A}$.

The properties stated in the theorem are, in fact, characteristic of maximal ideals and z-filters: if a z-filter $\mathscr{A}$ contains every zero-set that meets all members of $\mathscr{A}$, then, clearly, $\mathscr{A}$ is a z-ultrafilter.

z-IDEALS AND PRIME IDEALS

2.7. An ideal I in $C(X)$ is called a *z-ideal* if $\mathbf{Z}(f) \in \mathbf{Z}[I]$ implies $f \in I$—that is to say, if $I = \mathbf{Z}^{\leftarrow}[\mathbf{Z}[I]]$.

If $\mathscr{F}$ is a z-filter, then $\mathbf{Z}^{\leftarrow}[\mathscr{F}]$ is a z-ideal (since $\mathscr{F} = \mathbf{Z}[\mathbf{Z}^{\leftarrow}[\mathscr{F}]]$). Hence if J is any ideal in C, then $I = \mathbf{Z}^{\leftarrow}[\mathbf{Z}[J]]$ is a z-ideal; clearly, I is the smallest z-ideal containing J.

It is evident that *every maximal ideal is a z-ideal.*

The intersection of an arbitrary (nonempty) family of z-ideals is a z-ideal.

The mapping $\mathbf{Z}$ is one-one from the set of all z-ideals onto the set of all z-filters. The discussion in 2.4 shows that the principal ideal $(\mathbf{i})$ in $C(\mathbf{R})$ is not a z-ideal. If S is a nonempty set, in any space X, then the family of all functions in $C(X)$ that vanish everywhere on S is a z-ideal.

In $C(\mathbf{N})$, *every* ideal I is a z-ideal. For, suppose that $\mathbf{Z}(f) = \mathbf{Z}(g)$, where $g \in I$. Define h as follows: $h(n) = 0$ for $n \in \mathbf{Z}(g)$, and $h(n) = f(n)/g(n)$ for $n \notin \mathbf{Z}(g)$. Since $\mathbf{N}$ is discrete, h is continuous. Evidently, $f = hg$. Therefore $f \in I$. (Compare 1D.*1*.)

2.8. THEOREM. *Every z-ideal in $C(X)$ is an intersection of prime ideals.*

PROOF. $Z(f^n) = Z(f)$ for every $n \in \mathbf{N}$. Hence if I is any z-ideal, then $f^n \in I$ implies $f \in I$. But this property characterizes I as the intersection of all the prime ideals containing it (0.18).

It is not obvious from the definition whether z-ideals can be described as algebraic objects in the ring $C(X)$. It will turn out later that they can be (see 4A.*5*). Some algebraic information has already been obtained: every intersection of maximal ideals is a z-ideal, and every z-ideal is an intersection of prime ideals.

The converses are not true, however. As an example of a z-ideal that is not an intersection of maximal ideals, consider the set $\boldsymbol{O}_0$ of all functions f in $C(\mathbf{R})$ for which $Z(f)$ is a neighborhood of 0. Evidently, $\boldsymbol{O}_0$ is a z-ideal, and it is contained properly in the maximal ideal $\boldsymbol{M}_0$ of *all* functions that vanish at 0 (see 2.4). Note that every neighborhood of 0 contains a zero-set-neighborhood of 0, i.e., a member of $Z[\boldsymbol{O}_0]$. Now, if I is any ideal containing $\boldsymbol{O}_0$, then $Z[\boldsymbol{O}_0] \subset Z[I]$; hence every member of $Z[I]$ meets every neighborhood of 0, and therefore contains 0. It follows that $I \subset \boldsymbol{M}_0$. This shows that $\boldsymbol{M}_0$ is the *only* maximal ideal containing $\boldsymbol{O}_0$. Therefore $\boldsymbol{O}_0$ is not an intersection of maximal ideals.

Incidentally, since $\boldsymbol{O}_0$ is an intersection of prime ideals—all of which must be contained in $\boldsymbol{M}_0$—we have established the existence of non-maximal, prime ideals in $C(\mathbf{R})$.

In order to show that the converse of the theorem is not valid, it is enough to find a single prime ideal that is not a z-ideal. A construction is outlined in 2G.*1*.

The next theorem clarifies to some extent the relation between prime ideals and z-ideals.

2.9. THEOREM. *For any z-ideal I in C, the following are equivalent.*

(1) *I is prime.*

(2) *I contains a prime ideal.*

(3) *For all $g, h \in C$, if $gh = \mathbf{0}$, then $g \in I$ or $h \in I$.*

(4) *For every $f \in C$, there is a zero-set in $Z[I]$ on which f does not change sign.*

PROOF. (1) *implies* (2). Trivial.

(2) *implies* (3). If I contains a prime ideal P, and $gh = \mathbf{0}$, then $gh \in P$, whence either g or h is in P and hence in I.

(3) *implies* (4). It suffices to observe that $(f \vee \mathbf{0})(f \wedge \mathbf{0}) = \mathbf{0}$ for every $f \in C$.

(4) *implies* (1). Given $gh \in I$, consider the function $|g| - |h|$. By hypothesis, there is a zero-set Z of I on which $|g| - |h|$ is

nonnegative, say. Then every zero of g on Z is a zero of h. Hence

$$\mathbf{Z}(h) \supset Z \cap \mathbf{Z}(h) = Z \cap \mathbf{Z}(gh) \in \mathbf{Z}[I],$$

so that $\mathbf{Z}(h) \in \mathbf{Z}[I]$. Since I is a z-ideal, $h \in I$. Thus, I is prime.

2.10.

If J and J' are ideals, neither containing the other, then $J \cap J'$ is not prime.

In fact, this holds in any commutative ring. For, when $a \in J - J'$ and $a' \in J' - J$, then neither a nor a' belongs to $J \cap J'$, but $aa' \in J \cap J'$.

EXAMPLE. We saw in 2.4 that the set of all functions in $C(\mathbf{R})$ that vanish at 0 is a maximal ideal. A similar proof shows that the set of all functions vanishing at 1 is a maximal ideal. Let I denote the intersection of these ideals, i.e., I is the z-ideal of all functions that vanish at both 0 and 1. By the above, I is not prime. (In the proof, we may take, for example, $a = \mathbf{i}$, and $a' = \mathbf{i} - \mathbf{1}$.) By Theorem 2.9, the z-ideal I contains no prime ideal.

The next theorem generalizes this result to arbitrary maximal ideals in $C(X)$, for any X.

2.11. THEOREM. *Every prime ideal in $C(X)$ is contained in a unique maximal ideal.*

PROOF. We know that every ideal is contained in at least one maximal ideal. If M and M' are distinct maximal ideals, their intersection is a z-ideal (since M and M' are z-ideals), but it is not prime (2.10); by Theorem 2.9, $M \cap M'$ contains no prime ideal.

The corresponding theorem is valid for C^*, but we shall not prove it here. The conclusion will follow from the general result stated in 6.6(c).

2.12. By a *prime z-filter*, we shall mean a z-filter $\mathscr{F}$ with the following property: whenever the union of two zero-sets belongs to $\mathscr{F}$, then at least one of them belongs to $\mathscr{F}$.

THEOREM.

(a) *If P is a prime ideal in $C(X)$, then $\mathbf{Z}[P]$ is a prime z-filter.*

(b) *If $\mathscr{F}$ is a prime z-filter, then $\mathbf{Z}^{\leftarrow}[\mathscr{F}]$ is a prime z-ideal.*

PROOF. (a). Let $Q = \mathbf{Z}^{\leftarrow}[\mathbf{Z}[P]]$. Then $\mathbf{Z}[Q] = \mathbf{Z}[P]$, and Q is a z-ideal containing the prime ideal P. By Theorem 2.9, Q is prime. Suppose, now, that $\mathbf{Z}(f) \cup \mathbf{Z}(g) \in \mathbf{Z}[P]$. This implies that $\mathbf{Z}(fg) \in \mathbf{Z}[Q]$; therefore fg belongs to the z-ideal Q. Since Q is prime, it contains f, say. Then $\mathbf{Z}(f) \in \mathbf{Z}[Q] = \mathbf{Z}[P]$.

(b). We know that the ideal $P = \mathbf{Z}^{\leftarrow}[\mathscr{F}]$ is a z-ideal. Suppose that $fg \in P$. Then

$$\mathbf{Z}(fg) = \mathbf{Z}(f) \cup \mathbf{Z}(g) \in \mathbf{Z}[P] = \mathscr{F}.$$

By hypothesis, $\mathbf{Z}(f)$, say, belongs to $\mathbf{Z}[P]$. Then f belongs to the z-ideal P.

2.13. It follows that a prime z-filter is contained in a unique z-ultrafilter.

Since every maximal ideal in C is prime, every z-ultrafilter is a prime z-filter. This can be seen more directly. If zero-sets Z and Z' do not belong to a z-ultrafilter $\mathscr{A}$, then, by Theorem 2.6(b), there exist $A, A' \in \mathscr{A}$ such that $Z \cap A = Z' \cap A' = \emptyset$. Then $Z \cup Z'$ does not meet the member $A \cap A'$ of $\mathscr{A}$, and hence does not belong to $\mathscr{A}$.

In a discrete space X, there is no difference between prime and maximal, i.e., every prime filter $\mathscr{U}$ is an ultrafilter. For, if $A \notin \mathscr{U}$, then $X - A \in \mathscr{U}$; hence A cannot be adjoined to $\mathscr{U}$.

2.14. The correspondences between z-filters on X and ideals in $C(X)$ that have been established in this chapter are powerful tools in the study of $C(X)$. These correspondences, which also occur in a rudimentary form in C^* (e.g., in Theorem 2.3(b), $\mathbf{Z}^{\leftarrow}[\mathscr{F}] \cap C^*$ is an ideal in C^*), are inconsequential there, as many of the theorems of the chapter become false if C is replaced by C^*.

However, there is another correspondence, between a certain class of z-filters on X and ideals in $C^*(X)$, that leads to theorems quite analogous to those for C. The requisite information is outlined in 2L. It is worth noting that the theory is far more complicated for C^* than for C.

The development in 2L discloses a natural one-one correspondence between the maximal ideals in C and those in C^* (2L.*16*). In the text itself, we will not arrive at this correspondence until Chapter 7, at which time its significance will be clearer.

PROBLEMS

2A. BOUNDED FUNCTIONS IN IDEALS.

The functions f, $f \cdot (1 + f^2)^{-1}$, and $(-\mathbf{1} \vee f) \wedge \mathbf{1}$ belong to exactly the same ideals in C. [1E.*1*.] Hence every ideal in C has a set of bounded generators.

2B. PRIME IDEALS.

1. An ideal P in C is prime if and only if $P \cap C^*$ is a prime ideal in C^*.

2. If P and Q are prime ideals in C, or in C^*, then $PQ = P \cap Q$ (by definition, the product IJ of two ideals is the smallest ideal containing all products fg, where $f \in I$ and $g \in J$). [If $f \in P$, then $f^{1/3} \in P$.] In particular, $P^2 = P$. Hence $M^2 = M$ for every maximal ideal M in C or C^*.

3. An ideal I in a commutative ring is an intersection of prime ideals if and only if $a^2 \in I$ implies $a \in I$.

2C. FUNCTIONS CONGRUENT TO CONSTANTS.

1. Let I be an ideal in C; if $f \equiv \mathbf{r} \pmod{I}$, then $r \in f[X]$.

2. Let I be an ideal in C^*; if $f \equiv \mathbf{r} \pmod{I}$, then $r \in \mathrm{cl}_{\mathbf{R}} f[X]$.

2D. z-IDEALS.

1. Let I be a z-ideal in C, and suppose that $f \equiv \mathbf{r} \pmod{I}$. If $g(x) = r$ wherever $f(x) = r$, then $g \equiv \mathbf{r} \pmod{I}$.

2. If $f^2 + g^2$ belongs to a z-ideal I, then $f \in I$ and $g \in I$.

3. If I and J are z-ideals, then $IJ = I \cap J$. Compare 2B.*2*.

4. $\mathbf{Z}[(I, J)]$ is the set of all $Z_1 \cap Z_2$, where $Z_1 \in \mathbf{Z}[I]$ and $Z_2 \in \mathbf{Z}[J]$.

2E. PRIME z-FILTERS.

The following are equivalent for a z-filter $\mathscr{F}$.

(1) $\mathscr{F}$ is prime.

(2) Whenever the union of two zero-sets is all of X, at least one of them belongs to $\mathscr{F}$.

(3) Given $Z_1, Z_2 \in \mathbf{Z}(X)$, there exists $Z \in \mathscr{F}$ such that one of $Z \cap Z_1$, $Z \cap Z_2$ contains the other.

2F. FINITE SPACES.

Let X be a finite discrete space. In $C(X)$:

1. f is a multiple of g if and only if $\mathbf{Z}(f) \supset \mathbf{Z}(g)$.

2. Every ideal is a z-ideal.

3. Every ideal is principal, and, in fact, is generated by an idempotent.

4. Every ideal is an intersection of maximal ideals. The intersection of all the maximal ideals is $(\mathbf{0})$.

5. Every prime ideal is maximal.

2G. PRIME VS. z-IDEALS IN $C(\mathbf{R})$.

1. Select a function l in $C(\mathbf{R})$ such that $l(0) = 0$, while $\lim_{x\to 0}|l^n(x)/x| = \infty$ for all $n \in \mathbf{N}$. Apply 0.17 to construct a prime ideal in $C(\mathbf{R})$ that contains $\mathbf{i}$ but not l. This prime ideal is not a z-ideal (and hence is not maximal).

2. Let $\boldsymbol{O}_0$ denote the ideal of all functions f in $C(\mathbf{R})$ for which $\mathbf{Z}(f)$ is a neighborhood of 0. Define s in $C(\mathbf{R})$ as follows: $s(x) = x \sin(\pi/x)$ for $x \neq 0$, and $s(0) = 0$. Then $(\boldsymbol{O}_0, s)$ is not a z-ideal; and the smallest z-ideal containing $(\boldsymbol{O}_0, s)$ is not prime.

2H. THE IDENTITY FUNCTION $\mathbf{i}$ IN $C(\mathbf{R})$.

1. The principal ideal $(\mathbf{i})$ in $C(\mathbf{R})$ consists precisely of all functions in $C(\mathbf{R})$ that vanish at 0 and have a derivative at 0. Hence every nonnegative function in $(\mathbf{i})$ has a zero derivative at 0.

2. $(\mathbf{i})$ is not a prime ideal; in fact, $(\mathbf{i})^2 \neq (\mathbf{i})$. (See 2B.*2.*)

3. The ideal $(\mathbf{i}, |\mathbf{i}|)$ is not principal. [If $(\mathbf{i}, |\mathbf{i}|) = (d)$, there exist $g, h \in C$ such that $\mathbf{i} = gd$ and $|\mathbf{i}| = hd$. It follows that $g(0) = h(0) = 0$. Moreover, there exist $s, t \in C$ such that $s\mathbf{i} + t|\mathbf{i}| = d$. This implies that $sg + th = \mathbf{1}$, a contradiction.]

4. Exhibit a principal ideal containing $(\mathbf{i}, |\mathbf{i}|)$.

2I. $C(\mathbf{Q})$ AND $C^*(\mathbf{Q})$.

The set of all f in $C(\mathbf{Q})$ for which $\lim_{x \to \pi} f(x) = 0$ is not an ideal in $C(\mathbf{Q})$. But the bounded functions in this set do constitute an ideal in $C^*(\mathbf{Q})$.

2J. IDEAL CHAINS IN $C(\mathbf{R})$, $C(\mathbf{Q})$, AND $C(\mathbf{N})$.

1. Find a chain of z-ideals in $C(\mathbf{R})$ (under set inclusion) that is in one-one, order-preserving correspondence with $\mathbf{R}$ itself.

2. Find a chain of z-ideals in $C(\mathbf{Q})$ in one-one, order-preserving correspondence with $\mathbf{R}$.

3. Do the same for $C(\mathbf{N})$.

2K. z-FILTERS AND C^*.

If M is a maximal ideal in C^*, and $Z[M]$ is a z-filter, then $Z[M]$ is a z-ultrafilter.

2L. e-FILTERS AND e-IDEALS.

This problem contains an outline for a theory of z-filters applicable to C^*. For $f \in C^*$ and $\epsilon > 0$, we define

$$E_\epsilon(f) = f^{\leftarrow}[[-\epsilon, \epsilon]] = \{x : |f(x)| \leqq \epsilon\}.$$

Every such set is a zero-set; conversely, every zero-set is of this form: $Z(g) = E_\epsilon(\epsilon + |g|)$. For $I \subseteq C^*$, we write

$$E(I) = \{E_\epsilon(f) : f \in I, \epsilon > 0\},$$

i.e., $E(I) = \bigcup_\epsilon E_\epsilon[I]$. Finally, for any family $\mathscr{F}$ of zero-sets, we define

$$E^{\leftarrow}(\mathscr{F}) = \{f \in C^* : E_\epsilon(f) \in \mathscr{F} \text{ for all } \epsilon > 0\},$$

that is, $E^{\leftarrow}(\mathscr{F}) = \bigcap_\epsilon E_\epsilon^{\leftarrow}[\mathscr{F}]$.

1. $\mathscr{F} \supset E(E^{\leftarrow}(\mathscr{F})) = \bigcup_\epsilon \{E_\epsilon(f) : E_\delta(f) \in \mathscr{F} \text{ for all } \delta > 0\}$. Note that the inclusion may be proper, even when $\mathscr{F}$ is a z-filter. [Let $\mathscr{F}$ be the z-filter of all zero-sets in $\mathbf{R}$ that contain 0.]

2. A z-filter $\mathscr{F}$ is called an *e-filter* if $\mathbf{E}(\mathbf{E}^{\leftarrow}(\mathscr{F})) = \mathscr{F}$. Hence $\mathscr{F}$ is an e-filter if and only if, whenever $Z \in \mathscr{F}$, there exist f and ϵ such that $\mathbf{E}_\delta(f) \in \mathscr{F}$ for every $\delta > 0$, and $Z = \mathbf{E}_\epsilon(f)$.

3. $I \subset \mathbf{E}^{\leftarrow}(\mathbf{E}(I)) = \{f: \mathbf{E}_\epsilon(f) \in \mathbf{E}(I) \text{ for all } \epsilon > 0\}$. Note that the inclusion may be proper, even when I is an ideal. [Let I be the ideal of all functions in $C^*(\mathbf{R})$ that vanish on a neighborhood of 0, and consider any function that vanishes precisely at 0. Alternatively, take $I = (\mathbf{j}^2)$ in $C^*(\mathbf{N})$, and consider the function $\mathbf{j}$.]

4. An ideal I in C^* is called an *e-ideal* if $\mathbf{E}^{\leftarrow}(\mathbf{E}(I)) = I$. Hence I is an e-ideal if and only if, whenever $\mathbf{E}_\epsilon(f) \in \mathbf{E}(I)$ for all $\epsilon > 0$, then $f \in I$. Intersections of e-ideals are e-ideals.

5. If I is an ideal in C^*, then $\mathbf{E}(I)$ is an e-filter. [Verify (i), (ii′) and (iii) of 2.2. For (iii), let $Z(f') \supset \mathbf{E}_\epsilon(f)$, where $f' \in C^*$ and $f \in I$, with $f' \geqq \mathbf{0}$ and $f \geqq \mathbf{0}$. Define $g(x) = 1$ for $x \in \mathbf{E}_\epsilon(f)$, and

$$g(x) = f'(x) + \epsilon/f(x)$$

for $x \notin \mathbf{E}_\epsilon(f)$. Then $g \in C^*$, and $Z(f') = \mathbf{E}_\epsilon(fg)$.] The corresponding result holds in C.

6. If $\mathscr{F}$ is any z-filter, then $\mathbf{E}^{\leftarrow}(\mathscr{F})$ is an ideal in C^*. Note, however, that the corresponding result may fail in C, even if $\mathscr{F}$ is an e-filter. [Let $\mathscr{F}$ consist of the complements of the finite sets in $\mathbf{N}$, and consider the function $\mathbf{j}$.]

7. $I \subset J$ implies $\mathbf{E}(I) \subset \mathbf{E}(J)$, and $\mathscr{F} \subset \mathscr{G}$ implies $\mathbf{E}^{\leftarrow}(\mathscr{F}) \subset \mathbf{E}^{\leftarrow}(\mathscr{G})$.

8. If J is an e-ideal, then $I \subset J$ if and only if $\mathbf{E}(I) \subset \mathbf{E}(J)$. If $\mathscr{F}$ is an e-filter, then $\mathscr{F} \subset \mathscr{G}$ if and only if $\mathbf{E}^{\leftarrow}(\mathscr{F}) \subset \mathbf{E}^{\leftarrow}(\mathscr{G})$.

9. If $\mathscr{F}$ is any e-filter, then $\mathbf{E}^{\leftarrow}(\mathscr{F})$ is an e-ideal. If I is any ideal in C^*, then $\mathbf{E}^{\leftarrow}(\mathbf{E}(I))$ is the smallest e-ideal containing I. In particular, *every maximal ideal in C^* is an e-ideal.*

10. For any z-filter $\mathscr{G}$, $\mathbf{E}(\mathbf{E}^{\leftarrow}(\mathscr{G}))$ is the largest e-filter contained in $\mathscr{G}$.

11. If $\mathscr{A}$ is a z-ultrafilter, and a zero-set Z meets every member of $\mathbf{E}(\mathbf{E}^{\leftarrow}(\mathscr{A}))$, then $Z \in \mathscr{A}$. [Theorems 2.6(b) and 1.15.]

12. A maximal e-filter is called an *e-ultrafilter*. Every e-filter is contained in an e-ultrafilter.

13. If M^* is a maximal ideal in C^*, then $\mathbf{E}(M^*)$ is an e-ultrafilter; and if $\mathscr{E}$ is an e-ultrafilter, then $\mathbf{E}^{\leftarrow}(\mathscr{E})$ is a maximal ideal in C^*. [*9.*] Hence the correspondence $M^* \to \mathbf{E}(M^*)$ is one-one from the set of all maximal ideals in C^* onto the set of all e-ultrafilters.

14. The following property characterizes an ideal M^* in C^* as a maximal ideal: given $f \in C^*$, if every $\mathbf{E}_\epsilon(f)$ meets every member of $\mathbf{E}(M^*)$, then $f \in M^*$. [$(M^*, f) = C^*$ if and only if some $\mathbf{E}_\epsilon(f)$ fails to meet some member of $\mathbf{E}(M^*)$.]

15. If $\mathscr{A}$ is a z-ultrafilter, then it is the unique z-ultrafilter containing $\mathbf{E}(\mathbf{E}^{\leftarrow}(\mathscr{A}))$. [*11.*] Moreover, $\mathbf{E}(\mathbf{E}^{\leftarrow}(\mathscr{A}))$ is an e-ultrafilter, and it is the unique one contained in $\mathscr{A}$. [*10.*] Hence the correspondence $\mathscr{A} \to \mathbf{E}(\mathbf{E}^{\leftarrow}(\mathscr{A}))$ is one-one from the set of all z-ultrafilters onto the set of all e-ultrafilters.

16. If $\mathscr{A}$ is a z-ultrafilter, then $E^{-}(\mathscr{A})$ is the maximal ideal

$$E^{-}(E(E^{-}(\mathscr{A})))$$

in C^*. Hence *the correspondence*

$$M \rightarrow E^{-}(Z[M])$$

is one-one from the set of all maximal ideals in C onto the set of all maximal ideals in C^*. Its inverse is the correspondence $M^* \rightarrow Z^{\leftarrow}[\mathscr{A}]$, where $\mathscr{A}$ is the unique z-ultrafilter containing the e-ultrafilter $E(M^*)$.

2M. THE UNIFORM NORM TOPOLOGY ON C^*.

Let C' be a subring of $C(X)$ on which a topology has been defined. Then C' is called a topological ring if both subtraction and ring multiplication are continuous (from $C' \times C'$ into C'). If C' contains the constant functions, then it is a topological vector space if both addition and scalar multiplication (the latter being the mapping $(r, g) \rightarrow rg$ from $\mathbf{R} \times C'$ into C') are continuous. If C' is both a topological ring and a topological vector space, it is called a topological algebra.

By a norm is meant a mapping $f \rightarrow \|f\|$ into $\mathbf{R}$, satisfying: $\|f\| \geqq 0$, $\|f\| = 0$ if and only if $f = \mathbf{0}$, $\|f + g\| \leqq \|f\| + \|g\|$, and $\|rf\| = |r| \cdot \|f\|$. A metric d is defined from the norm, as usual, by: $d(f, g) = \|f - g\|$. A Banach algebra is a complete normed algebra whose norm satisfies: $\|fg\| \leqq \|f\| \cdot \|g\|$.

1. In any topological ring, the closure of an ideal is either an ideal or the whole ring.

2. A norm on C^* is given by: $\|f\| = \sup_{x \in X} |f(x)|$. The resulting metric topology is called the *uniform norm topology* on C^*. Convergence in this topology is uniform convergence of the functions. A base for the neighborhood system at g consists of all sets of the form

$$\{f : |g - f| \leqq \epsilon\} \qquad (\epsilon > 0).$$

Equivalently, a base at g is given by all sets

$$\{f : |g - f| \leqq u\},$$

where u is a positive unit of C^*.

3. C^* is a Banach algebra.

4. The closure of every ideal is a (proper) ideal. [If $\mathbf{1} \in \operatorname{cl} I$, then I contains a unit.] Hence every maximal ideal is closed.

5. Every e-ideal (2L) is closed. (Hence every maximal ideal is closed.) [Given $g \in \operatorname{cl} I$, and $\epsilon > 0$, there exists $f \in I$ such that $|g - f| \leqq \epsilon$. Then $E_{2\epsilon}(g) \supset E_{\epsilon}(f)$.] (It will be seen subsequently (6A.*2*) that every closed ideal is an intersection of maximal ideals. It follows that the closed ideals are precisely the e-ideals [2L.*4*].)

6. The topology of uniform convergence can also be defined on C, the neighborhood system at g being as described in *2*. However, C will not be either a topological ring or a topological vector space unless X is pseudocompact.

2N. THE m-TOPOLOGY ON C.

The *m-topology* is defined on $C(X)$ by taking as a base for the neighborhood system at g all sets of the form

$$\{f \in C: |g - f| \leqq u\},$$

where u is a positive unit of C. The same topology results if it is required further that u be a bounded function.

1. C is a topological ring.

2. The relative m-topology on C^* contains the uniform norm topology (2M), and the two coincide if and only if X is pseudocompact. In fact, when X is not pseudocompact, the set of constant functions in C^* is discrete (in the m-topology), so that C^* is not even a topological vector space.

3. The set of all units of C is open, and the mapping $f \to f^{-1}$ is a homeomorphism of this set onto itself.

4. The subring C^* is closed.

5. The closure of every ideal is a (proper) ideal. Hence every maximal ideal is closed. Every maximal ideal in C^* is closed.

6. Every closed ideal in C is a z-ideal. [Given $\mathbf{Z}(f) = \mathbf{Z}(g)$, with $g \in I$, and given u, define $h(x) = 0$ for $|f(x)| \leqq u(x)$, and

$$h(x) = \frac{f(x) \pm u(x)}{g(x)}$$

otherwise.]

7. In the ring $C(\mathbf{R})$, the z-ideal $\boldsymbol{O}_0$ of all functions that vanish on a neighborhood of 0 is not closed. [$\mathbf{i} \in \operatorname{cl} \boldsymbol{O}_0$.]

Chapter 3

COMPLETELY REGULAR SPACES

3.1. Up to this point in the text, we have not assumed any separation axioms for the topological space on which our ring of continuous functions is defined. Indeed, separation axioms were irrelevant to most of the subjects discussed. We have now reached the stage where separation properties of the space do enter in an essential way, so that we are forced to make a decision about what class or classes of spaces to consider. We have no desire to become involved in finding the weakest axiom under which each theorem can be proved, but prefer, if possible, to stick to a single class of topological spaces that is wide enough to include all of the interesting spaces, and, at the same time, restrictive enough to admit a significant theory of rings of continuous functions.

The class of *completely regular* spaces exactly fulfills this requirement. A space X is said to be completely regular provided that it is a Hausdorff space such that, whenever F is a closed set and x is a point in its complement, there exists a function $f \in C(X)$ such that $f(x) = 1$ and $f[F] = \{0\}$—in short, F and $\{x\}$ are completely separated. A simple but important consequence is that every subspace of a completely regular space is completely regular. Another is this: in a completely regular space, if $f(x) = f(y)$ for all $f \in C$, then $x = y$.

It is obvious that every metric space is completely regular. In particular, $\mathbf{R}$ and all its subspaces are completely regular.

Since complete regularity is defined in terms of the existence of continuous functions, it is not surprising that it should be a useful concept in our study. What is remarkable is that completely regular spaces have so many other desirable properties. For example, they are precisely the subspaces of compact spaces (Theorems 3.14 and 6.5); and they are precisely the spaces that admit Hausdorff uniform structures (Theorem 15.6).

Normal spaces have additional properties that are useful in the study of rings of continuous functions. In view of this, it is remarkable how little is gained by imposing upon a completely regular space X the stronger condition of normality. The contents of this book will attest to this fact—though it must be admitted that a good many of the results to be presented were first proved for normal spaces.

We can disclose here the system for bypassing the additional hypothesis that X be normal. The fundamental theorem about normal spaces is Urysohn's lemma (3.13), which states that any two disjoint closed sets in a normal space are completely separated. In our work, this result is applied essentially once—to yield the vital theorem that every *compact* space is completely regular.

In dealing with arbitrary completely regular spaces, the key device is Theorem 1.15, which states that in *any* space, disjoint *zero-sets* are completely separated. What makes this theorem so serviceable is that every closed set in a completely regular space is an intersection of zero-sets (Theorem 3.2).

It is well to point out two basic differences between Urysohn's lemma and Theorem 1.15. One concerns their content. Urysohn's lemma stands alone as a theorem whose conclusion asserts the existence of a continuous function, but whose hypothesis provides no functions to work from: in the proof, a function is constructed from "nothing."

The second point concerns our application of these theorems. As we have indicated, Urysohn's lemma is indispensable—but will rarely be referred to. For work with completely regular spaces, it is replaced by Theorem 1.15, which will be invoked over and over again (often without explicit mention). Notice that Urysohn's lemma is not replaced by Theorem 1.15 alone, but by the theorem *as applied to completely regular spaces*. Complete separation, in terms of existing functions, is provided by the theorem; existence of the functions in the first place is built into the *definition* of complete regularity.

Combination of Urysohn's lemma with Urysohn's extension theorem (1.17) yields the conclusion that in a normal space, every closed set is C^*-embedded. In the absence of normality, the last result can often be replaced by the fact that every *compact* set in a completely regular space is C^*-embedded (3.11(c)). The effectiveness of this device is enhanced by the existence of a compactification of an arbitrary completely regular space X, in which X is C^*-embedded (Theorem 6.5). Incidentally, section 3.11 provides a good illustration of how Theorem 1.15 is used.

That nothing can be achieved by considering a *wider* class than the completely regular spaces is the content of Theorem 3.9.

3.2. It turns out that most of our topological considerations will be expressed more conveniently in terms of closed sets than, as is more commonly the case, in terms of open sets. The next theorem shows why.

A collection $\mathscr{B}$ of closed sets is a *base* for the closed sets if every closed set in X is an intersection of members of $\mathscr{B}$. Equivalently, $\mathscr{B}$ is a base if whenever F is closed and $x \in X - F$, there is a member of $\mathscr{B}$ that contains F but not x.

THEOREM. *A Hausdorff space X is completely regular if and only if the family $\boldsymbol{Z}(X)$ of all zero-sets is a base for the closed sets.*

PROOF. *Necessity.* Suppose that X is completely regular. Then whenever F is a closed set and $x \in X - F$, there exists $f \in C(X)$ such $f(x) = 1$ and $f[F] = \{0\}$. Then $\boldsymbol{Z}(f) \supset F$, and $x \notin \boldsymbol{Z}(f)$. Consequently, $\boldsymbol{Z}(X)$ is a base.

Sufficiency. Suppose that $\boldsymbol{Z}(X)$ is a base. Then, whenever F is a closed set and $x \in X - F$, there is a zero-set, say $\boldsymbol{Z}(g)$, such that $\boldsymbol{Z}(g) \supset F$ and $x \notin \boldsymbol{Z}(g)$. Write $r = g(x)$. Then $r \neq 0$, and the function $f = gr^{-1}$ belongs to $C(X)$. Evidently, $f(x) = 1$ and $f[F] = \{0\}$. Therefore the Hausdorff space X is completely regular.

As a matter of fact, we have, as in Theorem 1.15:

(a) *Every closed set F in a completely regular space is an intersection of zero-set-neighborhoods of F.*

(b) *Every neighborhood of a point in a completely regular space contains a zero-set-neighborhood of the point.*

3.3. *Weak topology.* It is a triviality that the continuous functions on X to $\mathbf{R}$ are determined by the topology of X. The foregoing theorem says, in effect, that if X is completely regular, then the converse is also true: its topology is determined by the continuous real-valued functions.

This last statement can be made precise by introducing the notion of *weak topology.* Let X now be an abstract set, and consider an arbitrary subfamily C' of $\mathbf{R}^X$. The *weak topology induced by C' on X* is defined to be the smallest topology on X such that all functions in C' are continuous.

Let us see what this means. In order that a function f on X to $\mathbf{R}$ be continuous, it is necessary and sufficient that the preimage under f of each open set in $\mathbf{R}$ be open. Hence in order that every function in C' be continuous, it is necessary and sufficient that all such preimages, for all $f \in C'$, be open. Let $\mathscr{S}$ denote the collection of all these preimages: a subset U of X belongs to $\mathscr{S}$ if and only if there exist $f \in C'$, and an open set V in $\mathbf{R}$, such that $U = f^{\leftarrow}[V]$. In particular, we always have $X \in \mathscr{S}$ (if C' is not empty), since $X = f^{\leftarrow}[\mathbf{R}]$ for any $f \in C'$.

But $\mathscr{S}$ need not be a topology for X—or even a base for a topology. For example, if X is a three-element set $\{a, b, c\}$, and $C' = \{f, g\}$, where $f(a) = g(b) = 0$ and $f(b) = f(c) = g(a) = g(c) = 1$, then $\{c\}$ is the intersection of two members of $\mathscr{S}$ but contains no nonempty member of $\mathscr{S}$.

The weak topology generated by C' is the smallest topology containing the family $\mathscr{S}$. Therefore, it is the topology for which $\mathscr{S}$ is a subbase.

3.4. To obtain the weak topology, it is not necessary to consider the preimages of all the open sets in $\mathbf{R}$: the preimages of basic open sets—in fact, of subbasic open sets—already constitute a subbase for the weak topology.

If we consider the base for $\mathbf{R}$ consisting of all the ϵ-neighborhoods, then we see that a subbasic system of neighborhoods for a point x in X is given by all sets of the form

(a) $$\{y \in X : |f(x) - f(y)| < \epsilon\} \qquad (f \in C', \epsilon > 0).$$

Dually, we may work with closed sets. A family is a *subbase* for the closed sets if the finite unions of its members constitute a base. Since the closed rays form a subbase for the closed sets in $\mathbf{R}$, their preimages,

(b) $$\{x \in X : f(x) \geqq r\},$$

and

(b′) $$\{x \in X : f(x) \leqq r\} \qquad (f \in C', r \in \mathbf{R}),$$

form a subbase for the closed sets in X. In case $-f$ belongs to C' whenever f does—for example, if C' is an additive group—then the sets in (b′) are the same as those in (b).

3.5. Suppose, now, that X is given as a topological space. A natural undertaking is to compare its topology with the weak topology induced by some family C' of functions. When the weak topology turns out to coincide with the given one, we shall say that C' *determines* the topology of the space.

If $C' \subset C$, then every function in C' is continuous in the given topology (by definition of C). Therefore the weak topology is contained in the given one.

In case $C' = C$ or $C' = C^*$, the sets (b) coincide with the zero-sets of functions in C'. In fact,

$$\{x : f(x) \geqq r\} = \boldsymbol{Z}((f - \boldsymbol{r}) \wedge \mathbf{0}),$$

so that every such set is a zero-set. Conversely, every zero-set has this form:

$$Z(f) = \{x\colon -|f|(x) \geqq 0\}.$$

Since the union of two zero-sets is again a zero-set, the subbase $\mathbf{Z}[C] = \mathbf{Z}[C^*]$ is a *base.* We summarize all this in the following theorem, which includes a reformulation of Theorem 3.2.

3.6. THEOREM. *Let X be a topological space.*

The families $C(X)$ and $C^*(X)$ induce the same weak topology on X. A base for its closed sets is the family $\mathbf{Z}(X)$ of all zero-sets. A basic neighborhood system for a point x is given by the collection of all sets

$$\{y \in X\colon |f(x) - f(y)| < \epsilon\} \qquad (f \in C^*, \epsilon > 0).$$

Finally, if X is a Hausdorff space, then X is completely regular if and only if its topology coincides with the weak topology induced by C and C^* (i.e., its topology is determined by C and C^*).

3.7. This last result can be sharpened still further.

THEOREM. *If X is a Hausdorff space whose topology is determined by some subfamily C' of $\mathbf{R}^X$, then X is completely regular.*

PROOF. Clearly, every function in C' is continuous, i.e., $C' \subset C(X)$. Hence the weak topology induced by C' is contained in the weak topology induced by C. But the latter topology is always contained in the given topology on the space X. The hypothesis now implies that the two coincide, and so, by Theorem 3.6, X is completely regular.

3.8. Since the continuous functions determine the topology of a completely regular space, they determine the continuous mappings into the space. Precisely:

THEOREM. *Let C' be a subfamily of $C(Y)$ that determines the topology of Y. A mapping σ from a space S into Y is continuous if and only if the composite function $g \circ \sigma$ is in $C(S)$ for every $g \in C'$.*

PROOF. Necessity is obvious. To prove the sufficiency—that σ is continuous—we look at what happens to subbasic closed sets in Y under $\sigma^{\leftarrow}$. These are given, by hypothesis, as the sets of the form $g^{\leftarrow}[F]$, where $g \in C'$, and F is a closed set in $\mathbf{R}$. Now,

$$\sigma^{\leftarrow}[g^{\leftarrow}[F]] = (g \circ \sigma)^{\leftarrow}[F];$$

and this set is closed in S, since, by hypothesis, $g \circ \sigma$ is continuous. Therefore σ is continuous.

3.9. The next theorem eliminates any reason for considering rings of continuous functions on other than completely regular spaces.

THEOREM. *For every topological space X, there exist a completely regular space Y and a continuous mapping τ of X onto Y, such that the mapping $g \to g \circ \tau$ is an isomorphism of $C(Y)$ onto $C(X)$.*

PROOF. Define $x \equiv x'$ in X to mean that $f(x) = f(x')$ for every $f \in C(X)$. Evidently, this is an equivalence relation. Let Y be the set of all equivalence classes. We define a mapping τ of X onto Y as follows: τx is the equivalence class that contains x.

With each $f \in C(X)$, associate a function $g \in \mathbf{R}^Y$ as follows: $g(y)$ is the common value of $f(x)$ at every point $x \in y$. Thus, $f = g \circ \tau$. Let C' denote the family of all such functions g, i.e., $g \in C'$ if and only if $g \circ \tau \in C(X)$. Now endow Y with the weak topology induced by C'. By definition, every function in C' is continuous on Y, i.e., $C' \subset C(Y)$. The continuity of τ now follows from Theorem 3.8.

It is evident that if y and y' are distinct points of Y, then there exists $g \in C'$ such that $g(y) \neq g(y')$. This proves that Y is a Hausdorff space. Hence Y is completely regular, by Theorem 3.7.

Finally, consider any function $h \in C(Y)$. Since τ is continuous, $h \circ \tau$ is continuous on X. But this says that $h \in C'$. Therefore, $C' \supset C(Y)$. Thus, $C' = C(Y)$; and it is clear that the mapping $g \to g \circ \tau$ is an isomorphism. This completes the proof of the theorem.

It is equally clear that the mapping $g \to g \circ \tau$ is a lattice isomorphism as well, and that it carries $C^*(Y)$ onto $C^*(X)$. These conclusions also follow from Theorems 1.6 and 1.9.

We remark that τ is not necessarily a quotient mapping, i.e., the topology on Y need not be the largest such that τ is continuous; see 3I.*2* or 3J.*3*.

As a consequence of the foregoing theorem, algebraic or lattice properties that hold for all $C(X)$ [resp. $C^*(X)$], with X completely regular, hold just as well for all $C(X)$ [resp. $C^*(X)$], with X arbitrary. An example is the result that every residue class ring modulo a prime ideal is totally ordered (Theorem 5.5). We shall make no systematic attempt to distinguish between those results that are valid only for completely regular spaces and those of more general validity.

Beginning with Chapter 4, we shall impose the blanket assumption of complete regularity on all given spaces.

3.10. *Products of completely regular spaces.* We have defined the weak topology induced by a family of real-valued functions. More generally, let X be any set, and let Φ be an arbitrary family of mappings φ_α of X into topological spaces Y_α; then the weak topology induced by Φ on X is, by definition, the smallest topology in which each φ_α is continuous. Again, when the weak topology coincides with a given topology on X, we say that the latter is determined by Φ. Here, the spaces Y_α need not be all the same. In case each Y_α is completely regular, the family of all sets $\varphi_\alpha^{\leftarrow}[Z_\alpha]$, where Z_α is a zero-set in Y_α, and $\varphi_\alpha \in \Phi$, is a subbase for the closed sets in X.

A particular application is to product spaces. The product topology on $X = \bigtimes_\alpha X_\alpha$ may now be defined as the weak topology induced by the family of all projections $\pi_\alpha\colon X \to X_\alpha$. When each X_α is completely regular, the collection of all finite unions

$$\pi_{\alpha_1}^{\leftarrow}[Z_1] \cup \cdots \cup \pi_{\alpha_n}^{\leftarrow}[Z_n],$$

where Z_k is a zero-set in X_{α_k}, is a base for the closed sets in X. We notice that each such union is a zero-set, because

$$\pi_\alpha^{\leftarrow}[\mathbf{Z}_{X_\alpha}(f)] = \mathbf{Z}_X(f \circ \pi_\alpha).$$

We conclude that

(a) *An arbitrary product of completely regular spaces is completely regular*

(taking note of the simple fact that a product of Hausdorff spaces is a Hausdorff space).

If we examine the proof of Theorem 3.8, we find that it does not depend upon any special properties of **R**. Thus we have, more generally:

(b) *Let Φ be a family of mappings that determines the topology of a space X. A mapping σ from a space S into X is continuous if and only if $\varphi \circ \sigma$ is continuous for every $\varphi \in \Phi$.*

When X is given as a product space, this assumes the following familiar form:

(c) *A mapping σ from a space into a product $X = \bigtimes_\alpha X_\alpha$ is continuous if and only if $\pi_\alpha \circ \sigma$ is continuous for each projection π_α.*

3.11. *Complete separation of compact sets.* We recall that a Hausdorff space is said to be *compact* provided that every family of closed sets with the finite intersection property has nonempty intersection.

The separation properties in a completely regular space yield the following fundamental results.

(a) *In a completely regular space, any two disjoint closed sets, one of which is compact, are completely separated.*

Suppose that A and A' are disjoint closed sets, with A compact. For each $x \in A$, choose disjoint zero-sets Z_x and Z'_x, with Z_x a neighborhood of x, and $Z'_x \supset A'$. The cover $\{Z_x\}_{x \in A}$ of the compact set A has a finite subcover, say

$$\{Z_{x_1}, \cdots, Z_{x_n}\}.$$

Then A and A' are contained in the disjoint zero-sets

$$Z_{x_1} \cup \cdots \cup Z_{x_n} \quad \text{and} \quad Z'_{x_1} \cap \cdots \cap Z'_{x_n},$$

respectively.

(b) *In a completely regular space, every G_δ containing a compact set S contains a zero-set containing S.*

A G_δ-set A has the form $\bigcap_{n \in \mathbf{N}} U_n$, where each U_n is open. If $A \supset S$, then S is completely separated from $X - U_n$, and so there is a zero-set F_n satisfying $S \subset F_n \subset U_n$. Then

$$S \subset \bigcap_n F_n \subset A;$$

and $\bigcap_n F_n$, as a countable intersection of zero-sets, is a zero-set.

In particular, *every compact G_δ* in a completely regular space *is a zero-set.* Special case: every *G_δ-point* is a zero-set.

Compact is an absolute topological concept, not relative (like *closed*): a compact space is compact in any embedding. Let S be a compact subspace of a completely regular space X. Completely separated sets in S have disjoint closures in S. As these closures are compact, they are, by (a), completely separated *in* X. Urysohn's extension theorem (1.17) now yields:

(c) *Every compact set in a completely regular space is C-embedded.*

3.12. *Normal spaces.* A Hausdorff space X is said to be *normal* provided that any two disjoint closed sets have disjoint neighborhoods. Thus, X is normal if and only if every neighborhood of a closed set contains a closed neighborhood of the set. The crucial result about normal spaces is Urysohn's lemma, which states that disjoint closed sets are completely separated. We begin with the following preliminary result.

LEMMA. *Let X be an arbitrary space, and let R_0 be any dense subset of the real line $\mathbf{R}$. Suppose that open sets U_r of X are defined, for all $r \in R_0$, such that*

$$\bigcup_r U_r = X, \qquad \bigcap_r U_r = \emptyset,$$

and

$$\operatorname{cl} U_r \subset U_s \quad \textit{whenever} \quad r < s.$$

Then the formula

$$f(x) = \inf\{r \in R_0 : x \in U_r\} \qquad (x \in X)$$

defines f as a continuous function on X.

PROOF. The hypotheses regarding union and intersection imply that $f(x)$ is well defined as a real number. Evidently, $x \in U_r$ implies $f(x) \leqq r$, and $f(x) < r$ implies $x \in U_r$. Also, $x \in \text{cl}\, U_r$ implies $x \in U_s$ for all $s > r$, so that $f(x) \leqq r$. Now, fix a in X. Since R_0 is dense in $\mathbf{R}$, the intervals $[r, s]$, where $r, s \in R_0$ and $r < f(a) < s$, form a base for the neighborhoods of $f(a)$. The preceding remarks show that for any such r and s, $U_s - \text{cl}\, U_r$ is a neighborhood of a, and that $f(x) \in [r, s]$ for every x in that neighborhood. Therefore f is continuous at a.

3.13. URYSOHN'S LEMMA. *Any two disjoint closed sets in a normal space are completely separated. Hence every normal space is completely regular.*

PROOF. Let A and B be disjoint closed sets in a normal space X. We define open sets U_r, for all rational r, as follows.

First, take $U_r = \emptyset$ for all $r < 0$, and $U_r = X$ for $r > 1$.

Next, put $U_1 = X - B$; then U_1 is a neighborhood of A. Since X is normal, U_1 contains a closed neighborhood of A; we choose U_0 (open) so that $A \subset U_0$ and $\text{cl}\, U_0 \subset U_1$.

Now enumerate the rationals in $[0, 1]$ in a sequence $(r_n)_{n \in \mathbf{N}}$, with $r_1 = 1$ and $r_2 = 0$. Inductively, for each $n > 2$, we choose U_{r_n} (open) so that $\text{cl}\, U_{r_k} \subset U_{r_n}$ and $\text{cl}\, U_{r_n} \subset U_{r_l}$ whenever $r_k < r_n < r_l$ and $k, l < n$.

The sets U_r $(r \in \mathbf{Q})$ satisfy the hypotheses of Lemma 3.12. Clearly, the continuous function f provided by the lemma is equal to 0 on A, and to 1 on B.

In a nonnormal space, two closed sets with disjoint neighborhoods can fail to be completely separated; see 8J.*4* or 8L.*5*.

3.14. *Compact spaces; compactification.*

THEOREM. *Every subspace of a compact space is completely regular.*

PROOF. Corollary 0.11 shows that a compact space is normal. By Urysohn's lemma, it is completely regular, and therefore all its subspaces are completely regular.

By a *compactification* of a space X, we shall mean a compact space in which X is dense. Thus, if X is already compact, it is its only compactification.

We have referred to the fact that the completely regular spaces are precisely the subsets of compact spaces. What amounts to the same thing: they are just those spaces that have compactifications—for, if T is a compact space containing X, then $\text{cl}_T X$ is a compactification of X. We have just proved that every subspace of a compact space is completely regular. The converse result—that every completely regular space does have a compactification—will be proved in detail later (Chapter 6 or Chapter 11).

3.15. *Locally compact spaces.* A Hausdorff space is said to be *locally compact* provided that every point has a compact neighborhood; it follows that every neighborhood of a point contains a compact neighborhood of the point. Every locally compact, noncompact space X has a so-called *one-point compactification* X^*, defined as follows: one new point is adjoined to X, X is an open subspace of X^*, and the complements of compact subsets of X form a base of neighborhoods for the adjoined point. One verifies without difficulty that X^* is indeed a compactification of X. Thus, every locally compact space is completely regular.

Let X be a subspace of a Hausdorff space T.

(a) *If T is locally compact, and X is open in T, then X is locally compact.*

Indeed, for each $x \in X$, the neighborhood X of x contains a compact neighborhood of x.

(b) *If X is dense in T, then every compact neighborhood in X of a point $p \in X$ is a neighborhood in T of p.*

Let U be the interior of a compact neighborhood of p in X. Then $\text{cl}_X U$ is compact, hence closed in T, so that $\text{cl}_T U = \text{cl}_X U$. Let V be an open set in T such that $V \cap X = U$. Since X is dense, we have $\text{cl}_T V = \text{cl}_T U \subset X$, so that $V = U$.

This has the following corollaries.

(c) *If X is dense in T, and p is an isolated point of X, then p is isolated in T.*

(d) *If X is locally compact and dense in T, then X is open in T.*

CONVERGENCE OF z-FILTERS

3.16. The remainder of this chapter contains an outline of a theory of convergence of z-filters on a completely regular space. It is analogous to the standard theory of convergence of filters or filter bases on an arbitrary Hausdorff space.

Let X be a completely regular space. A point $p \in X$ is said to be a *cluster point* of a z-filter $\mathscr{F}$ if every neighborhood of p meets every member of $\mathscr{F}$. Thus, since the members of $\mathscr{F}$ are closed sets, p is a cluster point of $\mathscr{F}$ if and only if $p \in \bigcap \mathscr{F}$.

If S is a nonempty subset of X, then cl S is the set of all cluster points of the z-filter $\mathscr{F}$ of all zero-sets containing S, because the zero-sets in the completely regular space X form a base for the closed sets.

The z-filter $\mathscr{F}$ is said to *converge* to the *limit* p if every neighborhood of p contains a member of $\mathscr{F}$. Obviously, if $\mathscr{F}$ converges to p, then p is a cluster point of $\mathscr{F}$. We recall that in the completely regular space X, every neighborhood of p contains a zero-set-neighborhood of p (3.2(b)). Thus:

(a) *$\mathscr{F}$ converges to p if and only if $\mathscr{F}$ contains the z-filter of all zero-set-neighborhoods of p.*

Examples of z-filters on $\mathbf{R}$ that converge to 0 are provided by the families of all zero-sets Z in $\mathbf{R}$ satisfying the respective conditions: (i): Z is a neighborhood of 0; (ii): there exists $\epsilon > 0$ such that Z contains the interval $[0, \epsilon]$; (iii): there exists $\epsilon > 0$ such that Z contains $[-\epsilon, 0]$; (iv): $1/n \in Z$ for all but finitely many $n \in \mathbf{N}$; (v): $0 \in Z$. By 2.4 (or 3.18(b)), the last of these is a z-ultrafilter, and it is the only z-ultrafilter that converges to 0.

(b) *If p is a cluster point of $\mathscr{F}$, then at least one z-ultrafilter containing $\mathscr{F}$ converges to p.*

Let $\mathscr{E}$ denote the z-filter of all zero-set-neighborhoods of p. Then $\mathscr{F} \cup \mathscr{E}$ has the finite intersection property, and so it is embeddable in a z-ultrafilter $\mathscr{A}$. Since $\mathscr{A}$ contains $\mathscr{E}$, it converges to p.

In particular, a z-ultrafilter converges to any cluster point.

3.17. If $\mathscr{F}$ converges to p in a completely regular space, then $\bigcap \mathscr{F} = \{p\}$. (Thus, a z-filter has at most one limit.) The converse is not true. For example, let $\mathscr{F}$ consist of all subsets of $\mathbf{N}$ that contain the point 1, and whose complements are finite; then $\bigcap \mathscr{F} = \{1\}$, although $\mathscr{F}$ does not converge to 1.

The converse is valid, however, in the case of a z-ultrafilter, as we have seen. More generally, it holds for any prime z-filter:

THEOREM. *Let X be a completely regular space, let $p \in X$, and let $\mathscr{F}$ be a prime z-filter on X. The following are equivalent.*

(1) *p is a cluster point of $\mathscr{F}$.*

(2) *$\mathscr{F}$ converges to p.*

(3) $\bigcap \mathscr{F} = \{p\}$.

PROOF. It suffices to show that (1) implies (2). Let V be any zero-set-neighborhood of p. Since X is completely regular, V contains a neighborhood of p of the form $X - Z$, where Z is a zero-set. Since $V \cup Z = X$, either V or Z belongs to the prime z-filter $\mathscr{F}$. But Z cannot belong to $\mathscr{F}$, because $p \notin Z$. So $V \in \mathscr{F}$. Thus, $\mathscr{F}$ converges to p (3.16(a)).

3.18. It follows that in a completely regular space, a prime z-filter, and, in particular, a z-ultrafilter, can have at most one cluster point. It need not have any. For instance, the family of all zero-sets in **R** whose complements are bounded is a z-filter without a cluster point. Obviously, any z-ultrafilter containing it (and there are such, by the maximal principle) also has no cluster point.

The family of all zero-sets containing a given point p is denoted by A_p. Obviously, A_p is a z-filter. Because any zero-set not containing p is completely separated from $\{p\}$, A_p is actually a z-ultrafilter. It follows from Theorem 3.17 that the z-ultrafilters A_p $(p \in X)$ are precisely the convergent ones on X. We shall now prove this directly.

(a) *p is a cluster point of a z-filter $\mathscr{F}$ if and only if $\mathscr{F} \subset A_p$.*

For, p is a cluster point of $\mathscr{F}$ if and only if p belongs to every member of $\mathscr{F}$. Immediate consequences of this proposition are:

(b) *A_p is the unique z-ultrafilter converging to p.*

It is to be noted that $\{p\}$ need not be a zero-set and hence need not belong to A_p; see 4N.*1*.

(c) *Distinct z-ultrafilters cannot have a common cluster point.*

Any z-ultrafilter containing a z-filter converging to p also converges to p. Hence

(d) *If $\mathscr{F}$ is a z-filter converging to p, then A_p is the unique z-ultrafilter containing $\mathscr{F}$.*

By definition, a point p is a cluster point of a *filter* $\mathscr{F}$ if every neighborhood of p meets every member of $\mathscr{F}$; and $\mathscr{F}$ converges to the limit p if it contains the filter of all neighborhoods of p. In contrast to (c), distinct *ultrafilters* can have a common cluster point. On the one-point compactification

$$\mathbf{N}^* = \mathbf{N} \cup \{\omega\}$$

of **N**, let $\mathscr{F}$ be the filter of all sets that contain all but a finite number of the even integers, and $\mathscr{F}'$ those containing all but finitely many odd integers. Any ultrafilters $\mathscr{U}$ and $\mathscr{U}'$ containing $\mathscr{F}$ and $\mathscr{F}'$, respectively, are distinct, but both converge to ω. Note that $\mathscr{F}$ and $\mathscr{F}'$ are not z-filters on $\mathbf{N}^*$: their only members that are zero-sets are those containing ω.

PROBLEMS

3A. ZERO-DIVISORS, UNITS, SQUARE ROOTS.

Let X be a completely regular space containing more than one point.

1. $C^*(X)$, and hence $C(X)$, contains zero-divisors (i.e., it is not an integral domain).

2. C^*, and hence C, contains nonconstant units.

3. Let $\mathfrak{m}$ be an infinite cardinal, and let X be the one-point compactification of the discrete space of power $\mathfrak{m}$. In $C(X)$, $\mathbf{1}$ has just $\mathfrak{m}$ square roots.

3B. COUNTABLE SETS.

Let X be a completely regular space.

1. A countable set disjoint from a closed set F is disjoint from some zero-set containing F.

2. A C-embedded countable set S is completely separated from every disjoint closed set. [Theorem 1.18.] (This is false if S is uncountable (5.13) or if S is only C^*-embedded (4M), even if S is closed (8.20 and 6P).)

3. Any C-embedded countable set is closed. [Apply *2* to each point not in the set.] (An uncountable C-embedded set need not be closed (5.13); the appropriate generalization is in 8A.*1*.)

4. Any two countable sets, neither of which meets the closure of the other, are contained in disjoint cozero-sets. [Inductively, choose a suitable closed neighborhood of each point, alternating between the two sets.] (But the given sets need not be completely separated, even if they are closed; see 8J.*4*.)

5. A countable, completely regular space is normal. (More generally, see 3D.*4*.)

3C. G_δ-POINTS OF A COMPLETELY REGULAR SPACE.

Let p be a G_δ-point of a completely regular space X, and let $S = X - \{p\}$.

1. If $g \in C^*(S)$, $h \in C(X)$, and $h(p) = 0$, then $g \cdot (h|S)$ has a continuous extension to all of X.

2. If Z is a zero-set in S, then $\text{cl}_X Z$ is a zero-set in X. [Let $Z = \mathbf{Z}(f)$, with $\mathbf{0} \leqq f \leqq \mathbf{1}$. Let $\{p\} = \mathbf{Z}(h)$, with $\mathbf{0} \leqq h \leqq \mathbf{1}$ and $h[Z] = \{1\}$ in case $p \notin \text{cl } Z$. Consider $g = f$ or $g = \mathbf{1} - f$.]

3D. NORMAL SPACES.

1. The following are equivalent for any Hausdorff space X.

(1) *X is normal.*

(2) *Any two disjoint closed sets are completely separated.*

(3) *Every closed set is C^*-embedded.*

(4) *Every closed set is C-embedded.*

[Urysohn's extension theorem and Theorem 1.18.]

2. Every normal pseudocompact space is countably compact. (But a nonnormal, pseudocompact space need not be countably compact; see 5I or 8.20. And even a normal, pseudocompact space need not be compact; see 5.12.)

3. Every closed G_δ in a normal space is a zero-set.

4. Every completely regular space with the Lindelöf property (i.e., such that every open cover has a countable subcover) is normal. [Modify the proof of 3.11(a).] (It is well known, more generally, that every *regular* Lindelöf space is normal.)

5. Let X be a completely regular space. If $X = S \cup K$, where S is open and normal, and K is compact, then X is normal. [3.11(a).]

3E. NONNORMAL SPACES.

Let X be a nonnormal, Hausdorff space.

1. X contains a closed set that is not a zero-set.

2. X has a subspace S with the following property: any two completely separated sets in S have disjoint closures in X, yet S is not C^*-embedded in X. Compare Urysohn's extension theorem.

3F. T_0-SPACES.

In a topological space X, define $x \equiv x'$ to mean that $\text{cl}\,\{x\} = \text{cl}\,\{x'\}$. Let Y be the set of all equivalence classes thus defined, let τ map each point of X into its equivalence class, and provide Y with the quotient topology relative to τ.

1. For E closed in X, if $x \in E$ and $x \equiv x'$, then $x' \in E$; hence $\tau[X - E] = Y - \tau[E]$.

2. The weak topology induced by τ agrees with the given topology on X.

3. τ is both an open mapping and a closed mapping.

4. $\tau[\text{cl}\,\{x\}] = \text{cl}\,\{\tau x\}$, whence Y is a T_0-space.

5. In Theorem 3.2 (and hence in Theorems 3.6 and 3.7), it is enough to require that X be a T_0-space, rather than a Hausdorff space.

3G. WEAK TOPOLOGY.

If a family C' of real-valued functions on X is an additive group, contains the constant functions, and contains the absolute value of each of its members, then the collection of sets of the form 3.4(a) is a base of neighborhoods of x, for each $x \in X$, in the weak topology induced by C'.

3H. COMPLETELY REGULAR FAMILY.

Let X be a completely regular space. A subfamily C' of $C(X)$ is called a *completely regular family* if whenever F is closed and $x \notin F$, there exists $f \in C'$ such that $f(x) \notin \text{cl}\, f[F]$. For example, $\{\mathbf{i}\}$ is a completely regular family in $C(\mathbf{R})$. Every completely regular family determines the topology of X; in fact, C' is a completely regular family if and only if the collection of all sets of the form 3.4(a) $(x \in X)$ is a base for the topology.

3I. THEOREM 3.9.

1. Let A be a nonempty subfamily of $C(X)$. Define Y and τ as in 3.9, except that now f ranges only over A, instead of over all of $C(X)$. Then Y is completely regular, τ is continuous, and the mapping $g \to g \circ \tau$ is an isomorphism from $C(Y)$ *into* $C(X)$.

2. Let X be the set of real numbers, under the discrete topology, and take $A = \{\sigma\}$, where σ is the identity map of X onto $\mathbf{R}$. Then $\tau = \sigma$, and $Y = \mathbf{R}$. Hence τ is not a quotient mapping.

3J. A HAUSDORFF QUOTIENT SPACE THAT IS NOT COMPLETELY REGULAR.

Let S denote the subspace of $\mathbf{R} \times \mathbf{R}$ obtained by deleting $(0, 0)$ and all points $(1/n, y)$ with $y \neq 0$ and $n \in \mathbf{N}$. Define $\pi(x, y) = x$ for all $(x, y) \in S$; then π is a continuous mapping of S onto $\mathbf{R}$. Let E denote the quotient space of S associated with the mapping π; thus, E may be identified as the set of real numbers endowed with the largest topology for which the mapping π is continuous. (A set $A \subset E$ is open in E if and only if $\pi^{\leftarrow}[A]$ is open in S.)

1. E is a Hausdorff space. In fact, distinct points of E are completely separated. [$C(E) \supset C(\mathbf{R})$.]

2. E is not completely regular (in fact, it is not regular). [The set $\{1/n\}_{n \in \mathbf{N}}$ is closed.] Hence *a Hausdorff quotient space of a completely regular space need not be completely regular.*

3. $C(E) = C(\mathbf{R})$. Hence the completely regular space Y of Theorem 3.9 (with $X = E$) is $\mathbf{R}$; and the associated mapping τ is not a quotient mapping.

4. $E \cap [0, 1]$ is pseudocompact, but not countably compact (or completely regular).

3K. THE COMPLETELY REGULAR, NONNORMAL SPACE Γ.

Let Γ denote the subset $\{(x, y): y \geqq 0\}$ of $\mathbf{R} \times \mathbf{R}$, provided with the following enlargement of the product topology: for $r > 0$, the sets

$$V_r(x, 0) = \{(x, 0)\} \cup \{(u, v) \in \Gamma: (u - x)^2 + (v - r)^2 < r^2\}$$

are also neighborhoods of the point $(x, 0)$. Clearly, Γ satisfies the first countability axiom.

1. The subspace $D = \{(x, 0): x \in \mathbf{R}\}$ of Γ is discrete, and is a zero-set in Γ.

2. Γ is a completely regular space. [For $p \in D$, and any neighborhood $V_r(p)$ of p, define $f \in C(\Gamma)$ as follows: $f(p) = 0$; for every point q on the boundary of $V_r(p)$, $f(q) = 1$; and f is linear on the segment from p to q.]

3. The subspace $\Gamma \cap (\mathbf{Q} \times \mathbf{Q}) - D$ is dense in Γ. Hence $|C(\Gamma)| = \mathfrak{c}$.

4. The zero-set D is not C^*-embedded in Γ. [$|C^*(D)| > \mathfrak{c}$.] Hence Γ is not normal. [3D.*1*.]

5. Every closed set in Γ is a G_δ.

6. Γ contains a closed G_δ that is not a zero-set. [Γ has at most $|C(\Gamma)|$ zero-sets. The result also follows from 3E.*1*.]

3L. EXTENSION OF FUNCTIONS FROM A DISCRETE SET.

Let X be a completely regular space.

1. Let (V_α) be a family of disjoint sets in X with nonempty interiors, and such that for each index α, the set $\bigcup_{\sigma \neq \alpha} V_\sigma$ is closed. Any set D formed by selecting one element from the interior of each V_α is C-embedded in X. (Any attempt to weaken the hypothesis in a serious way runs into the pathology of the Tychonoff plank, **T**, described in detail in 8.20. **T** contains a *countable*, closed, discrete set that is not even C^*-embedded.)

2. Let $\{x_n\}_{n \in \mathbf{N}}$ be a discrete set (not necessarily closed) in X, and let $(r_n)_{n \in \mathbf{N}}$ be any convergent sequence of real numbers. Then there exists $f \in C^*(X)$ such that $f(x_n) = r_n$ for all $n \in \mathbf{N}$. [There exists a family (U_n) of disjoint open sets with $x_n \in U_n$.]

3. If X is infinite, then $C^*(X)$ contains a function with infinite range. [0.13.]

4. Let D be a countable discrete set in X. The following conditions are equivalent and imply that D is closed. (Cf. 3B.)

(1) D and any disjoint closed set are completely separated.

(2) D and any disjoint closed set have disjoint neighborhoods.

(3) D is C-embedded in X. [Enclose the points of D in disjoint neighborhoods, and apply *1.*]

5. If X is pseudocompact, then X is countably compact if and only if every countable, closed discrete subset D satisfies the conditions of *4.* (The interesting condition here is (2), which does not mention continuous functions. We remark that a countably compact space need not be normal; see 8L.)

3M. SUPREMA IN $C(\mathbf{R})$.

1. Construct a sequence of functions f_n in $C(\mathbf{R})$, with $f_n \leqq \mathbf{1}$, for which $\sup_n f_n$ does not exist in $C(\mathbf{R})$—that is, whenever $g \in C(\mathbf{R})$ satisfies $g \geqq f_n$ for all n, then there exists $h \in C(\mathbf{R})$ such that $h \leqq g$, $h \neq g$, and $h \geqq f_n$ for all n.

2. Construct a sequence of functions f_n in $C(\mathbf{R})$ for which $\sup_n f_n$ exists in $C(\mathbf{R})$, but is not the pointwise supremum—that is $(\sup_n f_n)(x) \neq \sup_n f_n(x)$ for at least one x.

3N. THE LATTICE $C(X)$.

Let X be a completely regular space.

1. Let $f \geqq \mathbf{0}$ in $C(X)$ be given. If

$$g = \sup_{n \in \mathbf{N}} (\mathbf{1} \wedge nf)$$

exists in $C(X)$, then g is 1 on $\operatorname{pos} f$ and 0 on $X - \operatorname{cl} \operatorname{pos} f$. [For each $x \in X - \operatorname{cl} \operatorname{pos} f$, there exists $h \geqq g$ such that $h(x) = 0$.]

2. Let V be an open set, and let B denote the family of all functions $\leqq \mathbf{1}$

in C that vanish on $X - V$. If $f = \sup B$ exists in C, then f is 1 on V and 0 on $X - \text{cl}\, V$.

Let (f_α) be a family of functions in C, and for $r \in \mathbf{R}$, define

$$U^r = \text{cl} \bigcup_\alpha \{x: f_\alpha(x) > r\}.$$

3. If $g \in C$, and $g \geqq f_\alpha$ for every α, then, for each x, $g(x) \geqq \sup \{r: x \in U^r\}$. [If $g(x) < s$, then $g(y) < s$ on a neighborhood of x, whence $x \notin U^s$, so that $\sup \{r: x \in U^r\} \leqq s$.]

4. If X is basically disconnected (1H), and if (f_α) is a countable family, then each U^r is open.

5. X is basically disconnected if and only if every countable family with an upper bound in C has a supremum in C (i.e., the lattice $C(X)$ is *conditionally σ-complete*). [Lemma 3.12.]

6. X is extremally disconnected if and only if every family with an upper bound in C has a supremum in C (i.e., the lattice $C(X)$ is *conditionally complete*).

3O. TOTALLY ORDERED SPACES.

Let X be a totally ordered set (of more than one element). We make X into a topological space by taking as a subbase for the open sets the family of all rays $\{x: x > a\}$ and $\{x: x < b\}$.

A nonempty subset S of X is called an *interval* of X if whenever an element x of X lies between two elements of S, then $x \in S$. When an interval is an open set, it is called an open interval. For example, the set of all positive rationals less than $\sqrt{2}$ is an open interval in the totally ordered space $\mathbf{Q}$.

The topology on X is called the *interval topology*, because the open intervals form a base.

1. Every open set is expressible in a unique way as a union of disjoint maximal open intervals.

2. X is a Hausdorff space.

3. For any nonempty subset A, if $\sup A$ exists, then $\sup A \in \text{cl}\, A$.

4. For $A \subset X$, the relative topology on A contains the interval topology, but the two need not be the same. [Let

$$A = \mathbf{R} - \{x: 0 < |x| \leqq 1\};$$

in the relative topology induced by $\mathbf{R}$, 0 is an isolated point of A, while in its interval topology, A is homeomorphic with $\mathbf{R}$. See also the example in *5*.]

5. If A is an interval of X, then the relative topology on A does coincide with the interval topology. The condition is not necessary: witness $\mathbf{N}$ in $\mathbf{R}$. Here $\mathbf{N}$ is a closed, discrete subspace; on the other hand, any nonvertical line is a closed, discrete subspace of the lexicographically ordered plane, yet, in its interval topology, is homeomorphic with $\mathbf{R}$.

6. X is connected if and only if X is Dedekind-complete (0.6) and has no consecutive elements.

7. X is compact if and only if it is lattice-complete (0.5). Thus, X is compact if and only if it is Dedekind-complete and has both a first element and a last element. [*Sufficiency.* (This is the Heine-Borel-Lebesgue theorem when $X \subset \mathbf{R}$.) Let $\mathscr{S}$ be a family of closed sets with the finite intersection property. Let A denote the set of all $a \in X$ such that $\{x \in X: x \geqq a\}$ meets the intersection of every finite subfamily of $\mathscr{S}$, and define $b = \sup A$; then $b \in \bigcap \mathscr{S}$.]

8. X has a totally ordered compactification. Hence X is completely regular.

9. X is normal. [Given disjoint closed sets H and K, define $f \in C(X)$ as follows. Let $f(x) = 0$ for all $x \in H$, and $f(y) = 1$ for all $y \in K$. Express the complement of $H \cup K$ as in *1*, and, using complete regularity, extend f to all of X by considering separately its definition on each maximal open interval.]

3P. CONVERGENCE OF z-FILTERS.

Let $\mathscr{F}$ be a z-filter on a completely regular space X, and let p be a cluster point of $\mathscr{F}$.

1. $\mathscr{F}$ converges to p if and only if $\mathscr{F}$ is contained in a unique z-ultrafilter. [Modify the proof of Theorem 3.17.]

2. If X is compact and p is the only cluster point of $\mathscr{F}$, then $\mathscr{F}$ converges to p.

Chapter 4

FIXED IDEALS. COMPACT SPACES

4.1. We have seen that in the study of rings of continuous functions there is no need to deal with spaces that are not completely regular. Accordingly,

IN THE SEQUEL, ALL GIVEN SPACES ARE ASSUMED TO BE COMPLETELY REGULAR.

Of course, when we *construct* a space, complete regularity must be checked.

Let I be any ideal in $C(X)$ or $C^*(X)$. If $\bigcap \boldsymbol{Z}[I]$ is nonempty, we call I a *fixed* ideal; if $\bigcap \boldsymbol{Z}[I] = \emptyset$, then I is a *free* ideal. Thus, I is free if and only if, for every point $x \in X$, there is a function in I that does not vanish at x.

The deeper relations among X, C and C^* depend upon an analysis of the set of all maximal ideals. The *fixed* maximal ideals are easy to describe, as we shall see in this chapter. Characterization of the *free* maximal ideals—the *sine qua non* of the entire theory—must await the development of further machinery.

But there are spaces that admit no free ideals—precisely, the compact spaces (Theorem 4.11). For these spaces, we can immediately present a result that represents one of the milestones in the development of the theory of rings of continuous functions: within the class of compact spaces, the ring structure of $C^*(X)$ determines X up to homeomorphism (Theorem 4.9).

4.2. Evidently the zero ideal is fixed. More generally, if $\boldsymbol{Z}(f)$ is not empty, then the principal ideal (f) is fixed, since, clearly, $\bigcap \boldsymbol{Z}[(f)] = \boldsymbol{Z}(f)$. Moreover, every free ideal I in C or C^* contains nonzero fixed ideals. In fact, if I contains a nonzero function h whose zero-set is nonempty, then I contains the nonzero fixed ideal (h). To see that I always does contain such an element (even when $I \subset C^*$), note, first of all, that since I is free, it contains a nonzero function f. In case $\mathbf{Z}(f) = \emptyset$, let r be any value assumed by f; then $h = f\cdot(f - \mathbf{r})$ is as required.

On the other hand, it is manifest that no fixed ideal can contain a free ideal.

As another example of a fixed ideal, let S be any nonempty subset of X. As we noticed in 2.7, the set

$$\{f \in C : f[S] = \{0\}\}$$

is an ideal in C (in fact, a z-ideal); plainly, this ideal is fixed, and its intersection with C^* is a fixed ideal in C^*. Furthermore, these ideals are nonzero so long as S is not dense in X.

4.3. EXAMPLES. Any ideal in $C^*(\mathbf{N})$ that contains $\mathbf{j}$ is an example of a free ideal. (Since $\mathbf{j}$ is not a unit of $C^*(\mathbf{N})$, such ideals exist.) In particular, every maximal ideal containing $\mathbf{j}$ is free.

The set $C_K(\mathbf{N})$ of all functions on $\mathbf{N}$ that vanish at all but a finite number of points is evidently a free ideal both in $C(\mathbf{N})$ and in $C^*(\mathbf{N})$. We can say more, namely, that C_K is the intersection of *all* the free ideals, again both in C and in C^*. Clearly, it suffices to prove that if I is any free ideal, then for each $n \in \mathbf{N}$, the function f defined as $f(n) = 1$, and $f(m) = 0$ for $m \neq n$ (in short, the characteristic function of $\{n\}$), belongs to I. Now, since I is free, there exists $g \in I$ such that $r = g(n) \neq 0$; and we have $f = \mathbf{r}^{-1}fg$. Hence $f \in I$.

For a generalization of this example, see 4D and 7E.

4.4. We turn our attention now to the *fixed maximal* ideals in the rings $C(X)$ and $C^*(X)$.

If I is a fixed ideal in C, then the set $S = \bigcap \mathbf{Z}[I]$ is nonempty, and the set

$$I' = \{f \in C : f[S] = \{0\}\}$$

is a fixed ideal containing I. Hence a fixed *maximal* ideal must be of this form. Moreover, since I' can be enlarged by making S smaller, the only candidates for fixed maximal ideals are the ideals I' for which S contains just one point. That these are indeed maximal will be shown below.

The corresponding statements hold for C^*.

The ideal I' considered above evidently contains the z-ideal $\mathbf{Z}^{\leftarrow}[\mathbf{Z}[I]]$. In general, the two are not the same: the set $S = \bigcap \mathbf{Z}[I]$ need not be a member of $\mathbf{Z}[I]$, even if S is a zero-set (see 4I.*1*). And since the zero-sets form a base for the closed sets, S itself can be an arbitrary closed set, not necessarily a zero-set.

4.5. *Residue class.* Let I be an ideal in an arbitrary ring A; the symbol $I(a)$ will denote the residue class of a. According to this functional notation, we are identifying the ideal I with the canonical homomorphism of A onto A/I.

4.6. THEOREM.

(a) *The fixed maximal ideals in $C(X)$ are precisely the sets*

$$\mathbf{M}_p = \{f \in C : f(p) = 0\} \qquad (p \in X).$$

The ideals $\mathbf{M}_p$ are distinct for distinct p. For each p, $C/\mathbf{M}_p$ is isomorphic with the real field $\mathbf{R}$; in fact, the mapping $\mathbf{M}_p(f) \to f(p)$ is the unique isomorphism of $C/\mathbf{M}_p$ onto $\mathbf{R}$.

(a*) *The fixed maximal ideals in $C^*(X)$ are precisely the sets*

$$\mathbf{M}^*_p = \{f \in C^* : f(p) = 0\} \qquad (p \in X).$$

The ideals $\mathbf{M}^*_p$ are distinct for distinct p. For each p, $C^*/\mathbf{M}^*_p$ is isomorphic with the real field $\mathbf{R}$; in fact, the mapping $\mathbf{M}^*_p(f) \to f(p)$ is the unique isomorphism of $C^*/\mathbf{M}^*_p$ onto $\mathbf{R}$.

PROOF. (a). $\mathbf{M}_p$ is the kernel of the homomorphism $f \to f(p)$ of $C(X)$ into $\mathbf{R}$. Since $\mathbf{r}(p) = r$ for each real r, the homomorphism is *onto* the field $\mathbf{R}$. Hence its kernel $\mathbf{M}_p$ is a maximal ideal. Uniqueness of p is an immediate consequence of the complete regularity of X.

On the other hand, if M is any fixed ideal in C, there exists a point p in $\bigcap \mathbf{Z}[M]$. Evidently, M is contained in $\mathbf{M}_p$, which has just been shown to be a (proper) ideal. Hence if M is maximal, we must have $M = \mathbf{M}_p$.

Since $\mathbf{M}_p$ is the kernel of a homomorphism onto $\mathbf{R}$, $C/\mathbf{M}_p$ is isomorphic with $\mathbf{R}$; and the isomorphism is unique, because the only automorphism of $\mathbf{R}$ is the identity (0.23).

The proof of (a*) is identical except for notation.

4.7. The foregoing theorem implies the existence of a one-one correspondence between the fixed maximal ideals in C and those in C^*. In addition, it yields, as an immediate corollary, a simple way of obtaining the correspondence, namely,

$$\mathbf{M}_p \to \mathbf{M}^*_p = \mathbf{M}_p \cap C^*.$$

Moreover, $\mathbf{M}_p$ is the only maximal ideal in C—fixed or free—whose intersection with C^* is $\mathbf{M}^*_p$. For, consider any maximal ideal M in C, distinct from $\mathbf{M}_p$. There exists $f \in M$ such that $f(p) \neq 0$. Let $g = |f| \wedge \mathbf{1}$. Then $g \in C^*$, and $\mathbf{Z}(g) = \mathbf{Z}(f)$. Hence $g(p) \neq 0$, so that $g \notin \mathbf{M}^*_p$; also, g belongs to the z-ideal M. (Alternatively, one could argue as in 2A.) Thus, g belongs to $M \cap C^*$, but not to $\mathbf{M}^*_p$.

If M is an arbitrary maximal ideal in C, then $M \cap C^*$ is always a prime ideal in C^* (2B.*1*). But, as we shall now show, (i) it need not be maximal, and (ii) the free maximal ideals in C^* need not be of this form. We obtain both of these conclusions by considering the function $\mathbf{j}$ in $C^*(\mathbf{N})$. Since $\mathbf{j}$ is a unit of $C(\mathbf{N})$, it belongs to no ideal in $C(\mathbf{N})$. We shall prove in the next paragraph that $\mathbf{j}$ belongs to *every* free maximal ideal in $C^*(\mathbf{N})$. Assuming this result for the moment, consider any free maximal ideal M in $C(\mathbf{N})$. (By 4.3, such ideals exist.) As was pointed out above, $M \cap C^*(\mathbf{N})$ cannot be a fixed maximal ideal in $C^*(\mathbf{N})$; and, since it does not contain $\mathbf{j}$, it cannot be a free maximal ideal either. This gives us (i). It also takes care of (ii): *no* free maximal ideal in $C^*(\mathbf{N})$ (again, by 4.3, such ideals exist) can assume the form $M \cap C^*(\mathbf{N})$, where M is a maximal ideal in $C(\mathbf{N})$.

To see that $\mathbf{j}$ does belong to every free maximal ideal in $C^*(\mathbf{N})$, consider any such ideal M^*, and suppose that $\mathbf{j} \notin M^*$. Then $(M^*, \mathbf{j}) = C^*(\mathbf{N})$; so there exists $f \in C^*(\mathbf{N})$ such that $f\mathbf{j} - \mathbf{1} \in M^*$. Since f is bounded, the set

$$A = \{n \in \mathbf{N}: f(n) > n/2\}$$

is finite. It follows from 4.3 that there exists a function g in M^* that has no zeros in A (specifically, the characteristic function of A). But then $g^2 + (\mathbf{1} - f\mathbf{j})^2$ is bounded away from zero, yet belongs to M^*—which is a contradiction.

The precise relationship between the maximal ideals M in $C(X)$ (for any X) and the ideals in $C^*(X)$ of the form $M \cap C^*$ is discussed in 7.9.

4.8. THEOREM. *If X is compact, then every ideal I in $C(X)$ is fixed.*

PROOF. $\mathbf{Z}[I]$ is a family of closed sets with the finite intersection property.

4.9. In the light of Theorem 4.6, we now have:

(a) *If X is compact, then the correspondence $p \to \mathbf{M}_p$ is one-one from X onto the set of all maximal ideals in $C(X)$.*

Since maximal ideals are algebraic invariants, this means that the points of a compact space can be recovered from the algebraic structure of the ring. Now, the zero-sets in X form a base for the closed sets; and the relation $p \in \mathbf{Z}(f)$ is equivalent to the purely algebraic relation $f \in \mathbf{M}_p$. Thus, the topology of X can also be recovered from $C(X)$.

THEOREM. *Two compact spaces X and Y are homeomorphic if and only if their rings $C(X)$ and $C(Y)$ are isomorphic.*

PROOF. Necessity is obvious. Since X and Y can both be recovered from their respective function rings, in the manner described above, the condition is also sufficient.

We have indicated how to recapture a compact space X from the ring $C(X)$. It is desirable to describe the process in more detail. Let

$$\mathfrak{M} = \mathfrak{M}(X)$$

denote the set of all maximal ideals in $C(X)$. (This notation will be used generally, not only for compact X.) We make $\mathfrak{M}$ into a *topological space* by taking, as a base for the closed sets, all sets of the form

$$\text{(b)} \qquad \{M \in \mathfrak{M} : f \in M\} \qquad (f \in C(X)).$$

For given f, M_p belongs to this set if and only if $f(p) = 0$. Hence the one-one correspondence $p \to M_p$ carries the zero-sets in X onto the family of all sets (b). Thus, $\mathfrak{M}$ is well defined as a topological space and is homeomorphic with X.

The topology thus defined is called the *Stone topology* on $\mathfrak{M}$. The set $\mathfrak{M}$, endowed with the Stone topology, is called the *structure space* of the ring C. For definition and properties of the structure space of an abstract ring, see 7M.

4.10. We shift our emphasis now from ideals to z-filters. (See Theorems 2.3 and 2.5.) We call a z-filter *free* or *fixed* according as the intersection of all its members is empty or nonempty. Thus, an ideal I in C is fixed if and only if $\mathbf{Z}[I]$ is fixed. It follows that every ideal in $C(X)$ will be fixed if and only if every z-filter on X is fixed.

LEMMA. *A zero-set Z is compact if and only if it belongs to no free z-filter.*

PROOF. Necessity is clear. Conversely, let $\mathscr{B}$ be any family of closed subsets of Z with the finite intersection property. The members of $\mathscr{B}$ are closed in X. The collection of all zero-sets in X that contain finite intersections of members of $\mathscr{B}$ is a z-filter $\mathscr{F}$; and, of course, $Z \in \mathscr{F}$. Since the zero-sets in the completely regular space X form a base for the closed sets, $\bigcap \mathscr{B} = \bigcap \mathscr{F}$. But this latter intersection is nonempty, by hypothesis; so $\bigcap \mathscr{B} \neq \emptyset$. Thus, Z is compact.

It is not true that if every member of a z-filter $\mathscr{F}$—or even a z-ultrafilter—is noncompact, then $\mathscr{F}$ is free (4N.*2*).

4.11. According to 3.16, to say that a z-filter is fixed is to say that it has a cluster point. Also, a *z-ultrafilter* converges to any cluster point.

THEOREM. *The following are equivalent.*

(1) *X is compact.*

(2) *Every ideal in $C(X)$ is fixed, i.e., every z-filter is fixed.*

(2*) *Every ideal in $C^*(X)$ is fixed.*

(3) *Every maximal ideal in $C(X)$ is fixed, i.e. every z-ultrafilter is fixed.*

(3*) *Every maximal ideal in $C^*(X)$ is fixed.*

PROOF. The equivalence of (1) with (2) is the special case $Z = X$ of the lemma. Likewise, (1) implies (2*), because $C = C^*$ when X is compact. Next, if I is a free ideal in C, then $I \cap C^*$ is a free ideal in C^*; therefore (2*) implies (2). Finally, (2) is equivalent with (3), and (2*) with (3*), because every free ideal is contained in a free maximal ideal.

It is possible to base a proof of Tychonoff's product theorem upon these criteria. While one can get the result itself more easily, *ab ovo*, by working with filters, rather than with z-filters, the proof to follow contains some instructive features and will be used as a model for a later proof (8.12).

4.12. *The mapping $\tau^{\#}$.* Let τ be a continuous mapping from X to Y, and let $\mathscr{F}$ be a z-filter on X. The image of $\mathscr{F}$ under τ is not, in general, a z-filter; in fact, the image of a zero-set need not even be closed. The total *preimage* of a zero-set, however, is a zero-set:

$$\tau^{\leftarrow}[\mathbf{Z}_Y(g)] = \mathbf{Z}_X(g \circ \tau).$$

It turns out that the collection of sets $\mathbf{Z}_Y(g)$ whose preimages belong to $\mathscr{F}$ is rich enough to reflect those properties of $\mathscr{F}$ in which we are interested. We denote it by $\tau^{\#}\mathscr{F}$:

$$\tau^{\#}\mathscr{F} = \{Z \in \mathbf{Z}(Y) : \tau^{\leftarrow}[Z] \in \mathscr{F}\}.$$

Clearly, $\tau^{\#}\mathscr{F}$ is a z-filter on Y. However, it need not be a z-ultrafilter, even when $\mathscr{F}$ itself is (4H.2). On the other hand, a straightforward check shows that when $\mathscr{F}$ is a z-ultrafilter, then $\tau^{\#}\mathscr{F}$ will be prime. More generally:

If a z-filter $\mathscr{F}$ is prime, then $\tau^{\#}\mathscr{F}$ is prime.

4.13. Let $X = \times_\alpha X_\alpha$ (where each X_α is completely regular). By definition, the product topology on X is the weak topology induced by the family of all projections π_α. A base for the closed sets in X is given as the collection of all zero-sets of the form

(a) $$\pi_{\alpha_1}^{\leftarrow}[Z_{\alpha_1}] \cup \cdots \cup \pi_{\alpha_n}^{\leftarrow}[Z_{\alpha_n}],$$

where $n \in \mathbf{N}$, and Z_{α_k} is a zero-set in X_{α_k} (see 3.10).

LEMMA. *Let $\mathscr{A}$ be a z-ultrafilter on $X = \times_\alpha X_\alpha$. If every z-filter $\pi_\alpha^{\#}\mathscr{A}$ is fixed, then $\mathscr{A}$ is fixed.*

PROOF. For each α, choose $x_\alpha \in \bigcap \pi_\alpha^{\#}\mathscr{A}$, and let x denote the point (x_α) of X. We shall show that $x \in \bigcap \mathscr{A}$. By its very definition, x belongs to every member of $\mathscr{A}$ of the form $\pi_\alpha^{\leftarrow}[Z_\alpha]$, where $Z_\alpha \in \boldsymbol{Z}(X_\alpha)$. Since $\mathscr{A}$ is prime, x belongs to every member of $\mathscr{A}$ of the form (a). Finally, an arbitrary member of $\mathscr{A}$ is an intersection of sets of the latter form; consequently, x belongs to every member of $\mathscr{A}$.

4.14. TYCHONOFF PRODUCT THEOREM. *An arbitrary product of compact spaces is compact.*

PROOF. Let $X = \times_\alpha X_\alpha$, where each X_α is compact. As a product of completely regular spaces, X is completely regular. Now consider any z-ultrafilter $\mathscr{A}$ on X. Since each space X_α is compact, each z-filter $\pi_\alpha^{\#}\mathscr{A}$ is fixed. By the lemma, $\mathscr{A}$ is fixed. Therefore X is compact.

PROBLEMS

4A. MAXIMAL IDEALS; z-IDEALS.

1. Maximal fixed ideal coincides with *fixed maximal* ideal, and *maximal free* ideal with *free maximal* ideal.

2. C and C^* are semi-simple, i.e., the intersection of all the maximal ideals is $(\mathbf{0})$.

3. Prove directly that $\boldsymbol{M}_p$ is a maximal ideal in C by showing that it is an ideal, and that if $f \notin \boldsymbol{M}_p$, then $(\boldsymbol{M}_p, f) = C$. Do the same for C^*.

4. Either in C or in C^*, if f belongs to every maximal ideal that g belongs to, then $\boldsymbol{Z}(f) \supset \boldsymbol{Z}(g)$.

5. The following algebraic condition is necessary and sufficient that an ideal I in C be a z-ideal: given f, if there exists $g \in I$ such that f belongs to every maximal ideal containing g, then $f \in I$.

4B. PRINCIPAL MAXIMAL IDEALS.

1. A point p of X is isolated if and only if the ideal $\boldsymbol{M}_p$ [resp. $\boldsymbol{M}^*_p$] is principal. [*Sufficiency.* If $\boldsymbol{M}_p = (f)$, then every element of $\boldsymbol{M}_p$ is divisible by f; and $\boldsymbol{Z}(f) = \{p\}$. Construct a function $g \in \boldsymbol{M}_p$ such that $g = hf$ implies that h is discontinuous at p unless p is isolated.] Hence X is discrete if and only if every fixed maximal ideal in C [resp. C^*] is principal.

2. X is finite if and only if *every* maximal ideal in C [resp. C^*] is principal. [For C, apply Theorem 4.11. For C^*, note that if a free maximal ideal is a principal ideal (f), then f has no zeros, and consider the function $f^{1/3}$.]

4C. FINITELY GENERATED IDEALS.

1. Every finitely generated ideal in C is fixed.

2. A necessary and sufficient condition that every finitely generated ideal in C^* be fixed is that X be pseudocompact. [*Necessity.* If $C \neq C^*$, then C contains an unbounded unit.]

3. If X is infinite, then both C and C^* contain fixed ideals that are not finitely generated. [Try it first for $\mathbf{N}$; see 4.3. In the general case, apply 0.13.]

4D. FUNCTIONS WITH COMPACT SUPPORT.

The *support* of f is, by definition, the closure of $X - \mathbf{Z}(f)$. Let $C_K(X)$ denote the family of all functions in C having compact support.

1. If X is compact, then $C_K = C$; otherwise, C_K is both an ideal in C and an ideal in C^*.

2. $C_K(\mathbf{Q}) = (\mathbf{0})$. [No point of $\mathbf{Q}$ has a compact neighborhood.]

3. C_K is a free ideal if and only if X is locally compact but not compact.

4. An ideal I in C or C^* is free if and only if, for every compact set A, there exists $f \in I$ having no zeros in A.

5. C_K is contained in every free ideal in C and in every free ideal in C^*. [1D.*1*.] (It will be seen in 7E that C_K is actually the intersection of these ideals.)

4E. FREE IDEALS.

1. Let $f \in C^*$. If f belongs to no free ideal in C^*, then $\mathbf{Z}(f)$ is compact. But the converse is false. Compare Lemma 4.10.

2. The intersection of all the free maximal ideals in C coincides with the set of all f in C for which $\mathbf{Z}(f)$ meets every noncompact zero-set.

3. A z-filter is a base for the closed sets if and only if it is free.

4. Let S be a compact set in X, I a free ideal in $C(X)$, and J the set of all restrictions $f|S$, for $f \in I$. Then $J = C(S)$. [Apply 4D.*4* to show that J is an improper ideal in $C(S)$.]

4F. z-ULTRAFILTERS ON $\mathbf{R}$ THAT CONTAIN NO SMALL SETS.

Let $\mathscr{F}$ denote the family of all closed subsets of $\mathbf{R}$ whose complements are of finite Lebesgue measure.

1. $\mathscr{F}$ is a free z-filter.

2. $\mathscr{F}$ is not a z-ultrafilter. [Theorem 2.6(b).]

3. Any z-ultrafilter containing $\mathscr{F}$ contains only sets of infinite measure.

4G. BASE FOR A FREE ULTRAFILTER.

1. A free ultrafilter cannot have a countable base. [Let $(U_n)_{n \in \mathbf{N}}$ be a countable base for a free filter $\mathscr{F}$. Inductively, pick two points from each U_n so as to obtain two disjoint sets, each of which meets every U_n.]

2. More generally, a free ultrafilter, each of whose members is of power $\geqq \mathfrak{m}$, cannot have a base of power $\leqq \mathfrak{m}$. [Use transfinite induction.]

4H. THE MAPPING $\tau^{\#}$.

Let X and Y be spaces with the same underlying point set, such that the identity mapping τ from X onto Y is continuous.

1. If $\mathscr{F}$ is a z-filter on X, then $\tau^{\#}\mathscr{F} = \mathscr{F} \cap \boldsymbol{Z}(Y)$.

2. If $\mathscr{U}$ is a z-ultrafilter on X, $\tau^{\#}\mathscr{U}$ need not be a z-ultrafilter on Y. [Take $Y = [0, 1]$, X discrete, and $\mathscr{U}$ a free ultrafilter.]

4I. THE IDEALS $\boldsymbol{O}_p$.

For $p \in X$, let $\boldsymbol{O}_p$ denote the set of all f in C for which $\boldsymbol{Z}(f)$ is a neighborhood of p.

1. $\boldsymbol{O}_p$ is a z-ideal in C, $\boldsymbol{O}_p \subset \boldsymbol{M}_p$, and $\bigcap \boldsymbol{Z}[\boldsymbol{O}_p] = \{p\}$.

2. $\boldsymbol{M}_p$ is the only maximal ideal, fixed or free, that contains $\boldsymbol{O}_p$. [3.18(d).]

3. If $\boldsymbol{O}_p \neq \boldsymbol{M}_p$, then $\boldsymbol{O}_p$ is contained in a prime ideal that is not maximal. [Theorem 2.8.]

4. If P is a prime ideal in C, and $P \subset \boldsymbol{M}_p$, then $P \supset \boldsymbol{O}_p$. [Let f vanish on a neighborhood U of p; by complete regularity of X, there exists $g \in C$ such that $g(p) = 1$ and $g[X - U] = \{0\}$; then $fg \in P$.] Compare Theorem 3.17.

5. If $f \in \boldsymbol{M}_p - \boldsymbol{O}_p$, then there exists a prime ideal, containing $\boldsymbol{O}_p$ and f, that is not a z-ideal (and hence not maximal). [Generalize the argument of 2G.*1*.]

6. If $f \in \boldsymbol{M}_p - \boldsymbol{O}_p$, then there exists a prime ideal containing $\boldsymbol{O}_p$, but not f, that is not a z-ideal. [Consider a function m in $C(\mathbf{R})$ such that $\lim_{x \to 0} m(x)/x^n = 0$ for every $n \in \mathbf{N}$.]

7. $\boldsymbol{O}_p$ is a countably generated ideal if and only if p has a countable base of neighborhoods.

8. There exists a countably generated ideal I containing $\boldsymbol{O}_p$ if and only if p is a G_δ-point. In fact, if p is a G_δ, then $\boldsymbol{O}_p$ is contained in a principal ideal. [*Necessity.* $\bigcap \boldsymbol{Z}[I]$ is a G_δ; apply *1* and *2*. *Sufficiency.* 1D.*1*.]

4J. P-SPACES.

For any X, every maximal ideal in $C(X)$ is prime. When, conversely, every prime ideal in $C(X)$ is maximal, we call X a *P-space*. (Recall our blanket assumption that X is completely regular.)

This definition is stated in terms of an algebraic property of $C(X)$; therefore, if $C(X)$ is isomorphic with $C(Y)$, and X is a P-space, then Y is a P-space.

The following assertions are equivalent. [Successive implications are suggested in the hints.] It will be obvious from several of them that every discrete space is a P-space.

(1) X is a P-space.

(2) For all $p \in X$, $M_p = O_p$, i.e., every function in C is constant on a neighborhood of p. [4I.*3*.]

(3) Every zero-set is open.

(4) Every G_δ is open. [3.11(b).]

(5) Every ideal in $C(X)$ is a z-ideal. [From (4), if $Z(f) \supset Z(g)$, then f is a multiple of g.]

(6) Every ideal is an intersection of prime ideals. [Theorem 2.8.]

(7) For every $f, g \in C$, the ideal (f, g) is the principal ideal $(f^2 + g^2)$. [From (5), $f \in (f^2 + g^2)$.]

(8) For every $f \in C$, there exists $f_0 \in C$ such that $f^2 f_0 = f$ (i.e., C is a *regular* ring). [Since f and f^2 belong to the same prime ideals, (6) implies (8). To obtain (8) from (7), set $g = \mathbf{0}$. And (8) implies (1) because every regular integral domain is a field; alternatively, for any f not belonging to a given prime ideal, consider the function $\mathbf{1} - f f_0$.]

(9) Every ideal is an intersection of maximal ideals.

(10) Every cozero-set in X is C-embedded. [This follows from (3) and implies (8).]

(11) Every principal ideal is generated by an idempotent. [(3) or (8).]

4K. FURTHER PROPERTIES OF P-SPACES.

1. Every countable subset of a P-space (4J) is closed and discrete. Hence every countable P-space is discrete, and every countably compact P-space is finite.

2. Every countable set in a P-space is C-embedded. [3L.] Hence every pseudocompact P-space is finite.

3. If X is a P-space, and every function in $C(X)$ is bounded on a subset S, then S is finite.

4. Every subspace of a P-space is a P-space.

5. Every (completely regular) quotient space of a P-space is a P-space.

6. Finite products of P-spaces are P-spaces. But no infinite product of spaces of more than one point is a P-space.

7. Every P-space is basically disconnected (1H). (The converse is false; see 4M.*3*,*4*.)

8. Every P-space—more generally, every basically disconnected space—has a base of open-and-closed sets.

9. Neither $C(\mathbf{R})$ nor $C(\mathbf{Q})$ is a homomorphic image of $C(\mathbf{N})$. [According to 4J(8), $C(\mathbf{N})$ is a regular ring.]

4L. P-POINTS.

If $M_p = O_p$, then p is called a *P-point* of X. Thus, X is a P-space if and only if every point is a P-point (4J(2)).

1. p is a P-point if and only if every G_δ containing p is a neighborhood of p. [Cf. 4J(4).]

2. A P-point of X is a P-point in any subspace containing it. [Cf. 4K.*4*.] Hence the set of all P-points of X is a P-space.

3. Let $f \in \mathbf{R}^X$. If f is continuous at a P-point p, then f is constant on a neighborhood of p.

4. If p is a P-point and M_p is countably generated, then p is isolated and M_p is principal. [Cf. 4I.7.]

4M. THE SPACE Σ.

Let $\mathscr{U}$ be a free ultrafilter on $\mathbf{N}$, let $\Sigma = \mathbf{N} \cup \{\sigma\}$ (where $\sigma \notin \mathbf{N}$), and define a topology on Σ as follows: all points of $\mathbf{N}$ are isolated, and the neighborhoods of σ are the sets $U \cup \{\sigma\}$ for $U \in \mathscr{U}$.

1. $\mathbf{N}$ is a dense subspace of Σ. Every set containing σ is closed; hence every subset of Σ is open or closed. Σ is a normal space (cf. 3B.*5* and 3D.*5*); in fact, every closed set is a zero-set.

2. The point σ does not have a countable base of neighborhoods. [4G.*1*.] Hence Σ is not metrizable.

3. Σ is extremally disconnected (1H), and so is every subspace.

4. *The z-ideal* O_σ *is prime but not maximal.* Hence Σ is not a P-space (4J).

5. The dense subspace $\mathbf{N}$ is C^*-embedded in Σ, but not C-embedded. [If $g \in C^*(\mathbf{N})$, then $g^\# \mathscr{U}$ is a prime z-filter on $\mathbf{R}$. Alternatively, use 1H.*6*.] Moreover, *every* subspace of Σ is C^*-embedded.

6. $C^*(\Sigma)$ is isomorphic with $C^*(\mathbf{N})$. But $C(\Sigma)$ is not isomorphic with $C(\mathbf{N})$. [*4*.]

7. If $\mathbf{0} \leqq h \leqq k$ in $C(\Sigma)$, then $h \in (k)$. [*5*.]

8. The ideal (f, g) in $C(\Sigma)$ is the principal ideal $(|f| + |g|)$. [*3* and *7*.] Hence every finitely generated ideal is principal.

9. The only z-ideals containing O_σ are O_σ and M_σ. [If Z belongs to $Z[M_\sigma]$, but not to $Z[O_\sigma]$, then $\Sigma - Z \in \mathscr{U}$.]

4N. A NONDISCRETE P-SPACE.

Let S be an uncountable space in which all points are isolated except for a distinguished point s, a neighborhood of s being any set containing s whose complement is countable.

1. The closed set $\{s\}$ is not a zero-set. S is a nondiscrete P-space (4J). (An example of a P-space with no isolated point is given in 13P.)

2. No member of $Z[M_s]$ is countably compact.

3. S is basically disconnected but not extremally disconnected (1H). (Obviously, every discrete space is an extremally disconnected P-space. The converse question is considered in 12H.)

4. Let T be the topological sum of Σ (4M) and S (i.e., Σ and S are disjoint open sets whose union is T). Then T is basically disconnected, but it is neither extremally disconnected nor a P-space.

4O. CLOSED IDEALS IN $C(X)$ FOR COMPACT X.

Let X be a *compact* space, and let I be an ideal in $C(X)$.

1. If $Z(f)$ is a neighborhood of $\bigcap Z[I]$, then $f \in I$. [Use compactness and 1D.*1*.]

2. Given $g \in C(X)$, and $\epsilon > 0$, there exists f such that $|g - f| \leqq \epsilon$ and $\boldsymbol{Z}(f)$ is a neighborhood of $\boldsymbol{Z}(g)$. [Define f so that $f(x) = 0$ wherever $|g(x)| \leqq \epsilon$.]

3. Define $\bar{I} = \bigcap \{M_p : M_p \supset I\}$. Then $\bar{I}$ is a closed ideal in the uniform norm topology on $C(X)$ [2M.*4*], and consists of all g for which $\boldsymbol{Z}(g) \supset \bigcap \boldsymbol{Z}[I]$. Furthermore, $\bigcap \boldsymbol{Z}[\bar{I}] = \bigcap \boldsymbol{Z}[I]$.

4. $\bar{I}$ is the closure of I in the uniform norm topology. [In *2*, if $g \in \bar{I}$, then $f \in I$, by *1*.] Hence *an ideal is closed if and only if it is an intersection of maximal ideals.*

Chapter 5

ORDERED RESIDUE CLASS RINGS

5.1. We saw in the preceding chapter that every residue class field of C or C^* modulo a *fixed* maximal ideal is isomorphic with the real field $\mathbf{R}$. The present chapter initiates the study of residue class fields modulo arbitrary maximal ideals. Each such field has the following properties, as will be shown: it is a totally ordered field, whose order is induced by the partial order in C, and the image of the set of constant functions is an isomorphic copy—necessarily order-preserving—of the real field.

More generally, for any ideal I, we shall wish to know when the residue class ring C/I can be ordered in such a way that the canonical mapping of C onto C/I will be order-preserving, and when, in addition, it will be a lattice homomorphism. The answers to these questions can be given for abstract partially ordered rings.

An ideal I in a partially ordered ring A is said to be *convex* if whenever $0 \leqq x \leqq y$, and $y \in I$, then $x \in I$. This agrees with the usual notion of convexity in partially ordered systems, because $a \leqq b \leqq c$ is equivalent to $0 \leqq b - a \leqq c - a$. In case A is a lattice-ordered ring, we have $-|x| \leqq x \leqq |x|$; hence if I is a convex ideal, $|x| \in I$ implies $x \in I$. Notice that in the ring of integers, the only convex ideal is (0).

An ideal I in a lattice-ordered ring is said to be *absolutely convex* if, whenever $|x| \leqq |y|$ and $y \in I$, then $x \in I$. Evidently, an absolutely convex ideal is convex. But the converse is not true; see 5B.*4* or 5E.*2*.

We recall that $I(a)$ denotes the residue class of a modulo I.

5.2. THEOREM. *Let I be an ideal in a partially ordered ring A. In order that A/I be a partially ordered ring, according to the definition:*

$$I(a) \geqq 0 \textit{ if there exists } x \in A \textit{ such that } x \geqq 0 \textit{ and } a \equiv x \pmod{I},$$

it is necessary and sufficient that I be convex.

PROOF. *Necessity*. If $0 \leqq x \leqq y$, and $y \in I$, then $0 \geqq I(x) - I(y) = I(x) \geqq 0$, whence $I(x) = 0$, i.e., $x \in I$.

Sufficiency. According to 0.19, we must verify (i): $I(a) \geqq 0$ and $I(-a) \geqq 0$ implies $I(a) = 0$, and (ii): $I(a) \geqq 0$ and $I(b) \geqq 0$ implies both $I(a) + I(b) \geqq 0$ and $I(a)I(b) \geqq 0$. In (i), there exist $x, y \in A$ such that $x \geqq 0$, $y \geqq 0$, $a \equiv x$, and $-a \equiv y$. Hence $x + y \equiv 0$. Since $0 \leqq x \leqq x + y$, we have $x \equiv 0$, by convexity. Therefore $I(a) = I(x) = 0$. Condition (ii) is an immediate consequence of the corresponding condition in A.

Whenever we say that a residue class ring is partially ordered, or totally ordered, we will always mean that it is an ordered ring whose order is induced as above. Trivially, the canonical mapping of A onto A/I is order-preserving, i.e., if $a \geqq b$, then $I(a) \geqq I(b)$.

5.3. THEOREM. *The following conditions on a convex ideal I in a lattice-ordered ring A are equivalent.*

(1) *I is absolutely convex.*

(2) *$x \in I$ implies $|x| \in I$.*

(3) *$x, y \in I$ implies $x \vee y \in I$.*

(4) *$I(a \vee b) = I(a) \vee I(b)$—whence A/I is a lattice.*

(5) *$I(a) \geqq 0$ if and only if $a \equiv |a| \pmod{I}$.*

PROOF. The pattern of proof will be (1) → (2) → (3) → (4) → (5) → (2) → (1).

(1) *implies* (2). Trivial.

(2) *implies* (3). If $x, y \in I$, then $|x| + |y| \in I$, and the result now follows from convexity and the relations

$$-(|x| + |y|) \leqq x \vee y \leqq |x| + |y|.$$

(3) *implies* (4). Obviously, $I(a \vee b) \geqq I(a)$, and $I(a \vee b) \geqq I(b)$. To prove that $I(a) \vee I(b)$ exists and equals $I(a \vee b)$, consider any element c of A such that $I(c) \geqq I(a)$ and $I(c) \geqq I(b)$. There exist $x, y \in I$ such that $c + x \geqq a$ and $c + y \geqq b$. Then

$$c + (x \vee y) = (c + x) \vee (c + y) \geqq a \vee b.$$

But $x \vee y \in I$, by hypothesis. Therefore $I(c) \geqq I(a \vee b)$.

(4) *implies* (5). Since $|a| \geqq 0$ (for the proof, see 5A), we have $I(|a|) \geqq 0$; hence if $a \equiv |a|$, then $I(a) \geqq 0$. Conversely, if $I(a) \geqq 0$, then

$$I(a) = I(a) \vee -I(a) = I(a) \vee I(-a) = I(a \vee -a) = I(|a|).$$

(5) *implies* (2). If $I(x) = 0$, then $0 \equiv x \equiv |x| \pmod{I}$.

(2) *implies* (1). If $|x| \leqq |y|$, and $y \in I$, then $|y| \in I$, by hypothesis, and convexity implies that $x \in I$.

REMARKS. Condition (3) states that I is a sublattice of A. Condition (4) implies that the canonical mapping of A onto A/I is a lattice homomorphism (0.5). Thus,

$$I(a \vee b) = I(a) \vee I(b), \quad I(a \wedge b) = I(a) \wedge I(b), \quad I(|a|) = |I(a)|.$$

The essential content of the theorem is that these are identities if and only if I is absolutely convex. The condition that A/I be a lattice is, by itself, not enough to guarantee that I be absolutely convex. See 5B.*4*.

5.4. *z-ideals.* Since every fixed maximal ideal in $C(X)$ or $C^*(X)$ is of the form $\boldsymbol{M}_p$ or $\boldsymbol{M}^*_p$ (Theorem 4.6), it is evident that each such ideal is absolutely convex. Furthermore, it is manifest that every z-ideal in C is absolutely convex, whence every maximal ideal in C, free as well as fixed, is absolutely convex. A similar argument holds for the e-ideals, and therefore for all maximal ideals, in C^* (2L). In 5.5, we shall present a proof, applicable to C and C^* simultaneously, that all maximal ideals—more generally, all prime ideals—are absolutely convex.

When M is a maximal ideal in $C(X)$, the order in C/M is intimately connected with the zero-sets in X. Let $f \in C$; we shall prove:

> $M(f) \geqq 0$ [resp. > 0] *if and only if f is nonnegative* [resp. *positive*] *on some zero-set of M.*

In fact, most of this statement is valid for any z-ideal.

(a) *If I is a z-ideal, then $I(f) \geqq 0$ if and only if f is nonnegative on some zero-set of I.*

Since I is absolutely convex, $I(f) \geqq 0$ is equivalent to $f - |f| \equiv 0 \pmod{I}$, and hence to $\boldsymbol{Z}(f - |f|) \in \boldsymbol{Z}[I]$, that is, f agrees with $|f|$ on some zero-set of I.

(b) *If f is positive on some zero-set of a z-ideal I, then $I(f) > 0$; and if I is maximal, the converse holds as well.*

If f is positive on $Z \in \boldsymbol{Z}[I]$, then $\boldsymbol{Z}(f)$ does not meet Z, and so $f \notin I$. By (a), $I(f) > 0$. On the other hand, if $I(f) > 0$, then f is nonnegative on some $Z \in \boldsymbol{Z}[I]$; and if I is maximal, then by Theorem 2.6(a), $\boldsymbol{Z}(f)$ is disjoint from some zero-set Z' of I. Thus, f is positive on the zero-set $Z \cap Z'$ of I.

This converse fails whenever I is not maximal. For, choose an ideal J containing I properly, and pick $f \in J - I$. Then $I(f^2) > 0$. Since $\boldsymbol{Z}(f^2)$ belongs to $\boldsymbol{Z}[J]$, it meets every member of $\boldsymbol{Z}[I]$, so that $\operatorname{pos} f^2$ cannot contain any such set.

The results are inapplicable to C^*. For example, the non-unit $\mathbf{j}$ of $C^*(\mathbf{N})$ belongs to some maximal ideal in C^* (in fact, to every free maximal ideal), in spite of the fact that $\mathbf{j}$ is positive everywhere on $\mathbf{N}$. See, however, 5Q.*2*.

We are also interested in knowing when a partially ordered residue class ring is *totally* ordered. To establish total order, it is enough to show that every element is comparable with 0. Hence if I is a z-ideal, then, by (a), C/I is totally ordered if and only if, for all f, there is a zero-set of I on which f does not change sign. By Theorem 2.9, this is the case precisely when I is prime. Thus:

(c) *If I is a z-ideal, then C/I is totally ordered if and only if I is prime.*

As an example, let I be the ideal of all functions in $C(\mathbf{N})$ that vanish at both 1 and 2; then I is a z-ideal, but not prime. Let f be any function having opposite signs at these points. Neither f nor $-f$ is nonnegative on any zero-set of I, and $I(f)$ is not comparable with 0 in C/I.

According to (c), C/I is totally ordered whenever I is a prime z-ideal in C (in particular, when I is a maximal ideal). The next theorem generalizes this result to arbitrary prime ideals in C or C^*.

5.5. THEOREM. *Every prime ideal P in $C(X)$* [resp. $C^*(X)$] *is absolutely convex, and the residue class ring C/P* [resp. C^*/P] *is totally ordered. Furthermore, the mapping $r \to P(\mathbf{r})$ is an order-preserving isomorphism of the real field $\mathbf{R}$ into the residue class ring.*

PROOF. To show, first, that P is absolutely convex, let $\mathbf{0} \leqq |f| \leqq |g|$, with $g \in P$. Define h as follows: $h(x) = f^2(x)/g(x)$ for $x \notin \mathbf{Z}(g)$, and $h(x) = 0$ for $x \in \mathbf{Z}(g)$. Then h is continuous, as one sees easily from the fact that $f(x)/g(x)$ is bounded on $X - \mathbf{Z}(g)$. Moreover, if g is bounded, then h is too, so that this construction is applicable to C^* as well as to C. Evidently, $f^2(x) = h(x)g(x)$ for every $x \in X$, i.e., $f^2 = hg$. (Compare 1D.*3*.) Hence $f^2 \in P$, and, since P is prime, $f \in P$.

Next, since $(f - |f|)(f + |f|) = \mathbf{0}$ and P is prime, we have $f \equiv |f|$ or $f \equiv -|f|$. Therefore C/P [resp. C^*/P] is totally ordered. Finally, the mapping of $\mathbf{R}$ into the residue class ring by means of the constant functions is clearly an isomorphism. As $r \geqq 0$ implies $\mathbf{r} \geqq \mathbf{0}$, and hence $P(\mathbf{r}) \geqq 0$, it also preserves order.

REMARK. As a matter of fact, *every* isomorphism of $\mathbf{R}$ to an ordered field is order-preserving. More generally, $\mathbf{R}$ may be replaced in this statement by any ordered field in which every positive element has a square root. For in such a field, the order is completely determined by the algebraic structure: an element is nonnegative if and only if it is a square.

RESIDUE CLASS FIELDS

5.6. *Real ideals; hyper-real ideals.* We know that every residue class field of C or C^* (modulo a maximal ideal M) contains a canonical copy of the real field $\mathbf{R}$: the set of images of the constant functions, under the canonical homomorphism. We shall identify this subfield with $\mathbf{R}$. Thus, in Theorem 4.6, we can write

$$M_p(f) = f(p), \qquad \text{for } f \in C(X),$$

and

$$M^*_p(f) = f(p), \qquad \text{for } f \in C^*(X).$$

When the canonical copy of $\mathbf{R}$ is the entire field C/M [resp. C^*/M], we shall refer to M as a *real* ideal. Thus, by Theorem 4.6, every fixed maximal ideal in C or C^* is real.

When the residue class field modulo M is not real, we call it a *hyper-real* field and refer to M as a *hyper-real* ideal.

Recall that a totally ordered field is said to be *archimedean* if for every element a, there exists $n \in \mathbf{N}$ such that $n \geqq a$ (0.21). Thus, a *non-archimedean* field is characterized (among all totally ordered fields) by the presence of *infinitely large* elements, that is, elements a such that $a > n$ for every $n \in \mathbf{N}$. An element b is *infinitely small* if it is positive but smaller than $1/n$ for every $n \in \mathbf{N}$. Hence b is infinitely small if and only if $1/b$ is infinitely large. Therefore, the presence of infinitely small elements also characterizes the nonarchimedean fields.

Every archimedean field is embeddable in $\mathbf{R}$ (0.21). It follows that every hyper-real field is nonarchimedean, since the only nonzero isomorphism of $\mathbf{R}$ into itself is the identity. Thus, the residue class field is archimedean if and only if it is real.

5.7. The following results relate unbounded functions on X with infinitely large elements modulo maximal ideals. Let $f \in C$.

(a) *For a given maximal ideal M in C, the following are equivalent.*

(1) $|M(f)|$ *is infinitely large.*

(2) f *is unbounded on every zero-set of* M.

(3) *For each* $n \in \mathbf{N}$, *the zero-set*

$$Z_n = \{x\colon |f(x)| \geqq n\}$$

belongs to $\mathbf{Z}[M]$.

By 5.4(a), $|M(f)| \leqq n$ if and only if there exists $Z \in \mathbf{Z}[M]$ such that $|f(x)| \leqq n$ for all $x \in Z$; thus, the negation of (1) is equivalent to the negation of (2). Again by 5.4(a), $M(|f|) \geqq n$ if and only if Z_n contains a member of $\mathbf{Z}[M]$; hence (1) is equivalent to (3).

(b) $|M(f)|$ *is infinitely large for some maximal ideal M in C if and only if f is unbounded on X.*

Necessity follows from (a). Conversely, if f is unbounded, then the family of sets Z_n has the finite intersection property, and hence is embeddable in a z-ultrafilter $\mathscr{A}$. The ideal $M = \mathbf{Z}^{\leftarrow}[\mathscr{A}]$ in C is maximal, and by (a), $|M(f)|$ is infinitely large.

(c) $|M(f)|$ *is infinitely large for every free maximal ideal M in C if and only if f is unbounded on every noncompact zero-set in X.*

This follows immediately from (a) and Lemma 4.10: every zero-set associated with a free ideal is noncompact, and, conversely, every noncompact zero-set belongs to $\mathbf{Z}[I]$ for some free ideal I, and hence to $\mathbf{Z}[M]$ for some free maximal ideal M.

The next theorem contains a complete description of the residue class fields of C^*.

5.8. THEOREM.

(a) *Every maximal ideal in C^* is real.*

(b) *Every maximal ideal in C is real when and only when X is pseudocompact.*

PROOF. (a). If $|f| \leqq \mathbf{n}$, then $|M(f)| \leqq n$. Therefore, C^*/M contains no infinitely large elements, and hence is real (see 5.6).

(b). This is immediate from 5.7(b).

REMARK. It is well to notice that if M is a hyper-real ideal in C, the theorem does *not* assert that $M(f)$ is real for every bounded f. On the contrary, there always exists a bounded function f with $M(f)$ infinitely small: simply take $f = (|g| + \mathbf{1})^{-1}$, where $M(g)$ is infinitely large.

5.9. *Realcompact spaces.* The problem of describing the residue class fields of C^* is now settled: *every* maximal ideal in C^*, fixed or free, is real. As for C, we know that every *fixed* maximal ideal is real. Hence the only candidates for hyper-real ideals in C or C^* are the free maximal ideals in C. According to the preceding theorem, hyper-real ideals will exist whenever there are unbounded functions in C.

But now another question presents itself: what are the spaces X for which *every* free maximal ideal in $C(X)$ is hyper-real? The ramifications of this question will occupy us throughout much of the book. Those spaces X that satisfy the condition will be termed ***realcompact***. We shall make a few remarks about realcompact spaces here, but their formal investigation must be postponed until Chapter 8.

If X is compact, then $C(X)$ contains no free ideals. Therefore every compact space is realcompact.

If X is pseudocompact but not compact, then every maximal ideal in C is real (Theorem 5.8), but not every maximal ideal in C is fixed (Theorem 4.11); hence X is not realcompact. An example of a noncompact, pseudocompact space is given in 5.12.

5.10. EXAMPLES. First, however, let us look at some other applications of the preceding theorems, by considering once more our favorite space $\mathbf{N}$, and the identity function $\mathbf{i}$ on $\mathbf{N}$. If M is any real maximal ideal in $C(\mathbf{N})$, then $M(\mathbf{i})$ has to be a real number r, i.e., we must have $\mathbf{i} \equiv \boldsymbol{r} \pmod M$. But then $\mathbf{Z}(\mathbf{i} - \boldsymbol{r})$—which contains at most one point—must belong to $\mathbf{Z}[M]$. If M is a free ideal, this is impossible. Therefore every real ideal is fixed, i.e., $\mathbf{N}$ is a realcompact space.

We can see that $\mathbf{N}$ is realcompact in another way. The function $\mathbf{i}$ is certainly unbounded on every noncompact set. Therefore, by 5.7(c), its image modulo any free maximal ideal in $C(\mathbf{N})$ is infinitely large. This implies, first of all, that every free maximal ideal in $C(\mathbf{N})$ is hyper-real; thus, $\mathbf{N}$ is realcompact.

Secondly, it implies that the image of $\mathbf{j}$ modulo any free maximal ideal in $C(\mathbf{N})$ is infinitely small. In contrast, we have seen that $\mathbf{j}$ is a member of every free maximal ideal in $C^*(\mathbf{N})$ (4.7). Here is another proof. For each $n \in \mathbf{N}$, let $f_n = \boldsymbol{n}^{-1} \wedge \mathbf{j}$. Then $f_n - \mathbf{j}$ vanishes at all but a finite number of points. By 4.3, $f_n - \mathbf{j}$ belongs to every free maximal ideal M in C or C^*, that is, $f_n \equiv \mathbf{j} \pmod M$. But since $\mathbf{0} \leqq f_n \leqq \boldsymbol{n}^{-1}$, we have $0 \leqq M(f_n) \leqq 1/n$. Hence $0 \leqq M(\mathbf{j}) \leqq 1/n$. As this is true for every n, $M(\mathbf{j})$ is infinitely small or zero. In the case of C, $\mathbf{j}$ is a unit, and so $M(\mathbf{j}) \neq 0$. In the case of C^*, the residue class field has no infinitely small elements (Theorem 5.8(a)), and so $M(\mathbf{j}) = 0$.

SPACES OF ORDINALS

5.11. For purposes of further illustration, both now and later, the space $\mathbf{W}$ of all countable ordinals is indispensable. Before examining its special properties, we have to gather some facts about spaces of ordinals in general. The set of all ordinals less than a given ordinal α is denoted by $W(\alpha)$:

$$W(\alpha) = \{\sigma : \sigma < \alpha\} \qquad (\alpha \text{ an ordinal}).$$

It is well-ordered, of course, and so it may be provided with the interval topology (3O). Thus, 0 is an isolated point (if $\alpha > 0$), and for any point $\tau > 0$, the set of all open-and-closed intervals

$$[\sigma + 1, \tau] = \{x : \sigma < x < \tau + 1\} \qquad (\sigma < \tau)$$

is a system of basic neighborhoods of τ. Evidently, a point of $W(\alpha)$ is an isolated point if and only if it is not a limit ordinal (i.e., it is 0 or has an immediate predecessor). The space $W(\omega)$ of all finite ordinals is homeomorphic with **N**. It is clear that:

(a) *For $\sigma < \alpha$, $W(\sigma)$ is a subspace of $W(\alpha)$.*

(Compare 3O.*4*,*5*.)

As one might expect, the properties of totally ordered spaces listed in 3O can be established more simply in the special case of well-ordered spaces. For the sake of completeness, we shall derive some essential properties of well-ordered spaces independently of the general theory of totally ordered spaces.

Every nonempty set of ordinals, being well-ordered, has a least element; and, according to the theory of ordinals, every set of ordinals has an upper bound, and hence has a supremum. If $A \subset W(\alpha)$, then $\sup A \in W(\alpha)$ if and only if A is bounded in $W(\alpha)$, i.e., there exists $\sigma < \alpha$ such that $x \leqq \sigma$ for all $x \in A$. Now, A is cofinal in $W(\alpha)$ provided that, for all $\sigma < \alpha$, there exists $x \in A$ such that $x \geqq \sigma$. Thus, if $W(\alpha)$ has a greatest element, then any subset containing this element is both bounded and cofinal; however, if α is a limit ordinal, then a subset is cofinal if and only if it is *un*bounded.

We now prove:

(b) *For every ordinal α, $W(\alpha)$ is a normal space.*

Verification of the Hausdorff axiom is trivial. Now let H and K be disjoint closed sets. For each $\tau \in H$, let U_τ be an open interval of the form $[\sigma, \tau]$ that does not meet K; correspondingly, define V_τ for $\tau \in K$. Then $\bigcup_{\tau \in H} U_\tau$ and $\bigcup_{\tau \in K} V_\tau$ are disjoint open sets containing H and K, respectively.

Finally, we have:

(c) *$W(\alpha)$ is compact if and only if α is a nonlimit ordinal.*

For the necessity, we observe that every *tail* in $W(\alpha)$, that is, every set of the form

$$W(\alpha) - W(\sigma) = \{x \in W(\alpha) : x \geqq \sigma\} \qquad (\sigma < \alpha),$$

is a closed set, and that the family of all tails has the finite intersection property. If α is a limit ordinal, then the intersection of this family is empty, so that $W(\alpha)$ is not compact.

For the sufficiency, we note first that $W(0)$ is empty, hence, trivially, compact. Now consider any nonlimit ordinal $\alpha' = \alpha + 1$, and assume, inductively, that $W(\tau)$ is compact for every nonlimit ordinal $\tau < \alpha + 1$. Let $\mathscr{U}$ be any cover of $W(\alpha + 1)$ by basic open sets. Since the point α

is covered, there exists $\sigma < \alpha$ such that $[\sigma + 1, \alpha] \in \mathscr{U}$. By the induction hypothesis, the subspace $W(\sigma + 1)$ (see (a)) is compact, so that a finite subcollection $\mathscr{F}$ of $\mathscr{U}$ covers $W(\sigma + 1)$. Then $\mathscr{F} \cup \{[\sigma + 1, \alpha]\}$ is a finite subfamily of $\mathscr{U}$ that covers $W(\alpha + 1)$.

It follows from this result that $W(\alpha)$ is always locally compact; and if $W(\alpha)$ is not compact, then $W(\alpha + 1)$ is its one-point compactification.

5.12. *The spaces* **W** *and* **W***. We define

$$\mathbf{W} = W(\omega_1) = \{\sigma : \sigma < \omega_1\}$$

and

$$\mathbf{W}^* = W(\omega_1 + 1) = \{\sigma : \sigma \leqq \omega_1\},$$

ω_1 denoting the first uncountable ordinal. As we have just seen, $\mathbf{W}^*$ is compact, while $\mathbf{W}$ is not; in fact, $\mathbf{W}^*$ is the one-point compactification of $\mathbf{W}$. Since ω_1 is the *smallest* uncountable ordinal, every uncountable set in $\mathbf{W}$ is cofinal. A more interesting feature is the converse:

(a) *No countable set in* **W** *is cofinal.*

For if S is cofinal, then $\mathbf{W} = \bigcup_{\sigma \in S} W(\sigma)$; then

$$\aleph_1 \leqq \sum_{\sigma \in S} |\sigma| \leqq |S| \cdot \aleph_0,$$

whence $|S| = \aleph_1$. (The symbol $|A|$ denotes the cardinal of A.)

It follows from (a) that every countable set A in $\mathbf{W}$ is contained in a compact subspace. For, let $\alpha = \sup A$; then $\alpha \in \mathbf{W}$, and A is contained in the compact subspace $W(\alpha + 1)$. As a consequence, every countable, closed set in $\mathbf{W}$ is compact. Furthermore, $\mathbf{W}$ *is countably compact*, since every countably infinite set has a limit point. (Compare 5H.*5*.)

Since $\mathbf{W}$ is countably compact, it is pseudocompact, and we have here an example of a space that is pseudocompact but not compact. It follows that all maximal ideals in $C(\mathbf{W})$ are real (Theorem 5.8) and that at least one of them is free (Theorem 4.11), so that $\mathbf{W}$ is not realcompact (5.9).

As is well known, every countably compact metric space is compact. (Hence, by 3D.*2*, every pseudocompact metric space is compact.) Therefore, $\mathbf{W}$ is not metrizable.

Next, we prove:

(b) *Of any two disjoint closed sets in* **W**, *one is bounded*

and hence countable and compact. For, if H and K are cofinal closed sets, we can choose an increasing sequence $(\alpha_n)_{n \in \mathbf{N}}$, where $\alpha_n \in H$ for n odd, and $\alpha_n \in K$ for n even. Then $\sup_n \alpha_n \in H \cap K$.

Finally:

(c) *Every function* $f \in C(\mathbf{W})$ *is constant on a tail* $\mathbf{W} - W(\alpha)$ (α depending upon f). To see this, observe that every tail $\mathbf{W} - W(\sigma)$ is countably compact (in fact, it is homeomorphic with $\mathbf{W}$ itself). Therefore each image set $f[\mathbf{W} - W(\sigma)]$ is a countably compact subset of $\mathbf{R}$, and hence compact, so that the intersection

$$\bigcap_{\sigma \in \mathbf{W}} f[\mathbf{W} - W(\sigma)]$$

of the nested family is nonempty. Choose any number r belonging to this intersection; then the closed set $f^{\leftarrow}(r)$ is cofinal in $\mathbf{W}$. Now, for every $n \in \mathbf{N}$, the closed set

$$\{x \in \mathbf{W}: |f(x) - r| \geqq 1/n\}$$

is disjoint from $f^{\leftarrow}(r)$; hence, by (b), it has an upper bound α_n in $\mathbf{W}$. For any countable ordinal $\alpha > \sup_n \alpha_n$, we have $f[\mathbf{W} - W(\alpha)] = \{r\}$.

5.13. *The rings* $C(\mathbf{W})$ *and* $C(\mathbf{W}^*)$. An immediate consequence of this last result is that $\mathbf{W}$ is C-embedded in $\mathbf{W}^*$: we extend $f \in C(\mathbf{W})$ to a function $f^\beta \in C(\mathbf{W}^*)$ by defining $f^\beta(\omega_1)$ to be the final constant value of f. (Note that $W(\omega)$ is not C^*-embedded in $W(\omega + 1)$; more generally, see 5N.*1*.) It is trivial that f^β is the *unique* continuous extension of f. In the other direction, given $g \in C(\mathbf{W}^*)$, the restriction of g to $\mathbf{W}$ belongs to $C(\mathbf{W})$. It follows that $C(\mathbf{W})$ is isomorphic with $C(\mathbf{W}^*)$, under the mapping $f \rightarrow f^\beta$.

We have here, then, two spaces that are topologically distinct, but whose rings of continuous functions are isomorphic. Accordingly, neither the algebraic structure of $C^*(X)$, nor even that of $C(X)$, is sufficient, in general, to determine X as a topological space.

Since $\mathbf{W}^*$ is compact, we already have a complete description of the maximal ideals in $C(\mathbf{W}^*)$: every ideal is fixed, and the maximal ideals assume the form

$$\boldsymbol{M}_\sigma = \{f^\beta \in C(\mathbf{W}^*): f^\beta(\sigma) = 0\} \qquad (\sigma \in \mathbf{W}^*)$$

(Theorem 4.6). By virtue of the isomorphism of $C(\mathbf{W}^*)$ with $C(\mathbf{W})$, the maximal ideals in $C(\mathbf{W})$ are in one-one correspondence with these. Moreover, the fixed maximal ideals in $C(\mathbf{W})$ correspond to the ideals $\boldsymbol{M}_\sigma$ in $C(\mathbf{W}^*)$ for which $\sigma \in \mathbf{W}$, leaving *just one free maximal ideal in* $C(\mathbf{W})$, namely, the one that corresponds to $\boldsymbol{M}_{\omega_1}$. (For an alternative proof, see 5L.) We observe immediately that *the concepts fixed and free are not algebraic invariants.*

Let $\boldsymbol{M}^{\omega_1}$ denote the (free, real) maximal ideal in $C(\mathbf{W})$ that corresponds to $\boldsymbol{M}_{\omega_1}$ in $C(\mathbf{W}^*)$. Then

$$\boldsymbol{M}^{\omega_1} = \{f \in C(\mathbf{W}): f^\beta(\omega_1) = 0\}.$$

The functions in this ideal are described in terms of their values on $\mathbf{W}$ in this way: $\boldsymbol{M}^{\omega_1}$ is the set of all functions in $C(\mathbf{W})$ that vanish on a tail.

For emphasis: if $f \in C(\mathbf{W})$ is equal to the constant r on a tail, then the function $f - \boldsymbol{r}$ vanishes on that tail, and therefore $f \equiv \boldsymbol{r} \pmod{\boldsymbol{M}^{\omega_1}}$; and, since $f^{\beta}(\omega_1) = r$, we have

$$\boldsymbol{M}^{\omega_1}(f) = \boldsymbol{M}_{\omega_1}(f^{\beta}) = f^{\beta}(\omega_1) = r.$$

Plainly $\boldsymbol{Z}[\boldsymbol{M}^{\omega_1}]$ has the countable intersection property; indeed, it is closed under countable intersection. These attributes are, in fact, characteristic of real maximal ideals, as we now show.

REAL MAXIMAL IDEALS

5.14. THEOREM. *The following are equivalent for any maximal ideal M in $C(X)$.*

(1) *M is real.*

(2) *$\boldsymbol{Z}[M]$ is closed under countable intersection.*

(3) *$\boldsymbol{Z}[M]$ has the countable intersection property.*

PROOF. (1) *implies* (2). Suppose that $(\boldsymbol{Z}(f_n))_{n \in \mathbf{N}}$ is a subfamily of $\boldsymbol{Z}[M]$ whose intersection does not belong to $\boldsymbol{Z}[M]$. Define

$$g(x) = \sum_{n \in \mathbf{N}} (|f_n(x)| \wedge 2^{-n}) \qquad (x \in X).$$

Then g is continuous (because of uniform convergence) and nonnegative, and $\boldsymbol{Z}(g) = \bigcap_n \boldsymbol{Z}(f_n) \notin \boldsymbol{Z}[M]$. Thus, $M(g) \geqq 0$ but $g \notin M$; therefore $M(g) > 0$.

On the other hand, for any $m \in \mathbf{N}$ and for every x belonging to the member

$$\boldsymbol{Z}(f_1) \cap \cdots \cap \boldsymbol{Z}(f_m)$$

of $\boldsymbol{Z}[M]$, we have

$$g(x) \leqq \sum_{n>m} 2^{-n} = 2^{-m}.$$

Therefore $M(g) \leqq 2^{-m}$ (5.4(a)); and this holds for each $m \in \mathbf{N}$. Hence $M(g)$ is infinitely small. Thus, C/M is nonarchimedean.

(2) *implies* (3). Since $\emptyset \notin \boldsymbol{Z}[M]$, this is trivial.

(3) *implies* (1). If M is hyper-real, there exists f for which $|M(f)|$ is infinitely large. The zero-sets

$$\{x\colon |f(x)| \geqq n\} \qquad (n \in \mathbf{N})$$

are all in $\boldsymbol{Z}[M]$ (5.7(a)), and, obviously, their intersection is empty.

REMARK. It is far less common for a z-filter to be closed under countable intersection than to have the countable intersection property. For instance, the family of all zero-set-neighborhoods of a point in **R** has the latter property but not the former. More generally, a non-maximal z-filter that is contained in a unique z-ultrafilter is *never* closed under countable intersection; see 7H.*4*.

5.15. *Real z-ultrafilters.* When M is a real maximal ideal in C, we refer to $\mathbf{Z}[M]$ as a *real z-ultrafilter*. Thus, the real z-ultrafilters are those with the countable intersection property.

(a) *If a countable union of zero-sets belongs to a real z-ultrafilter $\mathscr{A}$, then at least one of them belongs to $\mathscr{A}$.*

The proof is an exact analogue of the set-theoretic proof, given in 2.13, that any z-ultrafilter is prime.

It is perhaps tempting to conjecture that every z-filter with the countable intersection property is contained in a real z-ultrafilter. The proposition is false, however; see 8H.*5*.

A space X is realcompact (5.9) if and only if every real z-ultrafilter on X is fixed. It follows immediately that *every countable space is realcompact*: no free z-filter on a countable space can have the countable intersection property.

PROBLEMS

5A. ABSOLUTE VALUE.

For every element a of a lattice-ordered ring, $|a| \geqq 0$. [Put $x = |a| \wedge 0$, $y = a \wedge 0$, and $z = -a \wedge 0$. Then $y \leqq x$ and $z \leqq x$. Next, $z = y - a$. Hence

$$0 \leqq (x - y) \wedge (x - z) = (x - y) + (0 \wedge a) \leqq |a|.]$$

5B. CONVEX IDEALS.

1. An arbitrary intersection of [absolutely] convex ideals is [absolutely] convex. An arbitrary union of [absolutely] convex ideals, if it is an ideal, is [absolutely] convex.

2. Let I and J be convex ideals in A, with $I \subset J$. If A/I is totally ordered, then so is A/J.

3. The following condition characterizes a convex ideal I in a lattice-ordered ring as absolutely convex: $x \in I$ implies $x \vee 0 \in I$.

4. Let A be the direct sum of the ring **R** with itself, and let $(r, s) \geqq (0, 0)$ if and only if $r \geqq s \geqq 0$. Then A is a lattice-ordered ring. The set

$$I = \{(0, r) : r \in \mathbf{R}\}$$

is a convex maximal ideal, but is not absolutely convex. The ordered ring A/I

is totally ordered, hence a lattice, but the canonical mapping of A onto A/I, which is order-preserving, is not a lattice homomorphism.

5. If I is any nonmaximal, convex ideal in $C(X)$, then there exists f such that $I(f) > 0$, although f is not positive on any zero-set of I; and the latter assertion holds whenever $(I, f) \neq C$.

5C. HOMOMORPHISM INTO A FUNCTION RING.

Let $\mathfrak{t}$ be a homomorphism from $C(Y)$ (or $C^*(Y)$) into $C(X)$.

1. The kernel I of $\mathfrak{t}$ is absolutely convex. [Theorem 1.6.] (As a matter of fact, I is a z-ideal; see 10D.)

2. If $C(Y)/I$ is ordered as in Theorem 5.2, then the canonical isomorphism of $C(Y)/I$ into $C(X)$ is an order isomorphism.

5D. PRIME IDEALS.

Let P be a prime ideal in C or C^*.

1. If $f^2 + g^2 \in P$, then $f \in P$ and $g \in P$. Compare 2D.*2*.

2. If $f \geqq \mathbf{0}$, $r > 0$, and $f^r \in P$, then $f \in P$.

5E. IDEALS IN $C(\mathbf{R})$.

Let $\mathbf{i}$ denote the identity function on $\mathbf{R}$.

1. The ideal $(|\mathbf{i}|)$ in $C(\mathbf{R})$ is not convex. But $f \in (|\mathbf{i}|)$ implies $|f| \in (|\mathbf{i}|)$.

2. The ideal $(\mathbf{i})$ in $C(\mathbf{R})$ *is* convex [2H.*1*], but it is not absolutely convex. (Hence $(\mathbf{i})$ is neither prime nor a z-ideal. By 5G.*1*, $(\mathbf{i})$ contains no prime ideal.)

3. The set of all functions f in $C(\mathbf{R})$ such that $f(x)/x$ is bounded on a deleted neighborhood of 0 is an absolutely convex ideal, but it is neither prime nor a z-ideal.

5F. DIRECTED RING.

Let A be a partially ordered ring with unity and having the following property: for every $x \in A$, there exists $x' \in A$ such that $x' \geqq x$ and $x' \geqq 0$ (i.e., A is *directed*).

1. If M is a maximal ideal containing no element greater than 1, then M is convex. [If $0 \leqq a \leqq b$, $b \in M$, and $a \notin M$, then there exist $m \in M$ and $x \in A$ such that $1 \leqq m + x'b$.]

2. Suppose, in addition, that A is a lattice-ordered ring. M need not be absolutely convex. [5B.*4*.]

5G. TOTALLY ORDERED RESIDUE CLASS RING.

1. If a convex ideal J in C contains a prime ideal, then J is absolutely convex, and C/J is totally ordered.

2. If C/I is totally ordered, then I is contained in a unique maximal ideal. [5B.*2*, 5.4(c), and 2.10.]

3. Let J denote the principal ideal in $C(\Sigma)$ (see 4M) generated by the continuous extension of $\mathbf{j}$ to Σ. Then $O_\sigma \subset J \subset M_\sigma$ (properly), so that J is not a z-ideal. Furthermore, C/J is a totally ordered ring, although J is not prime.

4. Every ideal in $C(\Sigma)$ is absolutely convex.

5H. REALCOMPACT AND PSEUDOCOMPACT SPACES.

1. The arguments given in 5.10 show that $\mathbf{R}$ is realcompact.

2. X is compact if and only if it is both realcompact and pseudocompact.

3. Construct a normal space that is neither realcompact nor pseudocompact.

4. X is pseudocompact if and only if every z-filter has the countable intersection property. Contrast with countably compact (1.4).

5. X is countably compact if and only if every countable closed set is compact.

6. X is pseudocompact if and only if every countable, C-embedded subset is compact. [*Necessity.* Every C-embedded subset of X is pseudocompact; apply *2*. *Sufficiency.* Corollary 1.21.] (The space Λ of 6P is pseudocompact and has a countable, *C^*-embedded* subset that is not compact.)

7. If X is pseudocompact, then every countable zero-set is compact. [Lemma 4.10.] (The converse is false; see 6I.*3*.)

5I. THE SPACE Ψ.

1. There exists an infinite maximal family $\mathscr{E}$ of infinite subsets of $\mathbf{N}$ such that the intersection of any two is finite. [Start with infinitely many disjoint infinite sets, and apply the maximal principle.]

2. Let $D = \{\omega_E\}_{E \in \mathscr{E}}$ be a new set of distinct points, and define $\Psi = \mathbf{N} \cup D$, with the following topology: the points of $\mathbf{N}$ are isolated, while a neighborhood of a point ω_E is any set containing ω_E and all but a finite number of points of E. Thus, $E \cup \{\omega_E\}$ is the one-point compactification of E, and $\mathbf{N}$ is dense in Ψ. The space Ψ is completely regular.

3. D is an infinite, discrete zero-set in Ψ. Hence Ψ is not countably compact.

4. Ψ satisfies the first countability axiom, and every subset of Ψ is a G_δ.

5. Ψ is pseudocompact. [Maximality of $\mathscr{E}$ implies that an unbounded function on $\mathbf{N}$ is unbounded on some member of $\mathscr{E}$.] Hence Ψ is not normal. [3D.*2*.] Also, Ψ is not realcompact.

6. Each finite subset, but no countably infinite subset of D, is a zero-set in Ψ. [5H.*7*.] Hence D is not C^* embedded in Ψ. [1F.*3*.]

5J. CARDINALS OF RESIDUE CLASS RINGS.

If I is a z-ideal in C, and $\mathfrak{m}$ is the smallest of the cardinal numbers of all the dense subsets of Z, for all $Z \in \mathbf{Z}[I]$, then $|C/I| \leqq \mathfrak{c}^{\mathfrak{m}}$. [Two functions in C are congruent modulo I if they agree on a dense subset A of some $Z \in \mathbf{Z}[I]$. Consider the cardinal of $\mathbf{R}^A$.] (If I is not a z-ideal, the conclusion can fail; see 12F.*3*.)

5K. MAXIMAL IDEALS IN C AND C^*.

If M is a maximal ideal in C, then $M \cap C^*$ is a maximal ideal in C^* if and only if M is real. [*Necessity.* If M is not real, there exists $f \in C^*$ with $M(f)$ infinitely small; then $(M \cap C^*, f)$ is a proper ideal in C^*. *Sufficiency.* The canonical homomorphism of C onto $C/M = \mathbf{R}$ sends C^* onto $\mathbf{R}$.] (Another proof is given in 7.9, where the correspondence between maximal ideals in C and those in C^* is clarified.)

5L. UNIQUE FREE MAXIMAL IDEAL IN $C(\mathbf{W})$.

1. Any two cofinal, closed sets in $\mathbf{W}$ meet in a cofinal set. [5.12(b).]

2. Prove directly from *1* that the family $\mathscr{A}$ of all cofinal zero-sets is the unique free z-ultrafilter on $\mathbf{W}$. [Theorem 2.6(b).]

3. Prove directly from the fact that $\mathscr{A}$ is real that for every $f \in C(\mathbf{W})$, there exists $r \in \mathbf{R}$ such that $f^{\leftarrow}(r)$ is cofinal in $\mathbf{W}$. As in 5.12, this implies that f is constant on a tail.

5M. FURTHER PROPERTIES OF $\mathbf{W}$ AND $\mathbf{W}^*$.

1. The only free ideal in $C(\mathbf{W})$ is M^{ω_1}. [4D.]

2. $\mathbf{W}$ satisfies the first countability axiom. The set of all limit ordinals in $\mathbf{W}$ is a closed set, but not a G_δ. [3D.*3*.]

3. Characterize the closed sets in $\mathbf{W}$ that are zero-sets. [See 13B.*2*.]

4. Describe the finite zero-sets in $\mathbf{W}^*$.

5. In the one-point compactification of the topological sum of two copies of $\mathbf{W}$, each copy is C^*-embedded, but their union is not.

6. If X is compact, then $\mathbf{W} \times X$ is countably compact. [Every countable set is contained in a compact one.] (More generally, the product of a countably compact space with a compact space is countably compact; see 9J.)

5N. WELL-ORDERED SPACES.

1. If $W(\alpha)$ is C^*-embedded in $W(\alpha + 1)$, then it is C-embedded, and the conditions hold if and only if every countable set in $W(\alpha)$ is bounded. [Utilize the fact that a countably infinite, well-ordered set without last element contains a cofinal sequence; this result is proved in 13.6(a).]

2. Modify the proof of 5.11(b) to show, without using the axiom of choice, that every well-ordered space is normal.

5O. P-POINTS OF TOTALLY ORDERED SPACES.

1. It is plain that ω_1 is a P-point (4L) of $\mathbf{W}^*$. More generally, a point p of a totally ordered space (3O) is a P-point if and only if it is neither the limit of an increasing sequence nor the limit of a decreasing sequence. [4L.*1*.]

2. A point p of a totally ordered space is a P-point if and only if O_p (4I) is a prime ideal. [Theorem 2.9.]

5P. *P*-SPACES.

If X is a P-space (4J), then $C(X)/I$ is totally ordered if and only if I is maximal. [5.4(c).]

5Q. *e*-IDEALS.

Let I be an e-ideal (2L) in C^*, and let $f \in C^*$.

1. $f \equiv |f| \pmod I$ if and only if, for every $\epsilon > 0$, $f + \boldsymbol{\epsilon}$ is nonnegative on some member of $\boldsymbol{E}(I)$.

2. If there exists $\epsilon > 0$ such that $f - \boldsymbol{\epsilon}$ is positive on some member of $\boldsymbol{E}(I)$, then $f \equiv |f| \not\equiv \mathbf{0}$. And if I is maximal, the converse holds as well. [2L.*14*.] (For maximal ideals, these results also follow from 7.2 and 7R.*2*.)

5R. ALGEBRAIC INVARIANCE OF THE NORM.

The norm $\|f\| = \sup_{x \in X} |f(x)|$ on $C^*(X)$ satisfies

$$\|f\| = \sup_M |M(f)|,$$

M ranging over all the maximal ideals in C^*. [$|f| \leqq \boldsymbol{r}$ implies $|M(f)| \leqq r$. Alternatively, use 2C.*2*.] Compare 1J.*6*.

Chapter 6

THE STONE-ČECH COMPACTIFICATION

6.1. Among the major achievements in the theory of rings of continuous functions are the characterizations of the maximal ideals in $C(X)$ and in $C^*(X)$. We have already succeeded in obtaining characterizations in case X is compact—by attaching each maximal ideal to a point of the space (see 4.9(a)). The next step in our program is to extend this result, somehow, to the case of arbitrary (completely regular) X. In the general situation—when X is not pseudocompact—we will be faced with two distinct problems: that of C, and that of C^*.

In the case of C^*, we can be guided by the example of the noncompact space $\mathbf{W}$. Here we were able to find a compactification, $\mathbf{W}^*$, in which $\mathbf{W}$ is C^*-embedded. Thus, there is a natural isomorphism between $C^*(\mathbf{W})$ and $C^*(\mathbf{W}^*)$, and the characterization of the maximal ideals in $C^*(\mathbf{W})$ was accomplished via this isomorphism, by attaching each of these ideals to a point of the compactification $\mathbf{W}^*$.

In the case of $C(X)$, our guide will be the correspondence between its maximal ideals and the z-ultrafilters on X (Theorem 2.5). We wish to sharpen this characterization by associating each z-ultrafilter with a point of some suitable space, in a natural way. The fixed z-ultrafilters are associated with the points of X itself; roughly speaking, we wish to "fix the free z-ultrafilters." Once again, then, we seek a *compactification* of X.

It is a remarkable fact that these two problems not only have solutions, but a common one: the compactification, βX, that serves to characterize the maximal ideals in $C(X)$ is also a compactification in which X is C^*-embedded. Moreover, the space βX—known as the *Stone-Čech compactification* of X—is essentially unique.

Several different constructions of βX are known. Probably the simplest is the one to be presented in Chapter 11. This is also the construction in which the C^*-embedding of X shows up most transparently.

The procedure adopted in the present chapter is fitted to the problem of $C(X)$: we construct βX by adjoining to X one new point for each free z-ultrafilter, to serve as its limit (as defined below). The topological considerations involved in this process, and some of their ramifications, will occupy the entire chapter. The formal characterizations of the maximal ideals in $C(X)$ and in $C^*(X)$ are treated in Chapter 7.

6.2. In general, one compactifies a noncompact space X by adjoining new points to serve as limit points of sets that are closed in X but not compact. In considering this process, we may restrict our attention to the zero-sets, since they form a base for the closed sets. It will also be convenient, and instructive—though by no means necessary—to translate the problem into the language of convergence of z-filters. (Implicit in this transition is the circumstance that a noncompact zero-set is contained in a free z-filter (Lemma 4.10).) For this, the ideas set forth in 3.16 must first be cast in a more general setting.

Let X be dense in a space T, and consider a z-filter (or filter) $\mathscr{F}$ on X. We shall say that a point $p \in T$ is a *cluster point* of $\mathscr{F}$ if every neighborhood (in T) of p meets every member of $\mathscr{F}$—in other words,

(a) *p is a cluster point of $\mathscr{F}$ provided that*

$$p \in \bigcap_{Z \in \mathscr{F}} \operatorname{cl}_T Z.$$

Now, in any case, any such family of closed sets $\operatorname{cl}_T Z$ has the finite intersection property. Consequently, a necessary condition that T be *compact* is that every z-filter on X have a cluster point in T. It is not hard to see that the converse holds as well; in fact, it is sufficient that every *z-ultrafilter* on X have a cluster point in T (6F.4).

We shall say that $\mathscr{F}$ *converges* to the *limit* p if every neighborhood (in T) of p contains a member of $\mathscr{F}$. As in 3.16 ff., one sees immediately that limits are unique, and that a z-ultrafilter converges to any cluster point.

The crucial distinction between the situation in 3.16 (i.e., $T = X$) and the present one is this: *two different z-ultrafilters on X can have a common limit in T.* The counterexample given in 3.18 for ultrafilters leads to an example here: the traces of $\mathscr{U}$ and $\mathscr{U}'$ on $\mathbf{N}$ are distinct z-ultrafilters on $\mathbf{N}$, both of which converge to the point ω of $\mathbf{N}^*$. It is easy to generalize the construction in 3.18 so as to obtain infinitely many free z-ultrafilters on $\mathbf{N}$; and it is plain that *every* free z-ultrafilter on $\mathbf{N}$ converges to ω.

We have remarked that distinct free z-ultrafilters on X are to converge to distinct points of βX. Thus, the one-point compactification $\mathbf{N}^*$ of $\mathbf{N}$ is a hopelessly inadequate candidate for $\beta\mathbf{N}$. Even more obvious is the fact that $\mathbf{N}^*$ fails the test for C^*, i.e., $\mathbf{N}$ is not C^*-embedded in $\mathbf{N}^*$.

6.3. When will a given dense subspace X of a space T be C^*-embedded in T? To answer this question, we examine a more general one: when can every continuous mapping τ, from X into an arbitrary compact space Y, be extended to a continuous mapping $\bar{\tau}$ from T into Y?

The condition that $\bar{\tau}$ carry closures into closures will appear, in the language of convergence of z-filters, as the requirement that it take cluster points to cluster points—or limits to limits.

(a) *Let Z be a zero-set in X. If $p \in \mathrm{cl}_T Z$, then at least one z-ultrafilter on X contains Z and converges to p.*

Let $\mathscr{E}$ be the z-filter on T of all zero-set-neighborhoods (in T) of p, and let $\mathscr{B}$ be the trace of $\mathscr{E}$ on X. Since $p \in \mathrm{cl}\, Z$, $\mathscr{B} \cup \{Z\}$ has the finite intersection property, and so is contained in a z-ultrafilter $\mathscr{A}$. Clearly, $\mathscr{A}$ converges to p.

Notice that this result holds even if X is not dense in T.

(b) *If X is dense in T, then every point of T is the limit of at least one z-ultrafilter on X*

This follows from (a) (with $Z = X$).

6.4. THEOREM. *Let X be dense in T. The following statements are equivalent.*

(1) *Every continuous mapping τ from X into any compact space Y has an extension to a continuous mapping from T into Y.*

(2) *X is C^*-embedded in T.*

(3) *Any two disjoint zero-sets in X have disjoint closures in T.*

(4) *For any two zero-sets Z_1 and Z_2 in X,*

$$\mathrm{cl}_T (Z_1 \cap Z_2) = \mathrm{cl}_T Z_1 \cap \mathrm{cl}_T Z_2.$$

(5) *Every point of T is the limit of a unique z-ultrafilter on X.*

PROOF. (1) *implies* (2). A function f in $C^*(X)$ is a continuous mapping into the compact subset $\mathrm{cl}_{\mathbf{R}} f[X]$ of $\mathbf{R}$. Hence (2) is a special case of (1).

(2) *implies* (3). This follows from the simpler half of Urysohn's extension theorem (1.17).

(3) *implies* (4). If $p \in \operatorname{cl} Z_1 \cap \operatorname{cl} Z_2$, then for every zero-set-neighborhood V (in T) of p, we have

$$p \in \operatorname{cl}(V \cap Z_1) \quad \text{and} \quad p \in \operatorname{cl}(V \cap Z_2).$$

Hence (3) implies that $V \cap Z_1$ meets $V \cap Z_2$, i.e., V meets $Z_1 \cap Z_2$. Therefore

$$p \in \operatorname{cl}(Z_1 \cap Z_2).$$

Thus, $\operatorname{cl} Z_1 \cap \operatorname{cl} Z_2$ is contained in $\operatorname{cl}(Z_1 \cap Z_2)$. The reverse inclusion is trivial.

(4) *implies* (5). Since X is dense in T, each point of T is the limit of at least one z-ultrafilter. On the other hand, distinct z-ultrafilters contain disjoint zero-sets (Theorem 2.6(b)), and the hypothesis (in fact, its weaker form (3)) implies that a point p cannot belong to the closures of both these zero-sets. Hence the two z-ultrafilters cannot both converge to p.

We remark that the hypothesis that X is dense has been used, thus far, only to secure the existence of the z-ultrafilter in (5).

(5) *implies* (1). Given $p \in T$, let $\mathscr{A}$ denote the unique z-ultrafilter on X with limit p. As in 4.12, we write

$$\tau^{\#}\mathscr{A} = \{E \in \mathbf{Z}(Y) : \tau^{\leftarrow}[E] \in \mathscr{A}\}.$$

This is a z-filter on the compact space Y, and so it has a cluster point (Theorem 4.11). Moreover, since $\mathscr{A}$ is a prime z-filter, so is $\tau^{\#}\mathscr{A}$. Therefore, by Theorem 3.17, $\tau^{\#}\mathscr{A}$ has a limit in Y. Denote this limit by $\bar{\tau}p$:

(a) $$\bigcap \tau^{\#}\mathscr{A} = \{\bar{\tau}p\}.$$

This defines a mapping $\bar{\tau}$ from T into Y.

In case $p \in X$, we have $p \in \bigcap \mathscr{A}$, so that $\tau p \in \bigcap \tau^{\#}\mathscr{A}$. Therefore $\bar{\tau}$ agrees with τ on X.

For $F, F' \in \mathbf{Z}(Y)$, let us write $Z = \tau^{\leftarrow}[F]$ and $Z' = \tau^{\leftarrow}[F']$. If $p \in \operatorname{cl}_T Z$, then by 6.3(a), Z belongs to $\mathscr{A}$, and so $F \in \tau^{\#}\mathscr{A}$. Thus, $p \in \operatorname{cl} Z$ implies $\bar{\tau}p \in F$.

To establish continuity of $\bar{\tau}$ at the point p, we consider an arbitrary zero-set-neighborhood F of $\bar{\tau}p$ and exhibit a neighborhood of p that is carried by $\bar{\tau}$ into F. Let F' be a zero-set whose complement is a neighborhood of $\bar{\tau}p$ contained in F. Then $F \cup F' = Y$, so that $Z \cup Z' = X$, and therefore $\operatorname{cl} Z \cup \operatorname{cl} Z' = T$. Since $\bar{\tau}p \notin F'$, we have $p \notin \operatorname{cl} Z'$; therefore $T - \operatorname{cl} Z'$ is a neighborhood of p. And every point q in this neighborhood belongs to $\operatorname{cl} Z$, whence $\bar{\tau}q \in F$.

This completes the proof of the theorem. An alternative proof that (2) implies (1) will be presented in Theorem 10.7.

CONSTRUCTION OF βX

6.5. COMPACTIFICATION THEOREM. *Every (completely regular) space X has a compactification βX, with the following equivalent properties.*

(I) (STONE) *Every continuous mapping τ from X into any compact space Y has a continuous extension $\bar{\tau}$ from βX into Y.*

(II) (STONE-ČECH) *Every function f in $C^*(X)$ has an extension to a function f^β in $C(\beta X)$.*

(III) (ČECH) *Any two disjoint zero-sets in X have disjoint closures in βX.*

(IV) *For any two zero-sets Z_1 and Z_2 in X,*

$$\mathrm{cl}_{\beta X}(Z_1 \cap Z_2) = \mathrm{cl}_{\beta X} Z_1 \cap \mathrm{cl}_{\beta X} Z_2.$$

(V) *Distinct z-ultrafilters on X have distinct limits in βX.*

Furthermore, βX is unique, in the following sense: if a compactification T of X satisfies any one of the listed conditions, then there exists a homeomorphism of βX onto T that leaves X pointwise fixed.

The assertion that βX satisfies (I) will be referred to as *Stone's theorem*, and the mapping $\bar{\tau}$ will be called the *Stone extension* of τ into Y.

PROOF. We commence with the proof of uniqueness. By Theorem 6.4, if T satisfies one of (I)—(V), it satisfies *all* of them. By (I), the identity mapping on X, which is a continuous mapping into the compact space T, has a Stone extension from all of βX into T; similarly, it has a Stone extension from T into βX. It follows, as was pointed out in 0.12(a), that these extensions are homeomorphisms onto.

We turn now to the construction of βX. There is to be a one-one correspondence between the z-ultrafilters on X and the points of βX, each z-ultrafilter converging to its corresponding point. Now, we already have such a correspondence between the *fixed* z-ultrafilters and the points of X (3.18(b)); hence X constitutes a ready-made index set for the fixed z-ultrafilters. We increase it in any convenient way to an index set for the family of *all* z-ultrafilters.

(a) *The points of βX are defined to be the elements of this enlarged index set.*

The family of all z-ultrafilters on X is written

$$(A^p)_{p \in \beta X}$$

—with the understanding that for $p \in X$, A^p represents the (fixed) z-ultrafilter with limit p (i.e., the family of all zero-sets containing p). When emphasis is desirable, we shall denote A^p by A_p, for $p \in X$; thus,

$A_p = \boldsymbol{Z}[M_p]$. The topology on βX will be defined in such a way that p is the limit of the z-ultrafilter A^p for every $p \in \beta X$, not only for $p \in X$.

In what follows, Z will always stand for a zero-set in the given topological space X. Let us write

$$\bar{Z} = \{p \in \beta X: Z \in A^p\},$$

that is, $p \in \bar{Z}$ if and only if $Z \in A^p$. In particular, since X itself belongs to every z-ultrafilter, we have $\bar{X} = \beta X$.

We know that $Z_1 \cup Z_2 \in A^p$ if and only if $Z_1 \in A^p$ or $Z_2 \in A^p$; therefore

$$\bar{Z}_1 \cup \bar{Z}_2 = \overline{Z_1 \cup Z_2}.$$

And since $\emptyset$ belongs to no z-ultrafilter, $\bar{\emptyset} = \emptyset$. Thus, the family of sets $\bar{Z}$ is closed under finite union and contains the empty set.

(b) *βX is made into a topological space by taking the family of all sets $\bar{Z}$ as a base for the closed sets.*

Let us verify first that X is a subspace of βX. Evidently, $p \in \bar{Z} \cap X$ if and only if $Z \in A_p$, which is to say that $p \in Z$. So $\bar{Z} \cap X = Z$. Thus, the identity mapping on X carries the family of basic closed sets in the relative topology onto a family of basic closed sets in the original completely regular topology (see Theorem 3.2); therefore it is a homeomorphism.

Next, we show that X is dense in βX. In fact, we shall prove, more generally, that

$$\text{cl}_{\beta X} Z = \bar{Z},$$

from which the conclusion $\text{cl}_{\beta X} X = \bar{X} = \beta X$ follows. We know that $Z \subset \bar{Z}$, whence $\text{cl } Z \subset \bar{Z}$. On the other hand, for every basic closed set $\bar{Z}'$ containing Z, we have

$$Z' = \bar{Z}' \cap X \supset Z,$$

so that $\bar{Z}' \supset \bar{Z}$. Therefore, $\text{cl } Z \supset \bar{Z}$.

We now have:

(c) *$p \in \text{cl}_{\beta X} Z$ if and only if $Z \in A^p$.*

Since $Z_1 \cap Z_2 \in A^p$ if and only if $Z_1 \in A^p$ and $Z_2 \in A^p$, this immediately yields (IV). Accordingly, the proof will be complete as soon as we know that βX is compact—for then, as already mentioned, all five of the listed conditions will hold.

To see, first of all, that βX is a Hausdorff space, consider any two distinct points p and p'. Choose disjoint zero-sets $A \in A^p$ and $A' \in A^{p'}$

(Theorem 2.6(b)). By 1.15(a), there exist a zero-set Z disjoint from A, and a zero-set Z' disjoint from A', such that $Z \cup Z' = X$. Evidently, $Z \notin A^p$ and $Z' \notin A^{p'}$; that is to say, $p \notin \text{cl } Z$ and $p' \notin \text{cl } Z'$. Since

$$\text{cl } Z \cup \text{cl } Z' = \beta X,$$

the neighborhoods $\beta X - \text{cl } Z$ of p, and $\beta X - \text{cl } Z'$ of p', are disjoint.

Finally, consider any collection of basic closed sets cl Z with the finite intersection property, Z ranging over some family $\mathscr{B}$. By (IV), already established, $\mathscr{B}$ itself also has the finite intersection property. Consequently, $\mathscr{B}$ is embeddable in a z-ultrafilter A^p, and we have

$$p \in \bigcap_{Z \in A^p} \text{cl } Z \subset \bigcap_{Z \in \mathscr{B}} \text{cl } Z,$$

so that the latter intersection is nonempty. Therefore βX is compact.

6.6. REMARKS. The space βX is known as the *Stone-Čech compactification* of X. According to the theorem, it is characterized as *that compactification of X in which X is C^*-embedded.*

Incidentally, the equivalence of (II) with (III) is an immediate consequence of Urysohn's extension theorem (because disjoint closed sets in βX are completely separated).

In the construction of βX, nothing was said about *what* objects are to serve as the points of $\beta X - X$. In principle, of course, it makes no difference what these points are. It often happens, in a given discussion, that there are natural candidates for some or all of the points of $\beta X - X$; we shall then allow ourselves to regard these as the points in question, even without explicit mention. For example, we say that $\mathbf{W}^*$ is $\beta\mathbf{W}$. More significant instances occur in 6.7 and 6.9(a) below.

Let $\mathscr{F}$ be a z-filter on X, and let $p \in \beta X$; from 6.2(a) and 6.5(c), we get:

(a) *p is a cluster point of $\mathscr{F}$ if and only if $\mathscr{F} \subset A^p$.*

In particular, A^p itself converges to p.

From 6.4(a), we can read off the definition of the Stone extension $\bar{\tau}$:

$$\{\bar{\tau}p\} = \bigcap \tau^{\#}A^p \qquad (p \in \beta X).$$

From (II), it follows immediately that:

(b) *The mapping $f \to f^\beta$ is an isomorphism of $C^*(X)$ onto $C(\beta X)$.*

Thus, "every C^* is a C." Consequently:

(c) *Any purely algebraic result that is valid for every C holds as well for every C^*.*

By virtue of Theorem 3.9, this is true without restriction on the topology of the space.

The converse to the last assertion in Theorem 6.5, regarding uniqueness, is trivially true. (See, however, 6C.*2*.) The assertion itself may, of course, be expressed in terms of an arbitrary homeomorphism, not just the identity: if τ is a homeomorphism of X onto X', then $\bar{\tau}$ is a homeomorphism of βX onto $\beta X'$; and, evidently, $\bar{\tau}$ carries $\beta X - X$ onto $\beta X' - X'$.

6.7. THEOREM. *Let X be dense in T. The following are equivalent with the statements* (1) *to* (5) *of Theorem* 6.4.

(6) $X \subset T \subset \beta X$.

(7) $\beta T = \beta X$.

PROOF. We base the proof upon condition (2): X is C^*-embedded in T.

(2) *implies* (7). X is dense and C^*-embedded in βT. Therefore $\beta X = \beta T$.

(7) *implies* (6). $X \subset T \subset \beta T = \beta X$.

(6) *implies* (2). X is C^*-embedded in βX, and hence in T.

This result specifies all of the spaces in which X is dense and C^*-embedded: they are precisely the subspaces of βX that contain X. The only *compact* one among them is βX itself, of course.

APPLICATIONS OF THE COMPACTIFICATION THEOREM

6.8. The power of Stone's theorem is illustrated by the following proof.

TYCHONOFF PRODUCT THEOREM. *Any product of compact spaces is compact.*

PROOF. Let $X = \times_\alpha X_\alpha$, where each X_α is compact. Then X is completely regular.

Each projection $\pi_\alpha: X \to X_\alpha$ has a Stone extension $\bar{\pi}_\alpha: \beta X \to X_\alpha$. By 3.10(c), the mapping $p \to (\bar{\pi}_\alpha p)_\alpha$ from the compact space βX onto X is continuous.

6.9. Let S be a subspace—not necessarily dense— of X. Evidently, S is C^*-embedded in X if and only if it is C^*-embedded in βX. Since the compact set $\text{cl}_{\beta X} S$ is C^*-embedded in βX (3.11(c)), these conditions hold if and only if S is C^*-embedded in $\text{cl}_{\beta X} S$. Therefore, under these conditions, $\text{cl}_{\beta X} S$ satisfies the characteristic properties of βS: it is a compactification of S in which S is C^*-embedded. Thus:

(a) *S is C^*-embedded in X if and only if* $\text{cl}_{\beta X} S = \beta S$.

The following result, referred to above, is restated here for emphasis:

(b) *Every compact set in X is C^*-embedded in X.*

This fact is particularly transparent from (a): if S is compact, then $\mathrm{cl}_{\beta X} S = S = \beta S$, whence S is C^*-embedded in X.

(c) *If S is open-and-closed in X, then $\mathrm{cl}_{\beta X} S$ and $\mathrm{cl}_{\beta X} (X - S)$ are complementary open sets in βX.*

For, βX is the union of cl S with cl $(X - S)$, and, by (III), these sets are disjoint.

From 3.15, we have:

(d) *An isolated point of X is isolated in βX; and X is open in βX if and only if X is locally compact.*

The first assertion is also a special case of (c).

6.10. EXAMPLES. $\beta\mathbf{N}$, $\beta\mathbf{Q}$, *and* $\beta\mathbf{R}$. Let us try to get some idea of what these spaces must be like.

By 6.9(d), $\mathbf{N}$ is open in $\beta\mathbf{N}$, and $\mathbf{R}$ is open in $\beta\mathbf{R}$, but $\mathbf{Q}$ is not open in $\beta\mathbf{Q}$.

The space $\beta\mathbf{N}$. More specifically, every point of $\mathbf{N}$ is an isolated point of $\beta\mathbf{N}$. These are the only isolated points, of course, since $\mathbf{N}$ is dense in $\beta\mathbf{N}$. By 6.9(c), the closure in $\beta\mathbf{N}$ of every subset of $\mathbf{N}$ is open in $\beta\mathbf{N}$. The points p of $\beta\mathbf{N} - \mathbf{N}$ are in one-one correspondence with the free ultrafilters A^p on $\mathbf{N}$, with A^p converging to p. Hence every neighborhood of p meets $\mathbf{N}$ in a member of A^p. On the other hand, if $Z \in A^p$, then cl Z is an open neighborhood of p. Next, $\beta\mathbf{N}$ is totally disconnected: given distinct points p and q, choose $Z \in A^p - A^q$; then cl Z is an open-and-closed set containing p but not q.

The subset N_1 of odd integers is C^*-embedded in $\mathbf{N}$. Therefore $\mathrm{cl}\, N_1 = \beta N_1$ (6.9(a)), and hence $\mathrm{cl}\, N_1$ is homeomorphic with $\beta\mathbf{N}$. Similarly for the subset N_2 of even integers. Thus, $\beta\mathbf{N}$ is expressible as the union of two disjoint copies of itself.

We can also decompose $\mathbf{N}$ into *infinitely* many disjoint infinite sets A_n $(n \in \mathbf{N})$. The sets cl A_n are then disjoint open-and-closed subsets of $\beta\mathbf{N}$, and each is homeomorphic with $\beta\mathbf{N}$. Now, though, their union

$$T = \bigcup_n \mathrm{cl}\, A_n$$

is *not* all of $\beta\mathbf{N}$, as a compact space cannot be a union of infinitely many disjoint open sets. However, T is dense in $\beta\mathbf{N}$; in fact, $\mathbf{N} \subset T \subset \beta\mathbf{N}$, so that $\beta T = \beta\mathbf{N}$ (6.7).

Let us choose a point $p_n \in \mathrm{cl}\, A_n - \mathbf{N}$, and define

$$D = \{p_1, p_2, \cdots\}.$$

Trivially, D is a discrete subspace of $\beta\mathbf{N}$; so D is homeomorphic with $\mathbf{N}$. Moreover, D is C^*-embedded (in fact, C-embedded) in T: to extend a function f on D to a continuous function on T, simply assign the constant value $f(p_n)$ to each point of cl A_n. It follows that D is C^*-embedded in βT—which is $\beta\mathbf{N}$. By 6.9(a), $\text{cl}_{\beta\mathbf{N}} D = \beta D$. Now, since D is contained in the closed set $\beta\mathbf{N} - \mathbf{N}$, so is cl D; thus, $\beta\mathbf{N} - \mathbf{N} \supset \beta D$. But βD is homeomorphic with $\beta\mathbf{N}$. We have proved:

(a) $\beta\mathbf{N} - \mathbf{N}$ *contains a copy of* $\beta\mathbf{N}$.

Let f be a function in $C^*(\mathbf{N})$ that assumes all rational values in $[0, 1]$. Then every real number in $[0, 1]$ belongs to the closure of $f[\mathbf{N}]$. Consequently, the compact set $f^\beta[\beta\mathbf{N}]$ is all of $[0, 1]$. In particular, the cardinal of $\beta\mathbf{N}$ must be at least the cardinal $\mathfrak{c}$ of the continuum. (As a matter of fact, its cardinal is $2^\mathfrak{c}$ (9.3 or 9O).)

The space $\beta\mathbf{Q}$. This space, too, is totally disconnected. For, distinct points p and q are contained in disjoint closed neighborhoods U and V, respectively. Then $U \cap \mathbf{Q}$ and $V \cap \mathbf{Q}$ are disjoint closed sets in $\mathbf{Q}$. Consequently, there is an open-and-closed set E in $\mathbf{Q}$ containing $U \cap \mathbf{Q}$ and disjoint from $V \cap \mathbf{Q}$. (This is proved in 16.16, to which the reader may refer directly.) The open-and-closed set $\text{cl}_{\beta\mathbf{Q}} E$ in $\beta\mathbf{Q}$ (6.9(c)) then contains p but not q.

Any mapping τ of $\mathbf{N}$ onto $\mathbf{Q}$ is a continuous mapping into the compact space $\beta\mathbf{Q}$; as such, it has a Stone extension $\bar{\tau}$ from all of $\beta\mathbf{N}$ into $\beta\mathbf{Q}$. Since the range of $\bar{\tau}$ is a compact set in $\beta\mathbf{Q}$, and contains the dense set $\mathbf{Q}$, it must be all of $\beta\mathbf{Q}$. Thus, $\beta\mathbf{Q}$ is a continuous image of $\beta\mathbf{N}$. On the other hand, 6.9(a) implies that

$$\text{cl}_{\beta\mathbf{Q}} \mathbf{N} = \beta\mathbf{N}.$$

Therefore $\beta\mathbf{Q}$ is equipotent with $\beta\mathbf{N}$.

Since $\beta\mathbf{Q}$ is compact, every neighborhood of a point contains a compact neighborhood. But, clearly, no compact neighborhood can be contained entirely in $\mathbf{Q}$. It follows that $\beta\mathbf{Q} - \mathbf{Q}$ is dense in $\beta\mathbf{Q}$. (This argument is general and shows that if T is any locally compact space containing $\mathbf{Q}$, then $T - \mathbf{Q}$ is dense in T.)

The space $\beta\mathbf{R}$. As above, $\beta\mathbf{R}$ is a continuous image of $\beta\mathbf{N}$, and

(b) $$\text{cl}_{\beta\mathbf{R}} \mathbf{N} = \beta\mathbf{N}.$$

Hence $\beta\mathbf{R}$ is equipotent with $\beta\mathbf{N}$.

Let R^+ denote the subspace of all nonnegative reals, and R^- the subspace of nonpositive reals. Since the closure of a connected set is always connected, $\text{cl}_{\beta\mathbf{R}} R^+$ and $\text{cl}_{\beta\mathbf{R}} R^-$, as well as $\beta\mathbf{R} = \text{cl}\, \mathbf{R}$, are connected.

Trivially, R^+ is C^*-embedded in $\mathbf{R}$, so that cl $R^+ = \beta R^+$. Since R^+ is homeomorphic with R^-, cl R^+ is homeomorphic with cl R^-, and cl $R^+ - R^+$ with cl $R^- - R^-$. Every neighborhood of a point of cl $R^+ - R^+$ meets R^+ in an unbounded set. Since R^+ is locally compact, it is open in cl R^+, whence cl $R^+ - R^+$ is compact.

Obviously, cl $R^+ \cup$ cl $R^- = \beta\mathbf{R}$. The arctangent function in $C(\mathbf{R})$ has a continuous extension to all of $\beta\mathbf{R}$; clearly, this extension assumes the sole value $\pi/2$ on cl $R^+ - R^+$, and the sole value $-\pi/2$ on cl $R^- - R^-$. Therefore $\beta\mathbf{R} - \mathbf{R}$ is disconnected: cl $R^+ - R^+$ is disjoint from cl $R^- - R^-$, and their union is $\beta\mathbf{R} - \mathbf{R}$.

Finally, we show that cl $R^+ - R^+$ is connected. If not, it admits a continuous function that assumes precisely the values 0 and 1. This has an extension to a function $f \in C(\text{cl } R^+)$ (since cl $R^+ - R^+$ is compact), and f must assume values near 0, and values near 1, at arbitrarily large $x \in R^+$. Since R^+ is connected, f must assume the value $1/2$ on an unbounded set in R^+, and hence also at some point of cl $R^+ - R^+$. This contradiction shows that cl $R^+ - R^+$ is connected. Thus, $\beta\mathbf{R} - \mathbf{R}$ is the union of two disjoint, homeomorphic connected sets. For an alternative proof that cl $R^+ - R^+$ is connected, see 6L.*3*.

6.11. LEMMA. *If φ is a continuous mapping of a space E into a space Y, whose restriction to a dense set X is a homeomorphism, then φ carries $E - X$ into $Y - \varphi[X]$.*

PROOF. Suppose, on the contrary, that $\varphi p = \varphi x$, where $x \in X$ and $p \neq x$. Let V be a closed neighborhood of x in E that does not contain p. The homeomorphism $\varphi|X$ carries $V \cap X$ onto a neighborhood of φx in $\varphi[X]$, i.e., onto a set of the form $W \cap \varphi[X]$, where W is a neighborhood of φx in Y. Since X is dense, every neighborhood of p contains points of $X - V$. The homeomorphism $\varphi|X$ takes all such points into $Y - W$. Hence no neighborhood of p is carried by φ into W. Thus, φ is not continuous at p.

6.12. From Stone's theorem and the lemma, we obtain:

THEOREM. *Every compactification of X is a continuous image of βX. Moreover, if τ is any homeomorphism from X into a compact space Y, then its Stone extension $\bar{\tau}$ into Y carries $\beta X - X$ into $Y - \tau[X]$.*

6.13. For later use, we record here some elementary consequences of the lemma.

(a) *Let $X \subset T$. If there exists a continuous mapping from T to X whose restriction to X is the identity (i.e., X is a retract of T), then X is closed.*

According to the lemma (with $E = \text{cl}\, X$, and $Y = X$), the mapping carries $\text{cl}\, X - X$ into the empty set. (Alternative proof: The fixed points of any continuous map of a space into itself form a closed set.)

This yields

(b) *If T contains a product $X = \times_\alpha X_\alpha$, and each projection $\pi_\alpha: X \to X_\alpha$ has a continuous extension $\pi_\alpha^1: T \to X_\alpha$, then X is closed.*

By 3.10(c), the mapping $p \to (\pi_\alpha^1 p)_\alpha$ from T to X is continuous; and, obviously, its restriction to X is the identity.

In similar vein, we obtain the following familiar result.

(c) *The graph of a continuous mapping $\varphi: A \to B$ is a closed set in $A \times B$.*

Each of the mappings $(a, b) \to a$ and $(a, b) \to \varphi a$ is continuous. By 3.10(c), the mapping $(a, b) \to (a, \varphi a)$ onto the graph X of φ is continuous. Obviously, its restriction to X is the identity.

PROBLEMS

6A. C^*.

1. $C^*(X)$ is isomorphic with $C^*(Y)$ if and only if βX is homeomorphic with βY.

2. In the uniform norm topology (2M), *every closed ideal in $C^*(X)$ is an intersection of maximal ideals.* [2M and 4O.]

3. Find X and Y such that $C^*(X)$ is isomorphic with $C^*(Y)$, and X is a P-space (4J), while Y is not a P-space.

6B. THEOREM 6.4.

Consider the conditions (1), (2), and (3) of Theorem 6.4, where now the subspace X of T is not necessarily dense. It was pointed out in the proof of the theorem that (1) still implies (2), and (2) still implies (3). If T is normal, then (3) implies (2), but not otherwise. [3E.*2*.] However, even if T is compact, (2) does not imply (1). [Let X be a closed set but not a retract.]

6C. COMPACTIFICATION THEOREM.

1. If $f \in C^*(X)$ and $p \in \beta X$, then

$$f^\beta(p) = \inf_{Z \in A^p} (\sup f[Z]) = \sup_{Z \in A^p} (\inf f[Z]).$$

2. Let X be dense in T. Trivially, if there is a homeomorphism of βX onto T that leaves X pointwise fixed, then T satisfies the conditions listed in Theorem 6.5. The pointwise invariance of X is essential here. Let

$S = \mathbf{R} - \mathbf{N}$, and $X = S - \{0\}$. Then X is dense in βS, and βS is homeomorphic with βX, but X is not C^*-embedded in βS.

3. X is normal if and only if it has a compactification Y such that any two disjoint closed sets in X have disjoint closures in Y. Any such space Y may be identified with βX.

6D. LEMMA 6.11.

1. Lemma 6.11 does not carry through if the condition that $\varphi|X$ be a homeomorphism is relaxed to allow $\varphi|X$ to be an arbitrary one-one continuous mapping.

2. If τ is a continuous mapping of X into a compact space Y, with the property that $\bar{\tau}[\beta X - X] \subset Y - \tau[X]$, then τ, regarded as a mapping from X into $\tau[X]$, is closed.

6E. ZERO-SETS.

1. If X is dense in T, then the family of all sets $\mathrm{cl}_T\, Z$, for $Z \in \mathbf{Z}(X)$, is a base for the closed sets in T.

2. A zero-set in βX need not be of the form cl Z, for $Z \in \mathbf{Z}(X)$. [Consider $\mathbf{N}$.]

3. Every zero-set in βX is a countable intersection of sets of the form cl Z, for $Z \in \mathbf{Z}(X)$.

4. Any zero-set in $\beta\mathbf{N}$ that meets $\beta\mathbf{N} - \mathbf{N}$ has at least $\mathfrak{c}$ points. [Such a set contains a copy of $\beta\mathbf{N} - \mathbf{N}$.] Compare 6O.*6*.

5. For suitable choice of $\sigma \in \beta\mathbf{N} - \mathbf{N}$, the subspace $\Sigma = \mathbf{N} \cup \{\sigma\}$ of $\beta\mathbf{N}$ is the same as the space Σ of 4M. Evidently, $\{\sigma\}$ is a zero-set in Σ; but $\mathrm{cl}_{\beta\Sigma}\{\sigma\}$ is not a zero-set in $\beta\Sigma\ (= \beta\mathbf{N})$.

6. If X is discrete, $Z \subset X$, and $p \in \beta X$, then $Z \in A^p$ if and only if $\mathrm{cl}_{\beta X}\, Z$ is an open neighborhood of p.

7. If X is dense in T, and $\mathscr{B}$ is an arbitrary base for the closed sets in X, then the family of all sets $\mathrm{cl}_T\, B$, for $B \in \mathscr{B}$, need not form a base in T. [Consider $\mathbf{N}^*$.]

6F. CONVERGENCE OF z-FILTERS.

Let X be dense in T. The following generalize results in 3.16, 3.17, and 4.11.

1. For any nonempty set S in X, $\mathrm{cl}_T\, S$ is the set of all cluster points of the z-filter on X of all zero-sets containing S. [6E.*1*.]

2. Every z-filter on X with cluster point p is contained in a z-ultrafilter on X with limit p. Hence a z-ultrafilter converges to any cluster point.

3. A prime z-filter on X converges to any cluster point in T.

4. *T is compact* if and only if every z-filter on X has a cluster point in T, and *if and only if every z-ultrafilter on X has a limit in T.* [The z-filter of all zero-set-neighborhoods of members of a given free z-filter on T is also free; its trace on X generates a z-filter on X with no cluster point.]

6G. EXTENSION OF CONTINUOUS MAPPINGS.

Let X be dense in T, and let $\mathscr{F}$ be a z-filter on X.

1. Let σ denote the identity mapping of X into T. A cluster point or limit of $\mathscr{F}$ in T is precisely a cluster point or limit, respectively, of $\sigma^{\#}\mathscr{F}$. Thus, to "fix" $\mathscr{F}$ is to ensure that $\sigma^{\#}\mathscr{F}$ be a fixed z-filter on T.

Let τ be a continuous mapping from X into Y, where now Y is not necessarily compact.

2. $\bigcap \tau^{\#}\mathscr{F} = \bigcap_{Z \in \mathscr{F}} \mathrm{cl}_Y \tau[Z]$; thus, τ carries cluster points of $\mathscr{F}$ to cluster points of $\tau^{\#}\mathscr{F}$.

Let X_0 denote the set of all points p in T with the following property: for all the z-ultrafilters $\mathscr{A}$ on X that converge to p, the z-filters $\tau^{\#}\mathscr{A}$ all converge in Y, and to a common limit $\tau_0 p$.

3. $X_0 \supset X$.

4. The mapping τ_0 from X_0 into Y (defined by the above) is a continuous extension of τ.

5. *X_0 is the largest subspace of T to which τ has a continuous extension into Y.* [*2.*] Compare Theorem 10.13.

6H. CONTINUITY OF EXTENDED MAPPING.

Let φ be a mapping from T to Y, and let X be dense in T. If the restriction of φ to $X \cup \{p\}$ is continuous for each p, then φ is continuous. [Given a closed neighborhood V of φp_0, let U be an open neighborhood of p_0 in T such that $\varphi[U \cap X] \subset V$. Then $\varphi[U] \subset V$.]

6I. PSEUDOCOMPACT SPACES.

1. X is pseudocompact if and only if every nonempty zero-set in βX meets X.

2. If βX, in *1*, is replaced by an arbitrary compactification of X, then the sufficiency may fail. [Consider the one-point compactification of an uncountable discrete space.]

3. Every nonempty zero-set in $\mathbf{N} \times (\beta\mathbf{N} - \mathbf{N})$ is uncountable [6E.*4*], although the space is not pseudocompact. See 5H.7.

6J. ALMOST COMPACT SPACES.

The following are equivalent for any space X. By (2) and Corollary 1.21 (or by (1) and 1G.*4*), X must be pseudocompact. By (2), the space $\mathbf{W}$ satisfies the stated conditions. For additional equivalences, see 15R.

(1) Of any two disjoint zero-sets in X, at least one is compact.

(2) $|\beta X - X| \leqq 1$. [Establish (1) $\leftrightarrow$ (2) $\rightarrow$ (3) $\rightarrow$ (4) $\rightarrow$ (6) $\rightarrow$ (2).]

(3) $X \subset T$ implies $\beta X \subset \beta T$.

(4) Every embedding of X is a C^*-embedding.

(5) Every embedding of X is a C-embedding.

(6) The only compactification of X is βX.

(7) Every embedding of any continuous image of X is a C-embedding. [If X satisfies (2), so does its image.]

6K. HOMOMORPHISM OF $C(Y)$ ONTO $C^*(Y)$.

Let N_1 and N_2 denote the subsets of odd and even integers in $\mathbf{N}$, respectively, and let Y be the subspace $N_1 \cup \text{cl}_{\beta\mathbf{N}} N_2$ of $\beta\mathbf{N}$.

1. Y contains a copy of βY.

2. There exists a homomorphism of $C(Y)$ *onto* $C^*(Y)$. [$C(Y)$ is essentially the set of all functions in $C(\mathbf{N})$ that are bounded on N_2.] Contrast with Corollary 1.8.

3. There exists a subspace Y' of Y with the following properties:

(i) Y' is dense and C^*-embedded in Y;
(ii) Y' is not C-embedded in Y;
(iii) $C(Y')$ is isomorphic with $C(Y)$.

6L. CONNECTEDNESS.

1. βX is connected if and only if X is connected.

2. A point $x \in X$ has a base of open-and-closed neighborhoods in βX if and only if it has a base of open-and-closed neighborhoods in X.

3. It is well known that the intersection of any chain of compact, connected sets is connected (see Corollary 16.14). Use this result to prove that the complement of R^+ in any compactification is connected. Do the same for $\mathbf{R}^n$, where $n > 1$.

4. Every nonempty, proper zero-set in $\beta R^+ - R^+$ is disconnected. [For $f \in C^*(R^+)$, if f^β assumes both the values 0 and 1 on $\beta R^+ - R^+$, then the set

$$\{x \in R^+ : f(x) \leqq 1/2\}$$

is a union of two disjoint closed sets whose closures in βR^+ both meet $Z(f^\beta) - R^+$.] Contrast with 10N.*5*.

6M. EXTREMALLY DISCONNECTED SPACES.

1. βX is extremally disconnected (1H) if and only if X is extremally disconnected. βX is basically disconnected if and only if X is basically disconnected.

2. *A space T is extremally disconnected if and only if every dense subspace is C^*-embedded.* [*Necessity.* 1H.*4*, and Urysohn's extension theorem or Theorem 6.4. *Sufficiency.* For an open set U, consider the subspace $U \cup (T - \text{cl}\, U)$.] Hence, *a compact space T is extremally disconnected if and only if $T = \beta X$ for every dense subspace X.*

3. According to 3N.*6*, if X is extremally disconnected, then $C(X)$ is a conditionally complete lattice. Still, the supremum in C of a family of functions—even of a countable family—need not be the same as the pointwise supremum: find an example in $C(\beta\mathbf{N})$.

6N. COMPACTIFICATION OF A PRODUCT.

Let D be an infinite discrete space.

1. $\beta D \times \beta D$ is not extremally disconnected (1H). [The open set $\{(x, x): x \in D\}$ does not have an open closure.]

2. $\beta D \times \beta D$ *is not homeomorphic with* $\beta(D \times D)$. [6M.*1*.]

6O. $\beta\mathbf{N}$, $\beta\mathbf{Q}$, AND $\beta\mathbf{R}$.

It was shown in 6.10 that $\beta\mathbf{Q}$ and $\beta\mathbf{R}$ are continuous images of $\beta\mathbf{N}$.

1. $\beta\mathbf{N}$ and $\beta\mathbf{R}$ are continuous images of $\beta\mathbf{Q}$.

2. Neither $\beta\mathbf{N}$ nor $\beta\mathbf{Q}$ is a continuous image of $\beta\mathbf{R}$. [6L.*1*.]

3. $\beta\mathbf{N}$ is not homeomorphic with $\beta\mathbf{Q}$. (For generalizations, see Corollary 9.8 and section 10.10)

4. $\beta\mathbf{Q} - \mathbf{Q}$ is not C^*-embedded in $\beta\mathbf{Q}$. [The function sgn on $\mathbf{Q} - \{0\}$ has a continuous extension to $\beta\mathbf{Q} - \{0\}$, but not to $\beta\mathbf{Q}$.] Hence $\beta(\beta\mathbf{Q} - \mathbf{Q}) \neq \beta\mathbf{Q}$.

5. $\beta\mathbf{Q} - \mathbf{Q}$ has no P-points (4L). [Consider a homeomorphism of $\mathbf{Q}$ into $[0, 1]$.] Contrast with 6V.

6. Every countable set E in $\beta\mathbf{N}$ is C^*-embedded in $\beta\mathbf{N}$. [By 3B.*5*, $\mathbf{N} \cup E$ is normal; apply 3D.*1*.] Hence every infinite closed set in $\beta\mathbf{N}$ contains a copy of $\beta\mathbf{N}$. [0.13.]

6P. THE SPACE Λ.

Let $\Lambda = \beta\mathbf{R} - (\beta\mathbf{N} - \mathbf{N})$ (see 6.10(b)).

1. $\Lambda \supset \mathbf{R}$. Hence $\beta\Lambda = \beta\mathbf{R}$.

2. $\Lambda - \mathbf{R}$ is dense in $\beta\Lambda - \mathbf{R}$. [Every closed neighborhood of a point in $\beta\Lambda - \mathbf{R}$ meets $\mathbf{R} - \mathbf{N}$.] Hence the zero-set $\Lambda - \mathbf{R}$ in Λ belongs to A^p for every $p \in \beta\mathbf{N} - \mathbf{N}$.

3. Λ is pseudocompact. [An unbounded function on Λ would be unbounded on a closed subset of $\mathbf{R}$ disjoint from $\mathbf{N}$.] But Λ is not countably compact. Hence Λ is not normal. [3D.*2*.] Also, Λ is not realcompact.

4. $\mathbf{N}$ *is closed and* C^**-embedded in* Λ, *but it is not* C*-embedded.* Compare 3L.*4*,*5*.

5. $\mathbf{N}$ is a closed G_δ, but not a zero-set, in Λ. Thus, a closed, discrete set, each of whose points has a countable base of neighborhoods, need not be a zero-set. Nor need a closed, countable union of zero-sets be a zero-set.

6Q. THE SPACE Π.

1. Let φ be a one-one mapping of $\mathbf{N}$ onto $\mathbf{Q}$. With each *irrational* number r, select an increasing sequence (s_n) of rationals converging to r. For each such sequence, consider the subset

$$E = \{\varphi^{\leftarrow}(s_1), \quad \varphi^{\leftarrow}(s_2), \cdots\}$$

of $\mathbf{N}$, and let $\mathscr{E}$ denote the family of all sets E thus defined. Then $\mathscr{E}$ has $\mathfrak{c}$ members, and the intersection of any two is finite.

2. For $E \in \mathscr{E}$, define $E' = \text{cl}_{\beta\mathbf{N}} E - \mathbf{N}$. The $\mathfrak{c}$ sets E' are mutually disjoint, open-and-closed subsets of $\beta\mathbf{N} - \mathbf{N}$.

3. Construct a set D by selecting one point p_E from each set E', and define Π to be the subspace $\mathbf{N} \cup D$ of $\beta\mathbf{N}$. Then $\mathbf{N}$ is dense in Π, and every subspace of Π is extremally disconnected [6M.*2*]. The following properties of Π are to be contrasted with those of the space Ψ of 5I.

4. D is a discrete zero-set in Π, of power $\mathfrak{c}$.

5. Every subset of Π is a G_δ, but the first countability axiom is not satisfied. [4M.*2*.]

6. The zero-set D is not C^*-embedded in Π. [Cf. 3K.*4*.] Therefore Π is not normal.

7. Define f as follows: $f(n) = \varphi(n)$, $f(p_E) = \sup \varphi[E]$. Then f is a continuous, one-one mapping of Π onto $\mathbf{R}$. Hence Π is not pseudocompact.

8. Let φ^* denote the Stone extension of φ from $\beta\mathbf{N}$ into the one-point compactification of $\mathbf{R}$. Then $\varphi^*|\Pi = f$.

9. Even if $\mathscr{E}$ were enlarged to be a maximal family, as in 5I, the resulting space would not be pseudocompact. [Any member of the enlarged family contains an infinite set without a limit point.]

6R. EXTREMALLY DISCONNECTED SUBSPACES.

1. The closed subspace $\beta\mathbf{N} - \mathbf{N}$ of the extremally disconnected space $\beta\mathbf{N}$ is *not* extremally disconnected. [In 6Q, let A and B be disjoint subsets of D that cannot be separated by open subsets of Π. Consider the unions of the corresponding sets E', and apply 1H.*1* and the fact that $\beta\mathbf{N} - \mathbf{N}$ is normal.] (For another proof, see 6W.)

2. The following are equivalent.

(1) Every subspace of T is C^*-embedded. (Such a space need not be discrete: witness the space Σ of 4M.)

(2) T is normal and every subspace is extremally disconnected. [3D.*1*.] (The latter condition does not require normality: witness Π.)

(3) T is normal and every closed subspace is extremally disconnected. [To derive (1), use 6M.*2*.]

3. If every subspace of a compact space X is C^*-embedded, then X is finite. [Otherwise, by 0.13, X contains a copy of $\beta\mathbf{N}$.]

4. Every infinite compact space contains a subspace that is not extremally disconnected.

6S. OPEN-AND-CLOSED SETS IN $\beta\mathbf{N} - \mathbf{N}$.

For infinite $A \subset \mathbf{N}$, define

$$A' = \mathrm{cl}_{\beta\mathbf{N}} A - \mathbf{N}.$$

1. Every set of the form A' is homeomorphic with $\beta\mathbf{N} - \mathbf{N}$.

2. $B' \subset A'$ if and only if $B - A$ is finite. Thus, $B' = A'$ if and only if B differs from A in a finite set. Hence there are just $\mathfrak{c}$ distinct sets A'. [Only $\aleph_0$ sets differ from A in a finite set.] As a matter of fact, we have already found $\mathfrak{c}$ disjoint such sets (6Q.*1*).

3. Every open-and-closed set in $\beta\mathbf{N}$ is of the form $\mathrm{cl}_{\beta\mathbf{N}}\, S$ for some $S \subset \mathbf{N}$.

4. Every nonempty open-and-closed set in $\beta\mathbf{N} - \mathbf{N}$ is of the form A'. The sets A' form a base for the closed sets in $\beta\mathbf{N} - \mathbf{N}$, and a base for the open sets in $\beta\mathbf{N} - \mathbf{N}$.

5. If A' and B' are proper subsets of $\beta\mathbf{N} - \mathbf{N}$, there exists a homeomorphism of $\beta\mathbf{N} - \mathbf{N}$ onto itself carrying A' onto B'.

6. For any $p \in \beta\mathbf{N} - \mathbf{N}$, the set of all images of p under homeomorphisms of $\beta\mathbf{N} - \mathbf{N}$ is dense.

7. If a sequence of sets A_n' has the finite intersection property, then $\bigcap_n A_n'$ contains a nonempty, open subset of $\beta\mathbf{N} - \mathbf{N}$. [Choose distinct $x_n \in A_1 \cap \cdots \cap A_n$, and consider the set $A = \{x_n\}_{n \in \mathbf{N}}$.]

8. Every nonempty G_δ in $\beta\mathbf{N} - \mathbf{N}$ has a nonempty interior.

6T. NON-P-POINTS OF $\beta\mathbf{N} - \mathbf{N}$.

Let $f \in C^*(\mathbf{N})$, and $p \in \beta\mathbf{N} - \mathbf{N}$, with $f^\beta(p) = 0$.

1. If $f^\beta[U] = \{0\}$ for some neighborhood U of p in $\beta\mathbf{N} - \mathbf{N}$, then there exists a sequence (s_n) in $\mathbf{N}$ such that p belongs to the closure of $\{s_n\}_{n \in \mathbf{N}}$, and $\lim_{n\to\infty} f(s_n) = 0$. [$U$ contains a neighborhood of p of the form A' of 6S.]

2. Conversely, suppose that no such neighborhood U exists. If (s_n) is any sequence such that $\lim_{n\to\infty} f(s_n) = 0$, then p does *not* belong to the closure of $\{s_n\}_{n \in \mathbf{N}}$.

3. The situation described in *2* actually occurs. [4K.*1*.]

4. Specifically, let g be any function in $C(\beta\mathbf{N} - \mathbf{N})$ with infinite range. There exists $q \in \beta\mathbf{N} - \mathbf{N}$ such that g is not constant on any neighborhood of q. [Otherwise, $\beta\mathbf{N} - \mathbf{N}$ would be a union of infinitely many disjoint open sets.]

6U. z-ULTRAFILTERS THAT CONTAIN NO SMALL SETS.

A subset S of $\mathbf{N}$ is said to have density r if

$$\lim_{n\to\infty} \frac{|\{s \in S : s \leqq n\}|}{n} = r.$$

Let $\mathscr{F}$ denote the family of all subsets of $\mathbf{N}$ of density 1.

1. $\mathscr{F}$ is a free filter on $\mathbf{N}$.

2. If $\mathscr{U}$ is any ultrafilter containing $\mathscr{F}$, then no member of $\mathscr{U}$ has density 0.

3. Define $f \in C^*(\mathbf{N})$ as follows: $f(n) = k/2^m$ for $n = 2^m + k - 2$, where $m, k \in \mathbf{N}$, and $k \leqq 2^m$. If $\lim_{n\to\infty} f(s_n)$ exists, for some sequence (s_n) in $\mathbf{N}$, then $\{s_n\}_{n \in \mathbf{N}}$ has density 0.

4. No cluster point in $\beta\mathbf{N}$ of the filter $\mathscr{F}$ is a P-point (4L) of $\beta\mathbf{N} - \mathbf{N}$. [6T.*1* and 6.6(a).]

5. There exists a point p in $\beta\mathbf{R} - \mathbf{R}$ such that every member of the z-ultrafilter A^p on $\mathbf{R}$ is of infinite measure. [4F.] Hence every neighborhood of p meets $\mathbf{R}$ in a set of infinite measure.

6V. *P*-POINTS AND NONHOMOGENEITY OF $\beta\mathbf{N} - \mathbf{N}$.

Assume the continuum hypothesis ($\mathfrak{c} = \aleph_1$). Then the $\mathfrak{c}$ nonempty open-and-closed sets in $\beta\mathbf{N} - \mathbf{N}$ (see 6S.*2*,*4*) can be indexed $(V_\alpha)_{\alpha<\omega_1}$.

Define W_α inductively, for $\alpha < \omega_1$, as follows. Assume that $\bigcap_{\sigma \leqq \tau} W_\sigma$ is nonempty for each $\tau < \alpha$; then, by 6S.*4*,*7*, $\bigcap_{\sigma \leqq \alpha} W_\sigma$ contains a set V_γ. Put $W_\alpha = V_\gamma$ or $V_\alpha \cap V_\gamma$, according as the latter set is empty or not.

1. If $\alpha_n < \omega_1$ for $n \in \mathbf{N}$, and $\alpha = \sup_n \alpha_n$, then $\bigcap_n W_{\alpha_n} \supset W_\alpha$.

2. $\bigcap_{\alpha<\omega_1} W_\alpha$ contains at least one point p.

3. If $p \in V_\alpha$, then $V_\alpha \supset W_\alpha$. Hence p is the only point in $\bigcap_\alpha W_\alpha$.

4. p is a *P*-point (4L) of $\beta\mathbf{N} - \mathbf{N}$. Compare 6T.

5. $\beta\mathbf{N} - \mathbf{N}$ is not homogeneous, i.e., there exist two points such that no homeomorphism of the space onto itself takes one to the other. [A *P*-point must go to a *P*-point.]

6. $\beta\mathbf{N} - \mathbf{N}$ has a dense set of *P*-points.

6W. DISCONNECTEDNESS OF $\beta\mathbf{N} - \mathbf{N}$.

Recall that $\beta\mathbf{N}$ is extremally disconnected (6M.*1*).

Let A_n ($n \in \mathbf{N}$) and A be infinite sets in $\mathbf{N}$ such that $(A_n')_{n\in\mathbf{N}}$ (as in 6S) is strictly increasing, and $A' \supset \operatorname{cl} \bigcup_n A_n'$.

1. For each n, $A \cap A_n - \bigcup_{k<n} A_k$ is nonempty. [6S.*2*.]

2. Choose $x_n \in A \cap A_n - \bigcup_{k<n} A_k$, and define

$$B = A - \{x_n : n \in \mathbf{N}\}.$$

Then $B' \supset \operatorname{cl} \bigcup_n A_n'$, and B' is contained properly in A'.

3. In $\beta\mathbf{N} - \mathbf{N}$, the closure of the union of a strictly increasing sequence of open-and-closed sets is never open. Hence $\beta\mathbf{N} - \mathbf{N}$ is not basically disconnected (1H).

4. More generally, $\beta X - X$ is not basically disconnected for any infinite discrete X.

Chapter 7

CHARACTERIZATION OF MAXIMAL IDEALS

7.1. The promise made at the beginning of Chapter 6 that βX is to be used to characterize the maximal ideals in $C(X)$ and in $C^*(X)$, will be fulfilled in this chapter. The key to the description of the maximal ideals in $C^*(X)$ has already been given: $C^*(X)$ is isomorphic with $C(\beta X)$, and the maximal ideals in the latter ring are in one-one correspondence with the points of βX.

Whenever $C(X) \neq C^*(X)$, the ring $C(X)$ is definitely not isomorphic with $C(\beta X)$ (Corollary 1.8). But the maximal ideals in $C(X)$ are still in one-one correspondence with the points of βX. In fact, of the various ways of constructing βX, we chose to present first the one that brings out this correspondence most clearly. The bridge between the maximal ideals in $C(X)$ and the points of βX is provided by the z-ultrafilters on X.

7.2. THEOREM. *The maximal ideals in $C^*(X)$ are precisely the sets*

$$\boldsymbol{M}^{*p} = \{f \in C^*(X) : f^\beta(p) = 0\} \qquad (p \in \beta X),$$

and they are distinct for distinct p.

PROOF. The mapping $f \to f^\beta$ is an isomorphism from $C^*(X)$ onto $C(\beta X)$; and since βX is compact, the maximal ideals in $C(\beta X)$ are precisely the fixed ideals

$$\{f^\beta \in C(\beta X) : f^\beta(p) = 0\}.$$

Evidently, $\boldsymbol{M}^{*p}$ is free or fixed according as $p \in \beta X - X$ or $p \in X$; and, in the latter case, $\boldsymbol{M}^{*p}$ is the same as $\boldsymbol{M}^*_p$.

We recall (5.6) that each residue class field of C or C^* (modulo a maximal ideal M) contains a canonical copy of $\mathbf{R}$, and we have agreed to identify this copy with $\mathbf{R}$ itself: $M(\boldsymbol{r}) = r$ for all $r \in \mathbf{R}$. Thus, an immediate consequence of the theorem is the formula

$$f^\beta(p) = \boldsymbol{M}^{*p}(f) \qquad (p \in \beta X, f \in C^*(X)).$$

As another application of the theorem, consider the result (established in 4.7 and 5.10) that $\mathbf{j}$ belongs to every free maximal ideal in $C^*(\mathbf{N})$. Third proof: obviously, $\mathbf{j}^{\beta}(p) = 0$ for every $p \in \beta\mathbf{N} - \mathbf{N}$ (since $\mathbf{N}$ is dense in $\beta\mathbf{N}$); hence $\mathbf{j} \in \boldsymbol{M}^{*p}$ for all such p.

7.3. Now we look at $C(X)$. The mapping $p \to A^p$ is one-one from βX onto the set of all z-ultrafilters on X. But the latter set is in one-one correspondence with the set of all maximal ideals in $C(X)$ (Theorem 2.5). Therefore, corresponding to each point p of βX, there is a maximal ideal $\boldsymbol{M}^p$ in $C(X)$, determined by

$$\boldsymbol{Z}[\boldsymbol{M}^p] = A^p,$$

and the correspondence $p \to \boldsymbol{M}^p$ is one-one.

THEOREM (GELFAND-KOLMOGOROFF). *For the maximal ideals in* $C(X)$, *we have*

$$\boldsymbol{M}^p = \{f \in C(X) : p \in \mathrm{cl}_{\beta X} \boldsymbol{Z}_X(f)\} \qquad (p \in \beta X).$$

PROOF. Since the maximal ideal $\boldsymbol{M}^p$ is a z-ideal, $f \in \boldsymbol{M}^p$ if and only if $\boldsymbol{Z}(f) \in A^p$, hence if and only if $p \in \mathrm{cl}_{\beta X} \boldsymbol{Z}(f)$ (6.5(c)).

Again, $\boldsymbol{M}^p$ is free or fixed according as $p \in \beta X - X$ or $p \in X$; and, in the latter case, $\boldsymbol{M}^p = \boldsymbol{M}_p$.

7.4. Here, then, is an explicit one-one correspondence between the maximal ideals in $C(X)$ and the points of βX. One item that now shows up very clearly is the algebraic characterization of z-ideals stated in 4A.*5*. As another application, consider the purely algebraic proposition

$$f^2 + g^2 \in \boldsymbol{M}^p \textit{ if and only if } f \in \boldsymbol{M}^p \textit{ and } g \in \boldsymbol{M}^p,$$

which is a special case of 2D.*2*, and the purely topological statement

$$p \in \mathrm{cl}\,(Z \cap Z') \textit{ if and only if } p \in \mathrm{cl}\, Z \textit{ and } p \in \mathrm{cl}\, Z',$$

given in the compactification theorem (Z and Z' denoting zero-sets in X). At first, these may seem unrelated. But the Gelfand-Kolmogoroff theorem shows that the two propositions are identical (since $\boldsymbol{Z}(f^2 + g^2) = \boldsymbol{Z}(f) \cap \boldsymbol{Z}(g)$). Notice that in the proof of the first, it is the necessity that is nontrivial, while in the second, it is the sufficiency.

THE EXTENSION f^*

7.5. We wish now to consider the problem of extending functions in $C(X)$ to various points p of βX and to relate their values at p to the corresponding maximal ideals $\boldsymbol{M}^p$. The analogous problem for $C^*(X)$

has already been solved: every function f in $C^*(X)$ has a continuous extension f^β, defined over all of βX, and for every point $p \in \beta X$, we have $f^\beta(p) = M^{*p}(f)$.

In $C(X)$, the considerations are complicated by the possible presence of unbounded functions, and hence of hyper-real ideals (5.6 ff.). Certainly, no unbounded function in $C(X)$ can have a continuous, real-valued extension to all of βX. Moreover, even if f is bounded, $M^p(f)$ need not be a real number.

The situation is clarified in the next theorem and the discussion following it. Our simplifying device is to notice that each function f in $C(X)$ may be regarded as a continuous mapping of X into the one-point compactification

$$\mathbf{R}^* = \mathbf{R} \cup \{\infty\}$$

of $\mathbf{R}$; as such, it has a Stone extension

$$f^*: \beta X \to \mathbf{R}^*.$$

7.6. THEOREM. *Let $f \in C(X)$.*

(a) *$f^*(p) = \infty$ if and only if $|M^p(f)|$ is infinitely large.*

(b) *$f^*(p) = r \in \mathbf{R}$ if and only if $|M^p(f) - r|$ is either infinitely small or zero.*

PROOF. If $f^*(p) = \infty$, then for each $n \in \mathbf{N}$, p is in the closure of the set

$$Z_n = \{x \in X: |f(x)| \geqq n\}.$$

By the Gelfand-Kolmogoroff theorem, $Z_n \in \mathbf{Z}[M^p]$. By 5.4(a), $|M^p(f)| \geqq n$. Therefore $|M^p(f)|$ is infinitely large. (Cf. 5.7(a).) Similarly, if $f^*(p) = r$, then $|M^p(f) - r| \leqq 1/n$ for each n, so that $|M^p(f) - r|$ is infinitely small or zero. The converses follow from the fact that the possibilities considered are mutually exclusive and exhaustive.

Obviously, if f is bounded, then f^* is the same as f^β.

7.7. EXAMPLE. As an application of the theorem, let us consider again the result (established in 5.10) that the image of $\mathbf{j}$, modulo any free maximal ideal in $C(\mathbf{N})$, is infinitely small. For each $p \in \beta\mathbf{N} - \mathbf{N}$, we have $\mathbf{j}^*(p) = 0$; and, since $\mathbf{j}$ is a unit, $M^p(\mathbf{j}) \neq 0$. Since $\mathbf{j} = |\mathbf{j}|$, we have $M^p(\mathbf{j}) = |M^p(\mathbf{j})|$, and, by the theorem, this last is either infinitely small or zero, hence is infinitely small.

7.8. Let M^p be a hyper-real ideal in C. The totally ordered field C/M^p is nonarchimedean, and it contains the real field $\mathbf{R}$ in the form of the residue classes of the constant functions. The following remarks supply further details about the relation between $M^p(f)$ and $f^*(p)$, for $f \in C$.

Suppose that $|M^p(f)|$ is not infinitely large. Then the sets

$$A = \{s \in \mathbf{R}: s < M^p(f)\}, \quad B = \{s \in \mathbf{R}: s \geqq M^p(f)\}$$

form a Dedekind cut in $\mathbf{R}$. The unique real number $\sup A = \inf B$ determined by this cut is none other than $f^*(p)$. In fact, for every $n \in \mathbf{N}$, we have

$$\sup A - 1/n < M^p(f) < \sup A + 1/n,$$

so that $|M^p(f) - \sup A| < 1/n$. Consequently, $|M^p(f) - \sup A|$ is either infinitely small or zero. Therefore, $\sup A = f^*(p)$.

Any one of the three possibilities:

$$f^*(p) < M^p(f), \quad f^*(p) = M^p(f), \quad f^*(p) > M^p(f)$$

may occur. In fact, if $M^p(g)$ is infinitely small, then the three functions $f = \mathbf{r} + g$, $f = \mathbf{r}$, $f = \mathbf{r} - g$ satisfy the three conditions, respectively ($f^*(p) = r$ in each case).

If r is any real number, then $M^p(f) \leqq r$ implies $f^*(p) \leqq r$ (and, dually, $f^*(p) < r$ implies $M^p(f) < r$). The converse is false, a counterexample being provided by any function $f = \mathbf{r} + g$ (or $\mathbf{r} - g$), where $M^p(g)$ is infinitely small.

7.9. The correspondence $M^p \to M^{*p}$ is one-one between the maximal ideals in $C(X)$ and the maximal ideals in $C^*(X)$. The results obtained in this chapter enable us to describe this correspondence in terms of the functions on X. If $f \in M^p$, then $p \in \operatorname{cl} Z(f)$ (Theorem 7.3), whence $f^*(p) = 0$. In case f is bounded, this last states that $f^\beta(p) = 0$, and hence holds if and only if $f \in M^{*p}$ (Theorem 7.2). Thus, $M^p \cap C^*$ is contained in M^{*p}. The two are not the same, in general. Precisely:

(a) *M^{*p} is the set of all $f \in C^*(X)$ for which $|M^p(f)|$ is either infinitely small or zero.*

In fact, f satisfies this condition if and only if $f^*(p) = 0$ (Theorem 7.6(b)).

A striking distinction between real and hyper-real ideals is provided by

(b) *M^p is hyper-real if and only if M^{*p} contains a unit of C.*

If M^p is hyper-real, there exists $g \geqq \mathbf{1}$ such that $M^p(g)$ is infinitely large. Then $M^p(g^{-1})$ is infinitely small, and by (a), $g^{-1} \in M^{*p}$. The converse is obvious from (a).

The equality $M_p \cap C^* = M^*_p$ for *fixed* ideals was derived in 4.7. We are now in a position to state the general relation.

(c) $M^p \cap C^* = M^{*p}$ *if and only if* M^p *is real.*

The necessity follows at once from (b), while the sufficiency is obvious from (a).

STRUCTURE SPACES

7.10. As we have seen, the mapping $p \to M^{*p}$ is one-one from βX onto the set $\mathfrak{M}^* = \mathfrak{M}^*(X)$ of all maximal ideals in $C^*(X)$. Accordingly, it may be used to define a topology on $\mathfrak{M}^*$—simply by transferring that of βX. The image of X under this transfer is, of course, the subspace of all fixed maximal ideals. The topology on $\mathfrak{M}^*$ can be described intrinsically as follows. Define

$$\mathfrak{C}^*(f) = \{M \in \mathfrak{M}^* : f \in M\} \qquad (f \in C^*(X)).$$

Theorem 7.2 states that $f \in M^{*p}$ if and only if $f^\beta(p) = 0$—hence that $M^{*p} \in \mathfrak{C}^*(f)$ if and only if $p \in Z_{\beta X}(f^\beta)$. Consequently, the mapping $p \to M^{*p}$ carries the basic family of closed sets $Z_{\beta X}(f^\beta)$ in βX onto the family of sets $\mathfrak{C}^*(f)$ in $\mathfrak{M}^*$; therefore this latter family constitutes a base in $\mathfrak{M}^*$.

As in 4.9, the topology thus defined is called the *Stone topology* on $\mathfrak{M}^*$; and the set $\mathfrak{M}^*$, endowed with the Stone topology, is called the *structure space* of C^*. See 7M.

7.11. In the Stone topology on the set $\mathfrak{M} = \mathfrak{M}(X)$ of all maximal ideals in $C(X)$, the sets

$$\mathfrak{C}(f) = \{M \in \mathfrak{M} : f \in M\} \qquad (f \in C(X))$$

form a base for the closed sets. By definition, the sets

$$\{p \in \beta X : Z(f) \in A^p\}$$

form a base for the closed sets in βX (6.5(b)). Therefore the mapping $p \to M^p$ is a homeomorphism of βX onto $\mathfrak{M}$. Again, the image of X under the homeomorphism is the subspace of fixed maximal ideals.

It follows that the mapping $M^p \to M^{*p}$ is a homeomorphism between the structure spaces $\mathfrak{M}$ and $\mathfrak{M}^*$. It should be noted, however, that in spite of the formal similarity between the basic closed sets in the definitions of the respective Stone topologies, the homeomorphism does not lead, in general, to a correspondence between the bases. In fact, we have

$$M^{*p} \in \mathfrak{C}^*(f) \textit{ if and only if } p \in Z_{\beta X}(f^\beta),$$

as was pointed out in 7.10; by the Gelfand-Kolmogoroff theorem, on the other hand,

$$M^p \in \mathfrak{E}(f) \textit{ if and only if } p \in \mathrm{cl}_{\beta X} \boldsymbol{Z}_X(f).$$

While $\boldsymbol{Z}_{\beta X}(f^\beta) \supset \mathrm{cl}_{\beta X} \boldsymbol{Z}_X(f)$ for every $f \in C^*(X)$, it is easy to construct examples where the two sets are not the same. What is more to the point, the family of all sets $\boldsymbol{Z}_{\beta X}(f^\beta)$ may differ from the family of all sets $\mathrm{cl}_{\beta X} \boldsymbol{Z}_X(f)$ (for $f \in C^*(X)$) (see 6E.2). Observe that to obtain the family of all sets $\mathfrak{E}(f)$, it is enough to let f range over the bounded functions.

$\boldsymbol{O}^p$ *AND PRIME IDEALS*

The material in the remaining sections of this chapter will not be needed in the text except in Chapter 14.

7.12. For each point p in βX, we denote by $\boldsymbol{O}^p$ the set of all f in $C(X)$ for which $\mathrm{cl}_{\beta X} \boldsymbol{Z}(f)$ is a neighborhood of p. We observe, at the outset, that

(a) *$f \in \boldsymbol{O}^p$ if and only if there is a neighborhood V of p such that $\boldsymbol{Z}(f) \supset V \cap X$.*

The condition is surely sufficient. And since $\boldsymbol{Z}(f)$ is closed in X, we have $\boldsymbol{Z}(f) = \mathrm{cl}_{\beta X} \boldsymbol{Z}(f) \cap X$, which yields the converse.

It follows at once from this characterization that $\boldsymbol{O}^p$ is a z-ideal in $C(X)$. (Alternatively, (IV) of the compactification theorem (6.5) implies that $\boldsymbol{Z}[\boldsymbol{O}^p]$ is a z-filter.) Evidently, $\boldsymbol{O}^p$ is contained in $\boldsymbol{M}^p$. For $p \in X$, $\boldsymbol{O}^p$ is the same as the ideal $\boldsymbol{O}_p$ defined in 4I.

REMARK. If $f^*(p) = \infty$, then each set

$$\{x \in X: |f(x)| \geqq n\} \qquad (n \in \mathbf{N})$$

is a member of $\boldsymbol{Z}[\boldsymbol{O}^p]$. Similarly, if $g^*(p) = 0$, then each set $\{x: |g(x)| \leqq 1/n\}$ belongs to $\boldsymbol{Z}[\boldsymbol{O}^p]$.

(b) *$f \in \boldsymbol{O}^p$ if and only if $fg = \mathbf{0}$ for some $g \notin \boldsymbol{M}^p$.*

If $f \in \boldsymbol{O}^p$, there exists $g \in C^*(X)$ such that $g^\beta(p) = 1$, while $g^\beta(q) = 0$ for all $q \notin \mathrm{int\,cl}\, \boldsymbol{Z}(f)$. For this g, we have $g \notin \boldsymbol{M}^p$, and $fg = \mathbf{0}$. Conversely, suppose that g satisfies these conditions. By the Gelfand-Kolmogoroff theorem, $p \notin \mathrm{cl}\, \boldsymbol{Z}(g)$, and this implies quickly that $f \in \boldsymbol{O}^p$.

7.13. THEOREM. *An ideal I in C is contained in a unique maximal ideal $\boldsymbol{M}^p$ if and only if $I \supset \boldsymbol{O}^p$.*

PROOF. *Necessity.* Given $f \in \boldsymbol{O}^p$, let g be as in (b). Since $g \notin \boldsymbol{M}^p$, and M^p is the only maximal ideal containing I, we must have $(I, g) = C$. Consequently, there exist $h \in I$ and $s \in C$ such that $\mathbf{1} = h + sg$. Hence $f = hf \in I$.

Sufficiency. We show that M^p is the only maximal ideal containing O^p. If $q \neq p$, there exists $f \in C^*(X)$ such that f^β vanishes on a neighborhood of p, while $f^\beta(q) \neq 0$; then $f \in O^p - M^q$.

7.14. The Gelfand-Kolmogoroff theorem states that $p \in \text{cl}\, Z(g)$ if and only if $g \in M^p$. Hence $\text{cl}\, Z(f)$ is a neighborhood of $\text{cl}\, Z(g)$ if and only if $f \in O^p$ whenever $g \in M^p$.

THEOREM. *$\text{cl}_{\beta X} Z(f)$ is a neighborhood of $\text{cl}_{\beta X} Z(g)$ if and only if there exists h such that*

$$Z(f) \supset X - Z(h) \supset Z(g).$$

PROOF. *Necessity.* There exists $h \in C^*(X)$ such that h^β is equal to 1 on the compact set $\text{cl}\, Z(g)$, and 0 on $\beta X - \text{cl}\, Z(f)$.

Sufficiency. If h is as described, then $Z(h)$ is disjoint from $Z(g)$, and so their closures in βX are disjoint. Also, $Z(f) \cup Z(h) = X$, so that

$$\text{cl}\, Z(f) \cup \text{cl}\, Z(h) = \beta X.$$

Therefore,

$$\text{cl}\, Z(f) \supset \beta X - \text{cl}\, Z(h) \supset \text{cl}\, Z(g).$$

Thus, $\text{cl}\, Z(f)$ is a neighborhood of $\text{cl}\, Z(g)$.

The mere fact that $Z(f)$ is a neighborhood of $Z(g)$ does not imply that $\text{cl}\, Z(f)$ is a neighborhood of $\text{cl}\, Z(g)$. See 8K.*2*.

7.15. The ideals O^p are basic to the study of *prime* ideals. The following result suggests why.

THEOREM. *Every prime ideal P in C contains O^p for a unique p, and M^p is the unique maximal ideal containing P.*

PROOF. We include a second proof that P is contained in a unique maximal ideal (Theorem 2.11). Let M^p be any maximal ideal containing P. Given $f \in O^p$, let g be as in 7.12(b). Then $fg = \mathbf{0} \in P$, but $g \notin P$; so $f \in P$. Thus, $O^p \subset P$. The rest of the theorem follows from Theorem 7.13.

Regarding the incidence of nonmaximal, prime ideals, we have:

(a) *M^p contains a nonmaximal prime ideal if and only if $M^p \neq O^p$.*

For, the z-ideal O^p is the intersection of all the prime ideals contained in M^p (Theorem 2.8).

The following observation concerning the order in the residue class ring C/P will be useful.

(b) *Let P be a prime ideal contained in M^p. If f is nonnegative on a zero-set of O^p, then $P(f) \geqq 0$.*

For, $f - |f| \in O^p \subset P$.

7.16. As in a field, an element a of a totally ordered integral domain is said to be *infinitely large* if $a > n$ for all $n \in \mathbf{N}$; and a is *infinitely small* if $0 < na < 1$ for all $n \in \mathbf{N}$. The discussion in 7.8 concerning the order in a residue class field modulo a maximal ideal applies equally well to the case of a prime ideal P, and shows that for any $f \in C(X)$, either $|P(f)|$ is infinitely large, or else there exists a unique real number r such that $|P(f) - r|$ is infinitely small or zero.

THEOREM. *Let P be a prime ideal contained in M^p. For each $f \in C$:*

(a) *$|P(f)|$ is infinitely large if and only if $|M^p(f)|$ is infinitely large—or, equivalently, $f^*(p) = \infty$.*

(b) *$|P(f) - r|$ is infinitely small or zero (where $r \in \mathbf{R}$) if and only if $|M^p(f) - r|$ is infinitely small or zero—or, equivalently, $f^*(p) = r$.*

PROOF. If $P(f) \geqq n$, then

$$f - \mathbf{n} - |f - \mathbf{n}| \in P \subset M^p$$

(Theorem 5.5), whence $M^p(f) \geqq n$. This yields the necessity in (a); and a similar proof establishes the necessity in (b). The converses follow from the fact that the possibilities considered are mutually exclusive and exhaustive. The conditions in terms of f^* were derived in Theorem 7.6.

It follows that C/P contains infinitely large elements if and only if C/M^p does—that is, M^p is hyper-real. On the other hand, if P is not maximal, then C/P will always contain infinitely small elements, even when M^p is real: namely, all elements $|P(f)|$ for $f \in M^p - P$.

PROBLEMS

7A. ISOMORPHISM OF FUNCTION RINGS.

Use the structure spaces to prove that if $C(X)$ is isomorphic with $C(Y)$, then $C^*(X)$ is isomorphic with $C^*(Y)$. Compare 1.9.

7B. RESIDUE CLASS FIELDS.

1. $M^p(f) > 0$ if and only if there exists a neighborhood V of p such that $f(x) > 0$ for all $x \in V \cap X$. [5.4(b).]

2. If $|M^p(f)|$ is not infinitely large, then for every real $\epsilon > 0$, there exists a neighborhood V of p such that the inequality $|f(x) - M^p(f)| < \epsilon$ holds in the field C/M^p for all $x \in V \cap X$.

7C. REAL MAXIMAL IDEALS.

M^p is real if and only if, for every countable family of functions f_n in $C(X)$, there exists $x \in X$ such that $f_n(x) = f_n^*(p)$ for all n.

7D. CHARACTERIZATION OF M^p.

1. M^p coincides with the set of all f in C such that $(fg)^*(p) = 0$ for every g in C. [C/M^p is a field.]

2. Prove directly from the definition that the latter set is a maximal ideal.

7E. FUNCTIONS WITH COMPACT SUPPORT.

The family $C_K(X)$ of all functions in $C(X)$ with compact support (4D) is the same as the family of all f in $C^*(X)$ such that $\mathbf{Z}_{\beta X}(f^\beta)$ is a neighborhood of $\beta X - X$. Thus,

$$C_K(X) = \bigcap_{p \in \beta X - X} O^p.$$

Hence C_K *is the intersection of all the free ideals in* C, *and of all the free ideals in* C^*. [4D.*5*.]

7F. FUNCTIONS VANISHING AT INFINITY.

Let $C_\infty(X)$ denote the family of all functions f in $C(X)$ for which the set

$$\{x \in X : |f(x)| \geqq 1/n\}$$

is compact for every $n \in \mathbf{N}$. (Such functions are said to *vanish at infinity*.) Obviously, $C_\infty \supset C_K$ (7E).

1. $C_\infty(X)$ *is the intersection of all the free maximal ideals in* $C^*(X)$.

2. The intersection of all the free maximal ideals in $C(X)$ is always contained in the intersection of all the free maximal ideals in $C^*(X)$.

3. In case X is locally compact and σ-compact, but not compact, then the inclusion in *2* is proper. (A σ-compact space is a countable union of compact spaces.) [$\beta X - X$ is a compact G_δ.]

4. $\beta X - X$ is dense in βX if and only if either of the intersections in *2* is $(\mathbf{0})$.

5. $C_\infty(\mathbf{Q}) = (\mathbf{0})$. (Hence $\beta\mathbf{Q} - \mathbf{Q}$ is dense in $\beta\mathbf{Q}$.)

7G. C_K AND C_∞ FOR LOCALLY COMPACT SPACES.

Let X be a locally compact, noncompact space, and let $X^* = X \cup \{\infty\}$ denote its one-point compactification.

1. $C_\infty(X)$ (see 7F) is isomorphic with the ideal $\mathbf{M}_\infty$ in $C(X^*)$, and $C_K(X)$ (see 7E) is isomorphic with the ideal O_∞ in $C(X^*)$.

2. The subrings C_K and C_∞ coincide if and only if every σ-compact subset of X is contained in a compact set in X. [The complement in X^* of such a subset is a G_δ; apply 3.11(b). Cf. 4L.*1*.] Hence these rings are distinct in case X is not countably compact.

7H. THE IDEALS O^p.

1. The sets $\mathrm{cl}_{\beta X}\, Z(f)$, for $f \in O^p$, form a base for the neighborhoods of p.

2. If a zero-set Z meets every member of $Z[O^p]$, then $Z \in Z[M^p]$.

3. $Z[M^p]$ has the countable intersection property (i.e., M^p is a real ideal) if and only if $Z[O^p]$ has the countable intersection property. [*Sufficiency.* Theorem 7.6.]

4. A nonmaximal z-filter containing $Z[O^p]$ cannot be closed under countable intersection. [Every member of $Z[M^p]$ is a countable intersection of members of $Z[O^p]$.]

5. If $f \in M^p - O^p$, then there exists a prime ideal, containing O^p and f, that is not a z-ideal. [Argue as in 4I.*5*.]

6. If $f \in M^p - O^p$, then there exists a prime ideal containing O^p, but not f, that is not a z-ideal. [Argue as in 4I.*6*.]

7I. GENERATORS FOR M^p AND O^p.

1. If $M^p = (\cdots, f_\alpha, \cdots)$, then

$$\bigcap_\alpha \mathrm{cl}\, Z(f_\alpha) = \{p\}.$$

In case the f_α are bounded (see 2A), $\bigcap_\alpha Z(f_\alpha^\beta)$ can contain points other than p.

2. $O^p = (\cdots, f_\alpha, \cdots)$ if and only if the sets $\mathrm{cl}\, Z(f_\alpha)$ form a subbase for the neighborhoods of p. [Theorem 6.5(IV) and 1D.*1*.]

7J. PRIME IDEALS.

1. No free maximal ideal contains a fixed prime ideal.

2. In C^*, M^{*p} is the unique maximal ideal containing $O^p \cap C^*$; and if P is a prime ideal contained in M^{*p}, then $P \supset O^p \cap C^*$. Hence every prime ideal in C^* is contained in a unique maximal ideal. Notice that this also follows from Theorem 2.11 (or 7.15) and the fact that $C^*(X)$ is isomorphic with $C(\beta X)$; see 6.6(c).

3. Prove the analogues of 5.7(a, b, c) for arbitrary prime ideals.

4. Give a proof of Theorem 7.16 analogous to that of Theorem 7.6. [7.15(b).]

7K. RESIDUE CLASS RINGS MODULO PRIME IDEALS.

Let P be a prime ideal in C, define $P^* = P \cap C^*$, and let B denote the subring of C/P consisting of all elements a for which $|a|$ is not infinitely large.

1. The mapping $P^*(f) \to P(f)$ $(f \in C^*)$ defines an order-preserving isomorphism of C^*/P^* onto B. [The homomorphism $f \to P(f)$ $(f \in C^*)$ maps C^* onto B, and its kernel is P^*.]

2. C/P is isomorphic with C^*/P^* if and only if $C/P = B$, that is, the maximal ideal containing P is real. [Consider the image of M/P under the isomorphism, and apply the second isomorphism theorem.]

7L. *P*-SPACES.

1. The following are equivalent.

(1) X is a P-space (4J).

(2) $M^p = O^p$ for all $p \in \beta X$. Note that for the special case of a discrete space, (1) implies (2) by 6E.*6*.

(3) Every prime ideal in $C(X)$ is an intersection of maximal ideals.

(4) Every z-ideal in $C(X)$ is an intersection of maximal ideals.

2. Derive 5P from 5G.*2*.

3. If X is a P-space, then the intersection of all the free ideals in $C(X)$ is the same as the intersection of all the free maximal ideals, and consists of all functions that vanish everywhere except on a finite set. [7E and 4K.*3*.] (The corresponding result holds for all realcompact spaces (Theorem 8.19). But not every P-space is realcompact (9L).)

7M. THE STRUCTURE SPACE OF A COMMUTATIVE RING WITH UNITY.

Let A be an arbitrary commutative ring with unity element. Denote the set of all maximal ideals in A by $\mathfrak{S}$. For each $a \in A$, define

$$\mathfrak{C}(a) = \{M \in \mathfrak{S}: a \in M\}.$$

1. $\mathfrak{S}$ may be made into a topological space by taking the family of all sets $\mathfrak{C}(a)$ as a base for the closed sets. This space is called the ***structure space*** of A; its topology is called the *Stone topology*.

2. The closure of any subset $\mathfrak{T}$ of $\mathfrak{S}$ is the set

$$\{M \in \mathfrak{S}: M \supset \bigcap \mathfrak{T}\}.$$

Hence $\mathfrak{T}$ is dense if and only if $\bigcap \mathfrak{T} = \bigcap \mathfrak{S}$. The set

$$\{M \in \mathfrak{S}: M \supset I\}$$

is called the *hull* of I, and $\bigcap \mathfrak{T}$ is called the *kernel* of $\mathfrak{T}$. Thus, the closure of $\mathfrak{T}$ is the hull of the kernel of $\mathfrak{T}$. For this reason, the Stone topology is often referred to as the *hull-kernel* topology.

3. $\mathfrak{S}$ is a T_1-space.

4. $\mathfrak{S}$ is a Hausdorff space if and only if, for each pair of distinct maximal ideals M and M', there exist $a, a' \in A$ such that $a \notin M$, $a' \notin M'$, and $aa' \in \bigcap \mathfrak{S}$. In particular, the structure space of the ring of integers is not a Hausdorff space.

5. For $B \subset A$, the collection $\{\mathfrak{C}(b)\}_{b \in B}$ has empty intersection if and only if the ideal (B) generated by B is improper, i.e., all of A. In fact, if $(B) = A$, then some finite subfamily has empty intersection.

6. Every family of closed sets in $\mathfrak{S}$ with the finite intersection property has nonempty intersection. Thus, *if $\mathfrak{S}$ is a Hausdorff space, it is compact.*

7N. $\mathfrak{M}$ AND $\mathfrak{M}^*$ AS MODELS FOR βX.

The properties of the structure spaces $\mathfrak{M}$ and $\mathfrak{M}^*$, listed below, lead to independent proofs of the existence of a compactification in which X is C^*-embedded, i.e., of βX.

1. $\mathfrak{M}$ is compact. [7M.*6*.]

2. For $f \in C^*(X)$, and $r, s \in \mathbf{R}$, the set

$$\{M: r < M(f) < s\}$$

is open in $\mathfrak{M}$.

3. The set $\mathfrak{F}$ of all fixed maximal ideals in $C(X)$ is a dense subspace of $\mathfrak{M}$ [7M.*2*] and is homeomorphic with X [4.9].

4. $\mathfrak{F}$ is C^*-embedded in $\mathfrak{M}$. [Evidently, $C^*(\mathfrak{F})$ is in one-one correspondence with $C^*(X)$. Consider Dedekind cuts as in 7.8. *Caution:* Note the last sentence in 7.8.]

5. Similarly, the set of all fixed maximal ideals in $C^*(X)$ is homeomorphic with X and is a dense, C^*-embedded subspace of the compact space $\mathfrak{M}^*$.

6. The topology of $\mathfrak{M}^*$ is the weak topology induced by the family of functions $M \to M(f)$ on $\mathfrak{M}^*$ ($f \in C^*(X)$).

7O. PROPERTIES OF cl $\mathbf{Z}(f)$.

For an ideal I in C, define

$$\theta(I) = \{p \in \beta X: M^p \supset I\}.$$

1. $\theta(I) = \bigcap_{f \in I} \operatorname{cl} \mathbf{Z}(f)$, that is, $\theta(I)$ is the set of all cluster points of the z-filter $\mathbf{Z}[I]$.

2. If cl $\mathbf{Z}(f)$ is a neighborhood of $\theta(I)$, then $f \in I$. [Argue as in 4O.*1*.] Note that the necessity in Theorem 7.13 is a special case of this result.

3. Given $g \in C$, and a positive unit u, there exists f such that $|g - f| \leqq u$ and cl $\mathbf{Z}(f)$ is a neighborhood of cl $\mathbf{Z}(g)$. [Argue as in 4O.*2*, and apply Theorem 7.14.]

7P. IDEALS IN C AND C^*.

For any ideal I in C, $I \subset M^p$ if and only if $I \cap C^* \subset M^{*p}$. [7O.*2*.]

7Q. CLOSED IDEALS IN THE m-TOPOLOGY.

For an ideal I in C, define

$$\bar{I} = \bigcap \{M^p: M^p \supset I\},$$

that is, $\bar{I}$ is the kernel of the hull of I (7M.*2*).

1. $\bar{I}$ is a closed ideal in the m-topology [2N.*5*], and consists of all g for which cl $\mathbf{Z}(g) \supset \theta(I)$ (see 7O). Furthermore, $\theta(\bar{I}) = \theta(I)$.

2. $\bar{I}$ is the closure of I in the m-topology. [In 7O.*3*, if $g \in \bar{I}$, then $f \in I$, by 7O.*2*.] Hence *an ideal is closed if and only if it is an intersection of maximal ideals.* This generalizes 6A.*2*.

3. The closed ideals in C^* coincide with the intersections of maximal ideals in C^* if and only if X is pseudocompact. [Theorem 5.8(b) and 7.9(c).] It follows that if X is not pseudocompact, then the m-topology is not preserved under the isomorphism from $C^*(X)$ to $C(\beta X)$; compare 1J.*6*.

4. X is a P-space (4J) if and only if every ideal in C is closed.

7R. *e*-IDEALS.

1. $\boldsymbol{E}^{\leftarrow}(\boldsymbol{Z}[\boldsymbol{M}^p]) = \boldsymbol{M}^{*p}$, that is, the correspondence defined in 2L.*16* is the mapping $\boldsymbol{M}^p \to \boldsymbol{M}^{*p}$.

2. $\boldsymbol{E}(\boldsymbol{M}^{*p}) = \boldsymbol{Z}[\boldsymbol{O}^p] = \boldsymbol{Z}[\boldsymbol{O}^p \cap C^*] = \boldsymbol{E}(\boldsymbol{O}^p \cap C^*)$.

3. The only *e*-ideal containing $\boldsymbol{O}^p \cap C^*$ is $\boldsymbol{M}^{*p}$. Hence every prime *e*-ideal is maximal.

Chapter 8

REALCOMPACT SPACES

8.1. Hewitt introduced the notion of *realcompact* space, and showed that to a very large extent these spaces play the same role in the theory of $C(X)$ that the compact spaces do in the theory of $C^*(X)$. For each X, there exists a unique realcompact space υX in which X is dense and C-embedded (Theorem 8.7), and the space υX serves to characterize the *real* maximal ideals in $C(X)$.

Just as C^* distinguishes among compact spaces (Theorem 4.9), so does C distinguish among realcompact spaces (Theorem 8.3). Now, we have already seen that C is at least as sensitive as C^*: if $C(X)$ and $C(Y)$ are isomorphic, then $C^*(X)$ and $C^*(Y)$ are isomorphic (Theorem 1.9). Since C distinguishes between $\mathbf{N}$ and $\beta\mathbf{N}$, for example, while C^* does not, C is genuinely more sensitive than C^*.

Although C^* also distinguishes among other classes of spaces—e.g., metric spaces (Corollary 9.8)—the compact spaces form a *maximal* class; explicitly: for any X, the compact space βX has the property that $C^*(\beta X)$ is isomorphic with $C^*(X)$. Similarly, the realcompact spaces form a maximal class with respect to C: for each X, υX is a realcompact space such that $C(\upsilon X)$ is isomorphic with $C(X)$.

As we know, X is compact if and only if every maximal ideal in $C(X)$ is fixed (Theorem 4.11). By definition, X is realcompact if every real maximal ideal in $C(X)$ is fixed (5.9). Equivalently, X is compact if and only if every z-ultrafilter is fixed; and X is realcompact if and only if every real z-ultrafilter—i.e., every z-ultrafilter with the countable intersection property (5.15)—is fixed. Hence it is plain that realcompactness is a topological invariant. This can also be seen algebraically: every homeomorphism induces a ring isomorphism; and any ring isomorphism takes real ideals to real ideals.

While every compact space is realcompact, we have met realcompact spaces that are not compact—e.g., $\mathbf{N}$ and $\mathbf{R}$. Considerable effort was expended to produce a space that is *not* realcompact—e.g., **W** (5.12), Ψ (5I), or Λ (6P).

We shall see in Chapter 12 that for most "practical" purposes, the discrete spaces are all realcompact. More generally, "practically" all metrizable spaces are realcompact (Theorem 15.24). For the moment, we point out one elementary criterion. Others will be derived later in the chapter.

8.2. X is said to be a *Lindelöf* space provided that every family of closed sets with the countable intersection property has nonempty intersection—i.e., every open cover has a countable subcover.

THEOREM. *Every Lindelöf space is realcompact.*

PROOF. In a Lindelöf space, every z-filter with the countable intersection property is fixed.

As a matter of fact, this condition on z-filters is characteristic of Lindelöf spaces; see 8H.5.

When X is expressible as a countable union of compact subspaces, it is said to be *σ-compact*. Clearly, every σ-compact space is a Lindelöf space, and hence is realcompact. (Therefore, again, every countable space is realcompact.) Furthermore, every space with a countable base of open sets is a Lindelöf space, and hence is realcompact. Therefore every separable metric space is realcompact. Thus, every subspace of a euclidean space is realcompact.

In particular, **R** and all its subspaces are realcompact.

8.3. THEOREM. *Two realcompact spaces X and Y are homeomorphic if and only if $C(X)$ and $C(Y)$ are isomorphic.*

PROOF. Necessity is obvious.

The correspondence $p \to \boldsymbol{M}_p$ is one-one from the realcompact space X onto the set of all *real* maximal ideals in $C(X)$; and the property of being a real maximal ideal is an algebraic invariant. This means that the points of X can be recovered from the algebraic structure of the ring $C(X)$. Furthermore, the purely algebraic relation $f \in \boldsymbol{M}_p$ is equivalent to the relation $p \in \boldsymbol{Z}(f)$. Since the family of zero-sets is a base for the closed sets in X, this shows that the topology of X can also be recovered from $C(X)$. But the realcompact space Y can be extracted in the same way from the isomorphic ring $C(Y)$. Therefore Y is homeomorphic with X.

This generalization of Theorem 4.9 could have been presented at the time: only the *definition* of realcompact was lacking. The proof of the theorem amounts to the observation that the space of all real maximal ideals is determined algebraically, and, for realcompact X, is homeomorphic with X.

THE SPACE vX

8.4. By a *realcompactification* of X is meant a realcompact space in which X is dense. In particular, every compactification of X is such a space. Our next goal is to find a realcompactification in which X is C-embedded.

It follows from Theorem 6.7 that every space in which X is dense and C-embedded lies between X and βX. Here, then, is where we must look for the desired realcompactification.

THEOREM. *The following conditions on a point p of βX are equivalent.*

(1) $\boldsymbol{M}^p$ *is real—alternatively,* $\boldsymbol{A}^p$ *has the countable intersection property.*

(2) $f^*(p) \neq \infty$, *for all* $f \in C(X)$.

(3) $f^*(p) = \boldsymbol{M}^p(f)$ *for all* $f \in C(X)$.

(4) $f^*(p) = 0$ *implies* $\boldsymbol{M}^p(f) = 0$, *i.e.,* $f \in \boldsymbol{M}^p$, *for all* $f \in C(X)$.

PROOF. If $\boldsymbol{M}^p$ is real, then by Theorem 7.6, each of the conditions (2), (3), and (4) is satisfied. Conversely, if $\boldsymbol{M}^p$ is hyper-real, there exists a positive function g in $C(X)$ such that $\boldsymbol{M}^p(g)$ is infinitely large. Then g violates (2) (and (3)), while g^{-1} violates (3) and (4).

> *The set of all points in βX that satisfy the conditions in the theorem is denoted by vX.*

Since $X \subset vX \subset \beta X$, we have $\beta(vX) = \beta X$.

8.5. We show now that vX is precisely the space we are looking for.

COROLLARY.

(a) vX *is the largest subspace of* βX *in which* X *is* C*-embedded.*

(b) vX *is the smallest realcompact space between* X *and* βX. *In particular,* X *is realcompact if and only if* $X = vX$.

PROOF. Since vX consists of *all* points satisfying the conditions of the theorem, (2) implies (a), while (1) yields the second statement in (b).

Next, $X \subset vX \subset v(vX) \subset \beta X$. By transitivity of C-embedding, X is C-embedded in $v(vX)$, and so, by (a), $v(vX) \subset vX$. Therefore $v(vX) = vX$, i.e., vX is realcompact.

Finally, if $X \subset T \subset \beta X$, and T is realcompact, then by (a) again, $T\,(=vT)$ is the only subspace of $\beta X\,(=\beta T)$ in which T is C-embedded. But, clearly, T is C-embedded in $T \cup vX$. Hence, $vX \subset T$.

Experience impels us to point out that the Greek letter introduced above is *upsilon*, not *nu*.

8.6. THEOREM. *Let X be dense in T. The following statements are equivalent.*

(1) *Every continuous mapping τ from X into any realcompact space Y has an extension to a continuous mapping from T into Y.*

(2) *X is C-embedded in T.*

(3) *If a countable family of zero-sets in X has empty intersection, then their closures in T have empty intersection.*

(4) *For any countable family of zero-sets Z_n* in X,

$$\mathrm{cl}_T \bigcap_n Z_n = \bigcap_n \mathrm{cl}_T Z_n.$$

(5) *Every point of T is the limit of a unique, real, z-ultrafilter on X.*

(6) $X \subset T \subset \upsilon X$.

(7) $\upsilon T = \upsilon X$.

PROOF. We notice that each of the conditions listed here is stronger than the corresponding condition in Theorems 6.4 and 6.7. Therefore, if T satisfies any one of the present conditions, then

$$X \subset T \subset \beta X, \quad \text{and} \quad \beta T = \beta X.$$

The pattern of proof will be (1) → (2) → (7) → (5) → (4) → (3) → (6) → (1).

(1) *implies* (2). Since $\mathbf{R}$ is realcompact, (2) is just a special case of (1).

(2) *implies* (7). The hypothesis implies that X is C-embedded in υT; therefore $\upsilon T \subset \upsilon X$, by (a) of the corollary, and $\upsilon T \supset \upsilon X$, by (b).

(7) *implies* (5). By definition (8.4), if $p \in \upsilon X$, then A^p, which is the unique z-ultrafilter converging to p (6.6(a)), is real.

(5) *implies* (4). Evidently, the left member in (4) is contained in the right. Conversely, if $p \in \bigcap_n \mathrm{cl}\, Z_n$, then each Z_n belongs to the real z-ultrafilter A^p (6.5(c)); hence $\bigcap_n Z_n \in A^p$, by Theorem 5.14, and so $p \in \mathrm{cl} \bigcap_n Z_n$.

(4) *implies* (3). Obvious.

(3) *implies* (6). If $p \in \beta X - \upsilon X$, there exist $Z_n \in A^p$ ($n \in \mathbf{N}$) such that $\bigcap_n Z_n = \emptyset$. By hypothesis, $\bigcap_n \mathrm{cl}_T Z_n = \emptyset$. But $\bigcap_n \mathrm{cl}_{\beta X} Z_n$ contains p. Therefore, $p \notin T$.

(6) *implies* (1). Consider the Stone extension $\bar{\tau}$ from βX into βY. The extension sought must be $\bar{\tau}|T$. Accordingly, we are to show that $\bar{\tau}[T] \subset Y$. Let $p \in T$; then $p \in \upsilon X$. For every $g \in C(Y)$, the function $g \circ \tau$ is in $C(X)$; therefore $(g \circ \tau)^*(p)$ is a real number. But

$$(g \circ \tau)^* = g^* \circ \bar{\tau},$$

since all the extensions in question are unique; thus, $g^*(\bar{\tau}p)$ is real. As this holds for every $g \in C(Y)$, we have $\bar{\tau}p \in \upsilon Y = Y$.

This completes the proof of the theorem. An alternative proof that (2) implies (1) will be given in Theorem 10.7. Proofs of other implications are indicated in 8D.

REMARK. This theorem specifies all of the spaces in which X is dense and C-embedded: they are precisely the spaces between X and vX. The only realcompact one among them is vX itself, of course. Thus, a realcompact space cannot be dense and C-embedded in any other space.

8.7. THEOREM. *Every (completely regular) space X has a realcompactification vX, contained in βX, with the following equivalent properties.*

(I) *Every continuous mapping τ from X into any realcompact space Y has a continuous extension τ° from vX into Y. (Necessarily, $\tau^\circ = \bar{\tau}|vX$, where $\bar{\tau}$ is the Stone extension into βY.)*

(II) *Every function f in $C(X)$ has an extension to a function f^v in $C(vX)$. (Necessarily, $f^v = f^*|vX$.)*

(III) *If a countable family of zero-sets in X has empty intersection, then their closures in vX have empty intersection.*

(IV) *For any countable family of zero-sets Z_n in X,*

$$\mathrm{cl}_{vX} \bigcap_n Z_n = \bigcap_n \mathrm{cl}_{vX} Z_n.$$

(V) *Every point of vX is the limit of a unique z-ultrafilter on X, and it is a real z-ultrafilter.*

Furthermore, the space vX is unique, in the following sense: if a realcompactification T of X satisfies any one of the listed conditions, then there exists a homeomorphism of vX onto T that leaves X pointwise fixed.

PROOF. Only uniqueness remains to be proved. As in the case of βX, it follows from (I) and 0.12(a).

8.8. REMARKS. The space vX is called the *Hewitt realcompactification* of X. By the theorem, it is characterized as that realcompactification in which X is C-embedded (just as βX was characterized as that compactification in which X is C^*-embedded). Evidently,

(a) *The mapping $f \to f^v$ is an isomorphism of $C(X)$ onto $C(vX)$.*

The converse to the last assertion in Theorem 8.7, regarding uniqueness, is trivially true. (See, however, 6C.*2*.)

It is worth while to summarize some facts about zero-sets and vX. From Theorem 8.4(4), we have $f^v(p) = 0$ if and only if $f \in \boldsymbol{M}^p$, hence, by the Gelfand-Kolmogoroff theorem, if and only if $p \in \mathrm{cl}_{\beta X} \boldsymbol{Z}(f)$. Therefore

(b) $$\boldsymbol{Z}_{vX}(f^v) = \mathrm{cl}_{vX} \boldsymbol{Z}_X(f).$$

Thus, every nonvoid zero-set in vX meets X.

On the other hand, if $p \in \beta X - \upsilon X$, there exists $f \in C^*(X)$ such that $\mathbf{Z}(f) = \emptyset$, while $f^\beta(p) = 0$. Conversely, if a function f in $C^*(X)$ is a unit of C but not of C^*, then $f^\beta(p) = 0$ for some p ($\in \beta X - \upsilon X$).

We have seen that C is more sensitive than C^* in distinguishing among spaces. We have also pointed out, in considering z-filters and e-filters in Chapter 2, that the relations between $C(X)$ and the topology of X are simpler than those between $C^*(X)$ and X. These two facts are related, and, indeed, they reflect the same phenomenon. For $p \in \beta X - X$, C^* never distinguishes between X and $X \,\dot{\cup}\, \{p\}$, while C does precisely when $p \notin \upsilon X$. And the latter is the case exactly when $\mathbf{M}^{*p}$ contains a unit of C, i.e., when $\mathbf{Z}[\mathbf{M}^{*p}]$ is not a z-filter.

PROPERTIES OF REALCOMPACT SPACES

8.9. We look now at the problem of manufacturing new realcompact spaces from old. The results to be obtained will be analogues of various well-known results about compact spaces. In two outstanding cases, however—quotients, and finite unions—the analogues fail; see 8I and 8H.*6*.

THEOREM. *An arbitrary intersection of realcompact subspaces of a given space is realcompact.*

PROOF. Let (Y_α) be a family of realcompact subspaces of a space Y, and let $X = \bigcap_\alpha Y_\alpha$. For each α, the identity mapping τ from X into Y has a continuous extension from υX into the realcompact space Y_α (Theorem 8.7). As τ can have only one continuous extension from υX into Y, these extensions all coincide; hence this common extension carries υX into $\bigcap_\alpha Y_\alpha$, i.e., into X. By 6.13(a), X is closed in υX. Since X is also dense, it is all of υX.

8.10. THEOREM. *Every closed subspace of a realcompact space is realcompact.*

PROOF. Let X be a closed subspace of a realcompact space Y. The identity mapping τ of X into Y has a continuous extension τ° from υX into Y. By Lemma 6.11, the preimage of the closed set X, under τ°, is X; therefore X is closed in υX. Hence $X = \upsilon X$.

A corollary of this theorem is:

(a) *If S is C-embedded in X, then* $\mathrm{cl}_{\upsilon X} S = \upsilon S$.

For then, S is C-embedded in υX, and $\mathrm{cl}_{\upsilon X} S$ is a realcompactification of S in which S is C-embedded. In the analogue for C^*-embedding, the converse is also true, as we saw in 6.9(a). This stemmed from the fact that a closed subset of a compact space is C^*-embedded. But a closed subset of a realcompact space, even though realcompact, need not be C-embedded. This is shown in 8.18 below: the space Γ is a nonnormal, realcompact space; and a space is normal precisely when every closed subset is C-embedded (3D.*1*). Thus, only a partial converse to (a) is valid:

(b) *In case X or υX is normal, then* $\mathrm{cl}_{\upsilon X} S = \upsilon S$ *implies that S is C-embedded in X.*

Normality of X and normality of υX are independent. If X is nonnormal and pseudocompact, then υX ($= \beta X$) is normal. (Examples are Ψ (5I) and Λ (6P); another is the Tychonoff plank, described in 8.20 below.) A normal space X for which υX is not normal is described in the *Notes*.

8.11. THEOREM. *An arbitrary product of realcompact spaces is realcompact.*

PROOF. Let $X = \times_\alpha X_\alpha$, where each X_α is realcompact. Since X is a product of completely regular spaces, it is completely regular.

Each projection $\pi_\alpha: X \to X_\alpha$ has a continuous extension $\pi_\alpha{}^\circ: \upsilon X \to X_\alpha$. Therefore X is closed in υX (6.13(b)). Hence $X = \upsilon X$.

8.12. We now present an alternative proof of the product theorem. Although not so efficient as the one just given, it is more elementary, in that it invokes no material beyond Chapter 5.

LEMMA. *X is realcompact if and only if every prime z-filter with the countable intersection property is fixed.*

PROOF. Sufficiency is trivial. To prove necessity, let $\mathscr{F}$ be a free, prime z-filter, and let $\mathbf{Z}[M]$ (where M is a maximal ideal) be a z-ultrafilter that contains $\mathscr{F}$. (Actually, there is only one (2.13).) Since M is free and X is realcompact, there exists $f \in C(X)$ such that $|M(f)|$ is infinitely large. For each $n \in \mathbf{N}$, consider the zero-sets

$$Z_n = \{x: |f(x)| \geqq n\} \quad \text{and} \quad Z'_n = \{x: |f(x)| \leqq n\}.$$

By 5.7(a), $Z_{n+1} \in \mathbf{Z}[M]$; hence Z'_n, which is disjoint from Z_{n+1}, does not belong to $\mathscr{F}$. Since $Z_n \cup Z'_n = X$, and $\mathscr{F}$ is prime, we have $Z_n \in \mathscr{F}$. Obviously, $\bigcap_n Z_n = \emptyset$. Therefore $\mathscr{F}$ does not have the countable intersection property.

ALTERNATIVE PROOF OF THE PRODUCT THEOREM. Let $X = \times_\alpha X_\alpha$, where each X_α is realcompact. Then X is completely regular.

Let $\mathscr{A}$ be a real z-ultrafilter on X. For each α, the z-filter $\pi_\alpha{}^\#\mathscr{A}$ is prime (4.12). Since $\mathscr{A}$ has the countable intersection property, so has

$\pi_\alpha{}^\#\mathscr{A}$; by the lemma, then, $\pi_\alpha{}^\#\mathscr{A}$ is fixed. By Lemma 4.13, $\mathscr{A}$ is fixed. Therefore X is realcompact.

8.13. THEOREM. *Let τ be a continuous mapping from a realcompact space X into a space Y. Then the total preimage of each realcompact subset of Y is realcompact.*

PROOF. Let F be a realcompact set in Y, and let $S = \tau^{\leftarrow}[F]$. Because X is realcompact, the identity map σ on S has a continuous extension to a mapping $\sigma^\circ\colon \upsilon S \to X$. Also, $\tau|S$ has a continuous extension $(\tau|S)^\circ\colon \upsilon S \to F$. Since S is dense in υS, both these extensions are determined by their values on S. Now, $\tau|S = \tau \circ \sigma$, and therefore $(\tau|S)^\circ = (\tau \circ \sigma)^\circ = \tau \circ \sigma^\circ$. But by Lemma 6.11,

$$\sigma^\circ[\upsilon S - S] \subset X - S,$$

so that

$$(\tau \circ \sigma^\circ)[\upsilon S - S] \subset Y - F,$$

whereas

$$(\tau|S)^\circ[\upsilon S - S] \subset F.$$

Therefore, $\upsilon S - S = \emptyset$.

8.14. CORCLLARY. *Every cozero-set in a realcompact space is realcompact.*

PROOF. $X - \mathbf{Z}(f) = f^{\leftarrow}[\mathbf{R} - \{0\}]$.

An arbitrary open set need not be realcompact, however: witness $\mathbf{W}$ in $\mathbf{W}^*$.

8.15. COROLLARY. *If Y is realcompact, and each point of Y is a G_δ, then every subspace of Y is realcompact.*

PROOF. Each one-element set in Y is a zero-set (3.11(b)), and hence its complement is realcompact (Corollary 8.14). An arbitrary proper subset of Y is an intersection of such complements, and so is also realcompact (Theorem 8.9).

8.16. THEOREM. *In any space, the union of a compact set with a realcompact set is realcompact.*

PROOF. Let $X = S \cup K$, where K is compact, and suppose that X is not realcompact. Choose p in $\upsilon X - X$. Since K is compact, $p \in \operatorname{cl}_{\upsilon X} S$. We show that S is C-embedded in $S \cup \{p\}$, whence S is not realcompact (see 8.6, REMARK). Consider any function f in $C(S)$. There is a function g in $C(\upsilon X)$ that vanishes on a neighborhood of K and is 1 on a neighborhood of p. The function $(g|S)\cdot f$ may be extended to a continuous function h on X, by setting it equal to 0 on K. In turn, h can be extended continuously to p. Since f agrees with h on a deleted neighborhood of p, f can be extended likewise.

8.17. THEOREM. *The following conditions on a space Y are equivalent.*

(1) *For each space X, if there exists a continuous mapping $\tau: X \to Y$ such that $\tau^{\leftarrow}(y)$ is compact for each $y \in Y$, then X is realcompact.*

(2) *Every space of which Y is a one-one, continuous image is realcompact.*

(3) *Every subspace of Y is realcompact.*

(4) *For each point y, $Y - \{y\}$ is realcompact.*

PROOF. It is obvious that (1) implies (2) and (3) implies (4). Also, by Theorem 8.16, (4) implies that Y is realcompact. (And by Theorem 8.9, (4) implies that every *proper* subspace of Y is realcompact, so that (4) implies (3).)

(2) *implies* (3). Given a subspace F of Y, enlarge the topology of Y by making both F and $Y-F$ open. The new space X thus defined is completely regular and the relative topology on F is the same in X as in Y. Since the identity map from X onto Y is continuous, (2) implies that X is realcompact. Therefore F, which is a closed subset of X (though not necessarily of Y), is realcompact.

(4) *implies* (1). Let X and τ satisfy the hypotheses of (1). By (4), as already noted, Y is realcompact. Therefore τ has a continuous extension $\tau^{\circ}: \upsilon X \to Y$. Now consider any point $y \in Y$. By Theorem 8.13, the set

$$S = \tau^{\circ\leftarrow}[Y - \{y\}]$$

is realcompact, and so, by Theorem 8.16, $S \cup \tau^{\leftarrow}(y)$ is realcompact. Since this space lies between X and υX, it must be υX itself. In other words, τ° sends no point of $\upsilon X - X$ onto y. As this holds for every $y \in Y$, we must have $\upsilon X - X = \emptyset$.

Incidentally, we already know a class of spaces that satisfy (3): the Lindelöf spaces all of whose points are G_δ's—e.g., separable metric spaces.

8.18. COROLLARY. *If $\sigma: X \to Y$ is one-one and continuous, and if every subspace of Y is realcompact, then every subspace of X is realcompact.*

As a first application, we see that every discrete space of cardinal $\leqq \mathfrak{c}$ is realcompact: every such space can be mapped one-one into $\mathbf{R}$. It follows that not every realcompact space is a Lindelöf space.

In the one-point compactification of the discrete space of power $\mathfrak{c}$, every subspace is realcompact, but not every point is a G_δ.

Finally, every subspace of the nonnormal space Γ of 3K is realcompact, because the identity mapping of Γ into $\mathbf{R} \times \mathbf{R}$ is continuous.

8.19. *Intersection of free maximal ideals.* In 7E, the intersection of all the free ideals in $C(X)$, or in $C^*(X)$, was characterized as the set $C_K(X)$ of all functions with compact support (i.e., all f for which $\mathrm{cl}\,(X - \mathbf{Z}(f))$ is compact). In 7F, the intersection of the free *maximal* ideals in $C^*(X)$ was characterized as the set of all functions that vanish at infinity (i.e., all f such that $\{x: |f(x)| \geqq 1/n\}$ is compact for all $n \in \mathbf{N}$). For *realcompact* X, we can round out the picture in the following way.

THEOREM. *If X is realcompact, then the intersection of all the free maximal ideals in $C(X)$ is the family $C_K(X)$ of all functions with compact support.*

PROOF. If f has compact support, then, of course, it belongs to every free maximal ideal.

Conversely, if f has noncompact support, there exists a point p that belongs to $\mathrm{cl}_{\beta X}\,(X - \mathbf{Z}(f))$ but not to X. As X is realcompact, there is a function h in $C(X)$ for which $h^*(p) = \infty$. Then h is unbounded on $X - \mathbf{Z}(f)$, and so, by Corollary 1.20, this set contains a noncompact, closed set S that is C-embedded in X. By Theorem 1.18, the disjoint sets S and $\mathbf{Z}(f)$ are completely separated; hence they have disjoint closures in βX. For any point q in $\mathrm{cl}_{\beta X} S - X$, then, we have $q \notin \mathrm{cl}\,\mathbf{Z}(f)$. (Such a point exists, as S is closed in X but not compact.) Therefore, by the Gelfand-Kolmogoroff theorem, f does not belong to the free maximal ideal $\boldsymbol{M}^q$.

The condition that X be realcompact is not necessary. For example, the lone free maximal ideal $\boldsymbol{M}^{\omega_1}$ in $C(\mathbf{W})$ is precisely $C_K(\mathbf{W})$. Another example was given in 7L.*3*. On the other hand, the condition cannot be avoided altogether, as will be seen in the next section.

8.20. *The Tychonoff plank.* This is the name commonly given to Tychonoff's classic example of a nonnormal space, which has achieved considerable fame as a "universal counterexample." Consider the compact space

$$\mathbf{T}^* = \mathbf{W}^* \times \mathbf{N}^*,$$

where $\mathbf{N}^*$ denotes the one-point compactification $\mathbf{N} \cup \{\omega\}$ of $\mathbf{N}$. Let

$$t = (\omega_1, \omega).$$

The plank is the dense subspace

$$\mathbf{T} = \mathbf{T}^* - \{t\}$$

of $\mathbf{T}^*$. Evidently, $\mathbf{T}^*$ is the one-point compactification of $\mathbf{T}$.

It is convenient to use descriptive, geometric terminology in analyzing these spaces. Thus, we shall refer to the subspace

$$W = \mathbf{W} \times \{\omega\}$$

as the top edge of $\mathbf{T}$, and to

$$N = \{\omega_1\} \times \mathbf{N}$$

as the right edge of $\mathbf{T}$. Trivially, W is homeomorphic with $\mathbf{W}$, and N with $\mathbf{N}$. These edges are disjoint closed sets in $\mathbf{T}$. For $n \in \mathbf{N}$, the horizontal line $\mathbf{W}^* \times \{n\}$ is homeomorphic with $\mathbf{W}^*$, and for $\alpha \in \mathbf{W}$, the vertical line $\{\alpha\} \times \mathbf{N}^*$ is homeomorphic with $\mathbf{N}^*$. Since $\mathbf{T}$ contains a closed copy of $\mathbf{W}$, it is not realcompact (Theorem 8.10). Since N has no limit point in $\mathbf{T}$, $\mathbf{T}$ is not countably compact.

We shall now prove that $\mathbf{T}^* = \beta\mathbf{T}$, by showing that $\mathbf{T}$ is C^*-embedded in $\mathbf{T}^*$—in fact, C-embedded, from which it follows, in addition, that $\mathbf{T}$ is pseudocompact. Hence, again, $\mathbf{T}$ is not realcompact. (And, by 3D.*2*, $\mathbf{T}$ is not normal.)

Let $f \in C(\mathbf{T})$. For each $n \in \mathbf{N}$, there is a countable ordinal α_n such that f is constant on the tail

$$\{(\sigma, n) : \sigma \geqq \alpha_n\}$$

of the horizontal line $\mathbf{W}^* \times \{n\}$ (5.12(c)). Let $\alpha = \sup_n \alpha_n$; then $\alpha < \omega_1$ (5.12(a)), and for each $n \in \mathbf{N}$, f is constant on the tail

$$\{(\sigma, n) : \sigma \geqq \alpha\}$$

of $\mathbf{W}^* \times \{n\}$. The common limit r of f on each vertical line $\{\sigma\} \times \mathbf{N}^*$, for $\sigma \geqq \alpha$, is the final constant value of f on the top edge. Hence we extend f to the corner point t by assigning the value r at that point, and this gives us a continuous extension of f.

The fact that $\mathbf{T}^* = \beta\mathbf{T}$ enables us to conclude at once that $\mathbf{T}$ is not normal: W and N have a common limit point in $\beta\mathbf{T}$, and therefore these disjoint closed sets are not completely separated in $\mathbf{T}$. (The argument above also shows this directly.) We can look at this in another way: N is homeomorphic with $\mathbf{N}$, but its closure in $\beta\mathbf{T}$ is not homeomorphic with $\beta\mathbf{N}$; by 6.9(a), N is not C^*-embedded in $\mathbf{T}$ (although it is closed), whence $\mathbf{T}$ is not normal (3D.*1*).

The ring $C(\mathbf{T})$ has only one free maximal ideal, namely, $\boldsymbol{M}^t$. The function $(\alpha, n) \to 1/n$, $(\alpha, \omega) \to 0$, belongs to $\boldsymbol{M}^t$, but its support is the entire noncompact space $\mathbf{T}$. Therefore, in $C(\mathbf{T})$, the intersection of all the free maximal ideals is not the same as the set of all functions with compact support. What amounts to the same thing: the above function does not belong to $\boldsymbol{O}^t$, so that $\boldsymbol{M}^t \neq \boldsymbol{O}^t$.

8.21. N.B. A number of authors have fallen into the trap of assuming that every countable, closed, discrete subset of a completely regular space is C^*-embedded. We have just seen a counterexample: the right edge, N, of $\mathbf{T}$ is countable, closed, and discrete, but it is not C^*-embedded in $\mathbf{T}$.

At the same time, however, it is well to be aware of the possibility of extending certain special functions. Consider, for example, the function $s_n \to 1/n$, defined on a discrete set $S = \{s_n\}_{n \in \mathbf{N}}$ in a space X. (It is not assumed that S is closed.) We extend the function to $\mathrm{cl}_{\beta X} S$ by assigning it the value 0 at all limit points of S. The extended function is continuous, and, in turn, it can be extended from the compact set cl S to a function f in $C(\beta X)$. Then $f\,|X$ is continuous, and agrees with the original function on S. (A more elementary construction is indicated in 3L.*2*.)

Another common error is to assume that a closed, C^*-embedded subspace must be C-embedded. A counterexample is offered by the space Λ described in 6P: $\mathbf{N}$ is closed and C^*-embedded in Λ, but it is not C-embedded.

PROBLEMS

8A. REALCOMPACT SPACES.

1. If a realcompact space X is dense in T, where $T \neq X$, then X is not C-embedded in T: this property characterizes realcompact spaces. Hence a C-embedded realcompact subset is closed.

2. The ring structure of $C(X)$ determines X up to homeomorphism if and only if X is realcompact and, for each nonisolated $p \in X$, $X - \{p\}$ is not C-embedded in X.

3. Exhibit a space X such that $C(X)$ is not isomorphic to $C(Y)$ for any normal Y.

4. $\upsilon X = \beta X$ if and only if X is pseudocompact.

5. υX is a P-space (4J) if and only if X is a P-space.

8B. SPACES BETWEEN X AND βX.

1. $f^{\upsilon}[\upsilon X] = f[X]$ for all $f \in C(X)$. Compare 2C.*1*.

2. For $f \in C(X)$, the space

$$\upsilon_f X = \{p \in \beta X : f^*(p) \neq \infty\}$$

is locally compact and σ-compact.

3. The realcompact spaces between X and βX are precisely the spaces $\bigcap_{f \in C'} \upsilon_f X$, for $C' \subset C(X)$. (Hence υX is the smallest.)

4. Let τ be a continuous mapping from X into a realcompact space Y, and let $\bar{\tau}$ be its Stone extension into βY. By Theorem 8.13, $\bar{\tau}^{\leftarrow}[Y]$ is realcompact. Derive the same conclusion from *3.*

5. If X is pseudocompact, then $\mathrm{cl}_{\beta X} Z$ is a zero-set in βX whenever $Z \in \boldsymbol{Z}(X)$.

8C. STONE EXTENSION OF AN INJECTION.

Let S be a cozero-set in X, and let σ denote the identity map on S and $\bar{\sigma}$ its Stone extension into βX. For each $x \in \mathrm{cl}_X S - S$, $\bar{\sigma}^{\leftarrow}(x)$ is contained in $\beta S - \upsilon S$.

8D. THEOREM 8.6.

Let X be dense and C^*-embedded in T. Establish directly (i.e., without invoking Theorem 8.6):

1. If X is C-embedded, then $\mathrm{cl}_T \boldsymbol{Z}(f) = \boldsymbol{Z}(f^\circ)$, for $f \in C(X)$, and f° its continuous extension to T.

2. Conversely, if every zero-set in T is of the form cl Z, where $Z \in \boldsymbol{Z}(X)$, then X is C-embedded. [Theorem 1.18.]

3. (1) $\rightarrow$ (5) in Theorem 8.6. [Take $Y = \upsilon X$.]

4. (2) $\rightarrow$ (4). [*1.*]

5. (3) $\rightarrow$ (2). [Apply Theorem 1.18, noting that every zero-set is a countable intersection of zero-set-neighborhoods.]

6. (3) $\rightarrow$ (4). [If $p \notin \mathrm{cl} \bigcap_n Z_n$, there exists a zero-set, disjoint from $\bigcap_n Z_n$, whose closure contains p.]

7. (3) $\rightarrow$ (5).

8. (5) $\rightarrow$ (6).

8E. BOUNDED RESTRICTIONS.

1. For any subset S of a realcompact space X, if $f|S$ is bounded for all $f \in C(X)$, then cl S is compact. [Consider any point in $\mathrm{cl}_{\beta X} S - X$.]

2. By 4K.*3*, the assertion also holds in any P-space. (Hence, by 9L.*5*, it can hold when X is not realcompact.) If X is not realcompact, the assertion can fail.

8F. CONVERGENCE OF z-ULTRAFILTERS.

Let X be dense in T. If T is realcompact, then every real z-ultrafilter on X converges in T. [Theorem 8.7(I).] The converse fails, however. [Consider the space Ψ of 5I.]

8G. C-EMBEDDED SETS.

1. If V is an open set in υX, then $V \cap X$ is C-embedded in V. [Extend to each point of $V - X$, as in 8.16. Then apply 6H.]

2. If K is a compact set in X, then $X - K$ is C-embedded in $\upsilon X - K$.

3. If X is C-embedded in Y, but not dense, then $X - \{p\}$ need not be C-embedded in $Y - \{p\}$. [Consider $\mathbf{T}^*$.]

4. Corresponding statements are valid for βX.

8H. EXAMPLES OF REALCOMPACT SPACES.

1. $\beta\mathbf{Q} - \mathbf{Q}$ is realcompact. [Take $Y = [0, 1]$ in Theorem 8.17.]

2. The space Π of 6Q is realcompact.

3. The lexicographically ordered space $\mathbf{R} \times [0, 1]$ (see 3O) is realcompact, and so is every subspace.

4. **W** has a dense, realcompact subspace.

5. A realcompact space can have a free z-filter with the countable intersection property. In fact, a space has the Lindelöf property if and only if every z-filter with the countable intersection property is fixed. [*Sufficiency.* Pass from closed sets to zero-sets as in the proof of Lemma 4.10.]

6. The union of two realcompact subspaces of a space need not be realcompact—even if both are discrete and one is countable. [Consider the space Ψ of 5I.]

7. In Theorem 8.17, "$\tau^{\leftarrow}(y)$ is compact" may not be replaced by "$\tau^{\leftarrow}(y)$ is realcompact."

8. Prove that the discrete space of cardinal $\mathfrak{c}$ is realcompact by considering the free ultrafilters on [0,1]. [They all converge.]

8I. **W** AS A QUOTIENT SPACE OF A REALCOMPACT SPACE.

Let $Y = \mathbf{W}^* \times D$, where D is the set **W** with the discrete topology.

1. Y is realcompact.

2. The subspace $X = \{(\alpha, \sigma): \alpha \leqq \sigma\}$ is realcompact.

3. The mapping $(\alpha, \sigma) \to \alpha$ is continuous and open, but not closed, from X onto **W**.

8J. PROPERTIES OF THE PLANK.

1. If a subset of $\mathbf{W} \times \mathbf{N}^*$ contains points (α_n, n) for arbitrarily large $n \in \mathbf{N}$, then its closure meets W. [5M.*6*.] Hence W and N are not contained in disjoint open sets in **T**, so that **T** is not normal. Compare 3L.*4*,*5*.

2. $\mathbf{W} \times \mathbf{N}^*$ is normal, and $\mathbf{T}^* = \beta(\mathbf{W} \times \mathbf{N}^*)$. [Disjoint closed sets have disjoint closures in $\mathbf{T}^*$: by 5.12(b), they cannot have a common limit point on N; a like argument, with the help of *1*, shows that t cannot be a common limit point either.]

3. Every closed G_δ-set E in **T** is a zero-set and is C-embedded in **T**. [$E \cap (\mathbf{W} \times \mathbf{N}^*)$ is a closed G_δ in the normal space $\mathbf{W} \times \mathbf{N}^*$; apply 3D.*3*, and observe that $\mathbf{W} \times \mathbf{N}^*$ is C-embedded, in a trivial way, in **T**.]

4. Describe the class of real-valued functions on N that can be extended continuously to all of **T**. According to Urysohn's extension theorem (1.17), N must contain two disjoint sets that are not completely separated in **T**; exhibit such a pair. Note that any two such sets are closed and have disjoint neighborhoods in **T**.

5. Every noncompact zero-set in **T** meets W, although a noncompact closed set need not.

6. In $C(\mathbf{T})$, $\boldsymbol{O}^t$ is the smallest, and $\boldsymbol{M}^t$ the largest free ideal. The ideal $\boldsymbol{O}^t$ is not prime.

7. $\mathbf{T}^*$ contains two C^*-embedded subsets whose intersection is not C^*-embedded.

8K. AN ENLARGEMENT OF THE PLANK.

Enlarge $\mathbf{T}^*$ by letting each point (ω_1, n) of N serve as the point at infinity for a separate copy of $\mathbf{N}$ (yielding, in each case, a copy of $\mathbf{N}^*$). Let X denote the space obtained by deleting the point $t = (\omega_1, \omega)$.

1. $t \in \beta X$.

2. Construct functions $f, g \in C(X)$ such that $g \in \boldsymbol{M}^t$, and $\boldsymbol{Z}(f)$ is a neighborhood of $\boldsymbol{Z}(g)$, but $f \notin \boldsymbol{O}^t$, i.e., cl $\boldsymbol{Z}(f)$ is not a neighborhood of t. Compare Theorem 7.14.

8L. THE SPACE Ω.

Let $\Omega^* = \mathbf{W}^* \times \mathbf{W}^*$, denote the corner point (ω_1, ω_1) by w, and define $\Omega = \Omega^* - \{w\}$. Then Ω^* is the one-point compactification of Ω. In Ω, the top edge $\mathbf{W} \times \{\omega_1\}$, the right edge $\{\omega_1\} \times \mathbf{W}$, and the diagonal

$$\{(\alpha, \alpha): \alpha \in \mathbf{W}\},$$

are mutually disjoint closed sets, each homeomorphic with $\mathbf{W}$ (whence Ω is not realcompact).

1. Let E be a subset of Ω, disjoint from both the top edge and the right edge. If E meets every neighborhood of w, then $\mathrm{cl}_\Omega E$ meets the diagonal in every neighborhood of w. [Apply the interlacing technique as in 5.12(b).]

2. Every noncompact, closed set E in Ω has w as a limit point in Ω^*. Hence E must meet either the top edge, or the right edge, or the diagonal, in every neighborhood of w.

3. Every function in $C(\Omega)$ has the same constant value on tails of the top edge, the right edge, and the diagonal. [Apply *1* to suitable subsets of $f^{\leftarrow}(t)$ and $f^{\leftarrow}(r)$, where t and r are the final constant values on the edges.]

4. Every function in $C(\Omega)$ is constant on a deleted neighborhood of w. [Argue as in 5.12(c).] Hence Ω is C-embedded in Ω^*, so that Ω is pseudocompact, and $\Omega^* = \beta\Omega$. In addition, Ω is countably compact.

5. Ω is not normal. Observe, incidentally, that the top and right edges are contained in disjoint open sets, although they are not completely separated (cf. 8J.*4*). On the other hand, the diagonal and an edge are not contained in disjoint open sets.

8M. FURTHER PROPERTIES OF Ω.

Let Ω be as in 8L.

1. Let τ be a countable limit ordinal. If a subset of $\mathbf{W} \times W(\tau + 1)$ contains points (α_σ, σ) for arbitrarily large $\sigma < \tau$, then its closure meets $\mathbf{W} \times \{\tau\}$. [Compare 8J.*1*.]

2. $\mathbf{W} \times \mathbf{W}$ is normal, and $\Omega^* = \beta(\mathbf{W} \times \mathbf{W})$. [Compare 8J.*2.*]

3. $\mathbf{W} \times \mathbf{W}$ is countably compact.

4. $\mathbf{W} \times \mathbf{W}^*$, a product of two normal spaces (one of which, in fact, is compact), is not normal.

5. Every closed G_δ in Ω is a C-embedded zero-set. [Compare 8J.*3.*]

6. w is a P-point (4L) of Ω^*, and the only free ideal in $C(\Omega)$ is $\mathbf{M}^w$.

7. Each edge of Ω is C-embedded, although their union is not C^*-embedded.

8N. DISCRETE SUBSPACES OF PSEUDOCOMPACT SPACES.

Let $\mathfrak{m}$ be a cardinal $> \aleph_0$. Define $Y = \mathbf{W}^* \times W(\alpha + 1)$, where α is of power $\mathfrak{m}$, and form X by removing from Y all points (ω_1, λ) for which λ is a limit ordinal. Then X is pseudocompact, although it contains a closed, discrete subspace of cardinal $\mathfrak{m}$. [If f is unbounded on a sequence of points (α_n, σ_n), then f is unbounded on a set that is homeomorphic with a subspace of $\mathbf{T}$ containing $\mathbf{W} \times \mathbf{N}^*$.]

Chapter 9

CARDINALS OF CLOSED SETS IN βX

9.1. It was pointed out in 6.10 that the cardinal number of $\beta\mathbf{N}$ is at least $\mathfrak{c}$. On the other hand, since each ultrafilter on $\mathbf{N}$ is a subset of the set of all subsets of $\mathbf{N}$, and $\beta\mathbf{N}$ is in one-one correspondence with the set of all (z-)ultrafilters, we have $|\beta\mathbf{N}| \leqq 2^{\mathfrak{c}}$. This latter argument applies equally well to any completely regular space X and shows that $|\beta X| \leqq 2^{2^{|X|}}$. (See also 9A.*1*.)

In this chapter, we shall obtain sharper results about the cardinal numbers of sets in βX. The key to these results is the precise evaluation of the cardinal of $\beta\mathbf{N}$. (See also 9O.)

9.2. THEOREM. *For every infinite discrete space X,*

$$|\beta X| = 2^{2^{|X|}}.$$

PROOF. For simplicity of notation, put $\mathfrak{m} = 2^{2^{|X|}}$. The problem is to show that there are at least $\mathfrak{m}$ ultrafilters on X. We reduce the complexity of the notation by considering the following auxiliary sets: the set $\mathscr{F}$ of all finite subsets F of X, and the set Φ of all finite subsets φ of $\mathscr{F}$. We shall construct $\mathfrak{m}$ ultrafilters on $\mathscr{F} \times \Phi$; since $\mathscr{F} \times \Phi$ is equipotent with X (X being infinite), the results are equivalent.

With each (arbitrary) subset S of X, associate a subset $\mathfrak{b}_S$ of $\mathscr{F} \times \Phi$, as follows:

$$\mathfrak{b}_S = \{(F, \varphi) \in \mathscr{F} \times \Phi: S \cap F \in \varphi\}.$$

For simplicity of notation, we shall denote the complement of $\mathfrak{b}_S$ in $\mathscr{F} \times \Phi$ by $-\mathfrak{b}_S$.

Next, for each of the $\mathfrak{m}$ subsets $\mathscr{S}$ of the set of all subsets of X, define the family

$$\mathfrak{B}_{\mathscr{S}} = \{\mathfrak{b}_S: S \in \mathscr{S}\} \cup \{-\mathfrak{b}_S: S \notin \mathscr{S}\}$$

of subsets of $\mathscr{F} \times \Phi$. We shall show that each family $\mathfrak{B}_{\mathscr{S}}$ has the finite intersection property.

Let

$$\mathfrak{b}_{S_1}, \cdots, \mathfrak{b}_{S_k}, -\mathfrak{b}_{S_{k+1}}, \cdots, -\mathfrak{b}_{S_n}$$

be distinct members of $\mathfrak{B}_{\mathscr{S}}$. The indices $S_1, \cdots, S_n$ are distinct subsets of X. For $i < j$, choose a single element x_{ij} that belongs to exactly one of S_i, S_j. The selected elements x_{ij}, for $1 \leqq i < j \leqq n$, form a finite subset F of X. For $i < j$, the sets $S_i \cap F$ and $S_j \cap F$ are distinct, since exactly one of them contains x_{ij}. Now consider the finite set

$$\varphi = \{S_1 \cap F, \cdots, S_k \cap F\},$$

which is a member of Φ. Trivially, $S_i \cap F \in \varphi$ for $i \leqq k$, and $S_j \cap F \notin \varphi$ for $j > k$. Therefore $(F, \varphi) \in \mathfrak{b}_{S_i}$ for $i \leqq k$, and $(F, \varphi) \in -\mathfrak{b}_{S_j}$ for $j > k$. This shows that $\mathfrak{B}_{\mathscr{S}}$ has the finite intersection property.

Each family $\mathfrak{B}_{\mathscr{S}}$ is embeddable, therefore, in at least one ultrafilter $\mathfrak{U}_{\mathscr{S}}$. Furthermore, distinct families cannot be contained in the same ultrafilter; for, if $S \in \mathscr{S} - \mathscr{S}'$, then $\mathfrak{B}_{\mathscr{S}}$ contains $\mathfrak{b}_S$, while $\mathfrak{B}_{\mathscr{S}'}$ contains its complement. Since there are $\mathfrak{m}$ sets $\mathscr{S}$, there are $\mathfrak{m}$ ultrafilters $\mathfrak{U}_{\mathscr{S}}$.

9.3. As a special case of the theorem, we have

$$|\beta\mathbf{N}| = 2^{\mathfrak{c}}.$$

We proved in 6.10 that $\beta\mathbf{Q}$ and $\beta\mathbf{R}$ are equipotent with $\beta\mathbf{N}$. Therefore,

$$|\beta\mathbf{Q}| = |\beta\mathbf{R}| = 2^{\mathfrak{c}}.$$

Furthermore,

$$|\beta\mathbf{N} - \mathbf{N}| = |\beta\mathbf{Q} - \mathbf{Q}| = |\beta\mathbf{R} - \mathbf{R}| = 2^{\mathfrak{c}}.$$

We shall apply Theorem 9.2 to find lower bounds for the cardinals of various closed sets in βX, or in $\beta X - X$, for general X. The idea, in each case, is to show that the set in question contains a copy of $\beta\mathbf{N} - \mathbf{N}$; the cardinal of the set, then, is at least $2^{\mathfrak{c}}$.

9.4. LEMMA. *Let $E \subset \beta X$, and suppose that Z is a zero-set in βX that meets* cl E *but not $X \cup E$. Then E contains a copy N of* $\mathbf{N}$, *and Z contains $\beta N - N$.*

PROOF. Write $Z = \mathbf{Z}(f)$, and $Y = \beta X - Z$. Since $Z \cap X = \emptyset$, we have $Y \supset X$, and therefore $\beta Y = \beta X$. Also, $Y \supset E$. In $C(Y)$, $h = (f|Y)^{-1}$ exists; and, because Z meets cl E, h is unbounded on E. By Corollary 1.20, E contains a copy N of $\mathbf{N}$ that is C-embedded in Y and on which h goes to infinity. The first of these conclusions shows that N is C^*-embedded in βX, so that cl $N = \beta N$ (6.9(a)). The second conclusion implies that cl $N - N$ is contained in Z.

9.5. THEOREM. *Every nonempty zero-set in βX, if disjoint from X, contains a copy of $\beta\mathbf{N}$, and so its cardinal is at least $2^{\mathfrak{c}}$.*

PROOF. Take $E = X$ in the lemma; and recall (6.10(a)) that $\beta N - N$ contains a copy of $\beta\mathbf{N}$.

9.6. COROLLARY. *No point of $\beta X - X$ is a G_δ in βX.*

PROOF. Every G_δ-point is a zero-set (3.11(b)).

For example, a compact metric space cannot be the Stone-Čech compactification of any other space.

9.7. If x is a G_δ-point of X, then $\{x\}$ is a zero-set in X, and hence is also a zero-set in υX (8.8(b)). Thus, every G_δ-point of X is a G_δ *in* υX. On the other hand, no point of $\upsilon X - X$ is a G_δ in υX, because every zero-set in υX meets X (8.8(b)).

It is not true that a G_δ-point of X is always a G_δ *in* βX. (Consider, for example, the space Σ of 6E.5.) However, if x has a *countable base of neighborhoods* in X, then it also has a countable base in βX (or, more generally, in any space in which X is dense). In fact, if (U_n) is a base in X at x, then $(\mathrm{cl}_{\beta X} U_n)$ is a base in βX, as is easily seen. On the other hand, no point of $\beta X - X$ has a countable base of neighborhoods in βX, as follows from the corollary.

THEOREM.

(a) *If $C(X)$ is isomorphic with $C(Y)$, and if all points of X and Y are G_δ's, then X is homeomorphic with Y.*

(a*) *If $C^*(X)$ is isomorphic with $C^*(Y)$, and if X and Y satisfy the first countability axiom, then X is homeomorphic with Y.*

PROOF. (a). If $C(X)$ is isomorphic with $C(Y)$, then υX is homeomorphic with υY (Theorem 8.3). As noted above, no point of $\upsilon X - X$ or $\upsilon Y - Y$ is a G_δ in υX or υY, respectively. Since a homeomorphism takes G_δ-points to G_δ-points, X must be carried onto Y.

The proof of (a*) is similar.

9.8. COROLLARY. *Two metric spaces X and Y are homeomorphic if and only if $C^*(X)$ and $C^*(Y)$ are isomorphic.*

9.9. We look now at the cardinals of closed sets that are not necessarily zero-sets.

LEMMA. *Every nondiscrete, closed set S in $\beta X - \upsilon X$ has a countable, discrete subset D with a limit point in $\beta X - \upsilon X$.*

PROOF. Let p be a limit point of S. Then $p \notin \upsilon X$, and so there exists $f \in C(\beta X)$ such that $f(p) = 0$, while f vanishes nowhere on υX. In case $\mathbf{Z}(f) \cap S$ is finite, we choose D to be a discrete set $\{s_n\}_{n \in \mathbf{N}}$ in S, such that $f(s_n) \neq 0$ for every n, while $\lim_{n\to\infty} f(s_n) = 0$. In case $Z(f) \cap S$ is infinite, we choose D to be a countably infinite, discrete set in this intersection (0.13). In either case, D is not closed in βX, of course, and f vanishes on $\text{cl}\, D - D$. Hence $\text{cl}\, D - D$ must be contained in $\beta X - \upsilon X$, and therefore D is not closed in $\beta X - \upsilon X$ either.

9.10. LEMMA. *Let D be a countable, discrete subset of βX, having a limit point p in $\beta X - \upsilon X$. Then D contains a copy N of* $\mathbf{N}$, *with*

$$\beta N - N \subset \beta X - \upsilon X.$$

PROOF. Define $g \in C(\beta X)$ as follows. First, g is to map D one-one onto the set $\{1/n\}_{n \in \mathbf{N}}$. Next, g is to vanish at all limit points of D. So far, we have a nonnegative, continuous function defined on the compact set $\text{cl}\, D$. We choose g to be any nonnegative, continuous extension of this function to all of βX. (This is the construction described in 8.21.) Since p is a limit point of D, g vanishes at p.

Because $p \notin \upsilon X$, there is a nonnegative function f in $C(\beta X)$ that vanishes at p, but nowhere on υX. Then $\mathbf{Z}(f + g)$ meets $\text{cl}\, D$, but not $\upsilon X \cup D$. By Lemma 9.4, D contains a copy N of $\mathbf{N}$, and the subset $Z(f + g)$ of $\beta X - \upsilon X$ contains $\beta N - N$.

9.11. THEOREM. *Each nondiscrete, closed set in $\beta X - \upsilon X$ contains a copy of* $\beta\mathbf{N}$, *and so its cardinal is at least* $2^{\mathfrak{c}}$.

PROOF. By Lemma 9.9, any such set S has a countable, discrete subset D with a limit point in $\beta X - \upsilon X$. By Lemma 9.10, S contains a copy N of $\mathbf{N}$, with $\beta N - N$ contained in $\beta X - \upsilon X$. Since S is closed in $\beta X - \upsilon X$, it contains $\text{cl}_{\beta X - \upsilon X} N$, which is βN.

9.12. COROLLARY. *If X is locally compact and realcompact, then every infinite closed set in $\beta X - X$ contains a copy of* $\beta\mathbf{N}$ *(and so its cardinal is at least* $2^{\mathfrak{c}}$*).*

PROOF. Since X is locally compact, $\beta X - X$ is compact (6.9(d)), and so any closed, discrete subset is finite.

REMARKS. When X is not locally compact, $\beta X - \upsilon X$ can contain a countably infinite, closed discrete set; see 9C.*1*.

When X is not realcompact, $\beta X - X$ can be of cardinal 1, as for $X = \mathbf{W}$, and it can be countable but not discrete, as for $X = \mathbf{W} \times \mathbf{N}^*$ (8J.*2*). As a matter of fact, any space can be $\beta X - X$ for suitable X (9K.*6*).

For practical purposes, as will be shown in Chapter 12, it may be assumed that every discrete space is realcompact. The corollary applies, then, to these spaces. By 9H.*2*, the conclusion of the corollary holds for all discrete spaces without reservation. Thus, *if X is any discrete space, then every infinite closed set in $\beta X - X$ contains a copy of* $\beta\mathbf{N}$.

PSEUDOCOMPACT SPACES

9.13. LEMMA. *In order that a space Y be pseudocompact, it is necessary and sufficient that for any decreasing sequence $(V_n)_{n \in \mathbf{N}}$ of nonempty open sets, $\bigcap_n \operatorname{cl} V_n$ be nonempty.*

PROOF. *Necessity.* Suppose that $\bigcap_n \operatorname{cl} V_n = \emptyset$. Choose $y_n \in V_n$, distinct for distinct n. Then $\{y_n\}_{n \in \mathbf{N}}$ is a closed, discrete set, because any limit point of this set would belong to $\bigcap_n \operatorname{cl} V_n$. Inductively, choose closed neighborhoods V'_n of y_n, such that $V'_n \subset V_n$, and $V'_m \cap V'_n = \emptyset$ when $m \neq n$. For each n, let g_n be a function in $C(Y)$ such that $g_n(y_n) = n$, and $g_n[Y - V'_n] = \{0\}$. Finally, let

$$g(y) = \sum_{n \in \mathbf{N}} g_n(y) \qquad (y \in Y).$$

There is no convergence problem here, because for each y, at most one of the summands is different from 0. Furthermore, each point y has a neighborhood meeting at most one V'_n. For, any finite number of these sets that do not contain y can be subtracted from a neighborhood; and if every neighborhood of y were to meet infinitely many V'_n, and hence infinitely many V_n, then y would belong to $\bigcap_n \operatorname{cl} V_n$. Therefore, g agrees with one of the g_n on a neighborhood of y. This shows that g is continuous. Since g is unbounded, Y is not pseudocompact. (This proof shows, in effect, that $\{y_n\}_{n \in \mathbf{N}}$ is C-embedded in Y; compare 3L.*1*.)

Sufficiency. If g is an unbounded function in $C(Y)$, then the open sets $V_n = \{y : |g(y)| > n\}$ form a decreasing sequence for which $\bigcap_n \operatorname{cl} V_n$ is empty.

9.14. THEOREM. *If X is compact, and Y is pseudocompact, then $X \times Y$ is pseudocompact.*

PROOF. Let (W_n) be a decreasing sequence of nonempty, open sets in $X \times Y$. We shall exhibit a point (x_0, y_0) in $\bigcap_n \operatorname{cl} W_n$. Let π denote the projection of $X \times Y$ onto Y. Then $(\pi[W_n])$ is a decreasing sequence of nonempty, open sets in Y. By the lemma, there exists a point y_0 in $\bigcap_n \operatorname{cl} \pi[W_n]$. Every neighborhood V of y_0 meets $\pi[W_n]$, for all n, and hence $\pi^{\leftarrow}[V]$ meets every W_n.

For each neighborhood V of y_0, let $U_{n,V}$ denote the projection into X of the set $W_n \cap \pi^{\leftarrow}[V]$. Since a finite intersection of neighborhoods V of y_0 is a neighborhood, and since the sequence (W_n) is nested, the family of all sets $U_{n,V}$ has the finite intersection property. As X is compact, there exists a point x_0 in $\bigcap_{n,V} \operatorname{cl} U_{n,V}$.

Now, a basic neighborhood of (x_0, y_0) in $X \times Y$ has the form $U \times V$, where U and V are neighborhoods of x_0 and y_0, respectively. By definition of x_0, U meets $U_{n,V}$ for every n. This implies that $U \times V$ meets every W_n. Therefore $(x_0, y_0) \in \bigcap_n \operatorname{cl} W_n$. By the lemma, $X \times Y$ is pseudocompact.

9.15. EXAMPLE. In contrast to this theorem, we shall now present an example of *a pseudocompact space* G such that $G \times G$ *is not pseudocompact*. The space G will be a countably compact subspace of $\beta\mathbf{N}$.

First, we define a homeomorphism τ of $\beta\mathbf{N}$ onto itself, as follows. As we saw in 6.10, $\beta\mathbf{N}$ is the union of two disjoint copies of itself, βN_1 and βN_2. We choose any homeomorphism of βN_1 onto βN_2, and define τ to agree on βN_1 with this homeomorphism, and on βN_2 with its inverse. Thus, τ has no fixed point, and $\tau \circ \tau$ is the identity on $\beta\mathbf{N}$.

The subspace G will be defined inductively. Let $\mathscr{S}$ denote the family of all countably infinite subsets of $\beta\mathbf{N}$. Since $|\beta\mathbf{N}| = 2^{\mathfrak{c}}$, we have

$$|\mathscr{S}| = (2^{\mathfrak{c}})^{\aleph_0} = 2^{\mathfrak{c}}.$$

Let $\prec$ be a well-ordering of $\mathscr{S}$ according to the smallest ordinal of cardinal $2^{\mathfrak{c}}$. Consider any $S \in \mathscr{S}$, and suppose that for each $E \prec S$, we have chosen a limit point p_E of E, distinct from $\tau p_{E'}$ for all $E' \prec S$. Now, $|\operatorname{cl} S| = 2^{\mathfrak{c}}$, by 9.12, so that $|\operatorname{cl} S - S| = 2^{\mathfrak{c}}$. Since the set of all predecessors of S is of smaller cardinal, we can select p_S in $\operatorname{cl} S - S$ so as to differ from τp_E for all $E \prec S$. We now define

$$G = \mathbf{N} \cup \{p_S\}_{S \in \mathscr{S}}.$$

By construction, every countably infinite subset of G (in fact, of $\beta\mathbf{N}$) has a limit point in G. Therefore G is countably compact, and hence pseudocompact.

We wish to show that there exists an unbounded, continuous function on $G \times G$. It suffices to prove that there exists an infinite, discrete set that is open-and-closed. Consider the infinite set

$$D = \{(n, \tau n) : n \in \mathbf{N}\}.$$

Since τ carries $\mathbf{N}$ into $\mathbf{N}$, each point of D is isolated, and hence D is open and discrete. On the other hand, if $p \notin \mathbf{N}$, then G, by construction, does not contain both p and τp. Therefore D is the intersection of $G \times G$ with the subset

$$\{(p, \tau p) : p \in \beta\mathbf{N}\}$$

of $\beta\mathbf{N} \times \beta\mathbf{N}$. Since the latter set is the graph of a continuous mapping, it is closed (6.13(c)). Therefore D is closed in $G \times G$.

PROBLEMS

9A. THE CARDINAL OF βX.

1. Let X be dense in T. Prove directly from the fact that T is a Hausdorff space that

$$|T| \leqq 2^{2^{|X|}}.$$

[Associate with each point p of T the family of all sets of the form $V \cap X$, where V is a neighborhood of p.] (Hence $|\beta\mathbf{Q}| \leqq 2^{\mathfrak{c}}$ and $|\beta\mathbf{R}| \leqq 2^{\mathfrak{c}}$.)

2. Derive the same conclusion from Stone's theorem.

9B. THE SPACE $\mathbf{R}$.

1. There are exactly $2^{\mathfrak{c}}$ z-filters on $\mathbf{R}$, and $2^{2^{\mathfrak{c}}}$ filters.

2. If S is an unbounded set in $\mathbf{R}$, then $|\mathrm{cl}_{\beta\mathbf{R}}\, S| = 2^{\mathfrak{c}}$.

9C. THE SPACE $\beta\mathbf{Q}$.

1. $\beta\mathbf{Q} - \mathbf{Q}$ contains a countably infinite, closed, discrete set. [Map $\beta\mathbf{Q}$ onto [0, 1], and consider a sequence of irrationals converging to a rational.]

2. $\beta(\beta\mathbf{Q} - \mathbf{Q})$ is not homeomorphic with $\beta\mathbf{Q}$. [Which points have countable bases?] This result improves 6O.*4*.

9D. $\beta X - \upsilon X$.

1. If $p \in \beta X - \upsilon X$, then every neighborhood of p in βX contains a copy of $\mathbf{N}$ that is C-embedded in X, and hence contains a copy of $\beta\mathbf{N}$.

2. If $C(X)$ contains a hyper-real ideal, then it contains at least $2^{\mathfrak{c}}$ hyper-real ideals. In other words, if $\beta X - \upsilon X$ is nonempty, then $|\beta X - \upsilon X| \geqq 2^{\mathfrak{c}}$.

3. If $|\beta X - X| < 2^{\mathfrak{c}}$, then X is pseudocompact. (Compare with 6I.*1* and 6J.) The converse is not true. [Consider the space Λ of 6P and 9E.] Notice that $\beta\mathbf{N}$ has $2^{\mathfrak{c}}$ infinite, pseudocompact subsets.

9E. THE SPACE Λ.

Let $\Lambda = \beta\mathbf{R} - (\beta\mathbf{N} - \mathbf{N})$. Every zero-set in Λ that meets $\Lambda - \mathbf{R}$ is of cardinal $2^{\mathfrak{c}}$.

9F. ULTRAFILTERS.

If X is infinite, there exist $2^{2^{|X|}}$ ultrafilters on X all of whose members are of cardinal $|X|$. [In the proof of Theorem 9.2, observe that every finite intersection of members of $\mathfrak{B}_{\mathscr{S}}$ is of cardinal $|X|$. Adjoin to each family $\mathfrak{B}_{\mathscr{S}}$ all subsets of $\mathscr{F} \times \Phi$ with complement of power less than $|X|$.] Compare 12I.

9G. ULTRAFILTERS ON N THAT CONTAIN NO SMALL SETS.

Let $\mathscr{F}$ be the filter of all sets in $\mathbf{N}$ of density 1 (6U).

1. Every set of positive density meets every member of $\mathscr{F}$.

2. $\mathscr{F}$ is contained in infinitely many ultrafilters. [Find infinitely many disjoint sets of positive density.]

3. $\mathscr{F}$ is contained in $2^{\mathfrak{c}}$ ultrafilters. [The set of all cluster points of $\mathscr{F}$ in $\beta\mathbf{N}$ is closed.]

9H. BASICALLY DISCONNECTED SPACES.

1. Every countable set S in a basically disconnected space X (1H) is C^*-embedded. [3B.*4* and Urysohn's extension theorem.] If S is not countable, the conclusion may fail, even when S is discrete and X is extremally disconnected. [6Q.]

2. If X is basically disconnected, then every infinite closed set E in βX contains a copy of $\beta\mathbf{N}$, and so its cardinal is at least $2^{\mathfrak{c}}$. [By 0.13, E contains a countably infinite discrete subset D. By 6M.*1*, βX is basically disconnected.] In particular, this holds for discrete X.

9I. REALCOMPACTIFICATION OF A PRODUCT.

For the space G of 9.15, $vG \times vG$ is compact, while $v(G \times G)$ is not. [8A.*4*.] Hence *$vG \times vG$ is not homeomorphic with $v(G \times G)$.* A similar result for βX was given in 6N.*2*.

9J. COUNTABLY COMPACT PRODUCT SPACE.

The product of a compact space with a countably compact space is countably compact. [A space is countably compact if and only if every decreasing sequence of nonempty closed sets has nonempty intersection.]

9K. GENERALIZATIONS OF W AND T.

The smallest ordinal of cardinal $\aleph_\alpha$ is denoted by ω_α. (Thus, $\omega_0 = \omega$.)

Let α be a *nonlimit* ordinal > 0. Proofs of the statements below may be modeled after those given for the case $\alpha = 1$ in 5.12, 5.13, and 8.20. See also 5N.*1*.

1. No subset of $W(\omega_\alpha)$ of cardinal $< \aleph_\alpha$ is cofinal. [$\aleph_\xi^2 = \aleph_\xi$.]

2. Of any two disjoint closed sets in $W(\omega_\alpha)$, one is bounded.

3. Every continuous function on the totally ordered space $W(\omega_\alpha)$ into $\mathbf{R}$ is constant on a tail.

4. $\beta W(\omega_\alpha) = W(\omega_\alpha + 1)$.

5. If X is a compact space, with $|X| < \aleph_\alpha$, then

$$\beta(X \times W(\omega_\alpha)) = X \times W(\omega_\alpha + 1).$$

6. Any space S is homeomorphic with a space $\beta Y - Y = \upsilon Y - Y$, for suitable Y. [In *5*, let X be a compactification of S.]

9L. A P-SPACE THAT IS NOT REALCOMPACT.

Let X be the subspace of $W(\omega_2)$ obtained by deleting all nonisolated points having a countable base.

1. As an ordered subset of $W(\omega_2)$, X is of order type ω_2, and hence no subset of X of power $< \aleph_2$ is cofinal. [9K.*1*.] However, the topology on X is not the interval topology (3O).

2. Every subset of X of type ω_1 has a supremum in X. [5.12(a).] (More generally, see 13I.*2*.)

3. Of any two disjoint closed sets in X, one is bounded.

4. Every function in $C^*(X)$ is constant on a tail. [Choose

$$r \in \bigcap_\alpha \operatorname{cl}\{f(\sigma): \sigma > \alpha\},$$

and modify the argument of 5.12(c).] Hence every function in $C(X)$ is constant on a tail.

5. X is a P-space (4J), but is not realcompact.

9M. P-POINTS AND NONHOMOGENEITY OF $\beta X - X$.

Recall that X contains a C-embedded copy of $\mathbf{N}$ if and only if it is not pseudocompact (Corollary 1.21).

1. If $\mathbf{N}$ is C-embedded in X and U is any neighborhood of $\mathbf{N}$, then $\operatorname{cl}_{\beta X}(X - U)$ is disjoint from $\operatorname{cl}_{\beta X}\mathbf{N}$ $(= \beta\mathbf{N})$. [3B.*2*.]

2. If $\mathbf{N}$ is C-embedded in a locally compact space T, then every P-point (4L) of $\beta\mathbf{N} - \mathbf{N}$ is a P-point of $\beta T - T$. [Let $g \in C(\beta T)$ vanish at the P-point p of $\beta\mathbf{N} - \mathbf{N}$. Let (s_n) be a sequence in $\mathbf{N}$ such that $\lim_n g(s_n) = 0$, as provided in 6T.*1*. For each n, take a compact neighborhood on which $|g(t) - g(s_n)| < 1/n$, and apply *1* to their union.] Local compactness is critical: see 6O.*5*.

3. Assume the continuum hypothesis. If T is locally compact but not pseudocompact, then $\beta T - T$ has P-points as well as non-P-points [6V.*4*]. If, in addition, T is realcompact, then both sets of points are dense in $\beta T - T$ [9D.*1*].

4. Let $\mathbf{N}$ be C-embedded in X, and let φ be a homeomorphism of $\beta X - X$ onto itself. If p is a P-point of $\beta\mathbf{N} - \mathbf{N}$, and if $\varphi p \in \beta\mathbf{N} - \mathbf{N}$, then φp is a P-point of $\beta\mathbf{N} - \mathbf{N}$. [Let $f \in C(X)$, with $f(n) = n$ for $n \in \mathbf{N}$, and define

$$T = \upsilon_f X - \varphi^{\leftarrow}[\beta\mathbf{N} - \mathbf{N}]$$

(8B.*2*). Apply *2* (justified by Theorem 1.19) to show that p is a P-point of $\varphi^{\leftarrow}[\beta\mathbf{N} - \mathbf{N}]$.]

5. Assume the continuum hypothesis. *If X is not pseudocompact, then $\beta X - X$ is not homogeneous.* [Apply *4*.]

9N. THEOREM 9.7.

1. If Y is C^*-embedded in T, V is open in T, and $V \cap Y$ is dense in V, then $V \cap Y$ is C^*-embedded in V. [Given $t \in V - Y$, there exists $h \in C^*(T)$ that vanishes outside V and is equal to 1 on a neighborhood of t. Given $g \in C^*(V \cap Y)$, the function $g \cdot (h|V \cap Y)$ has a continuous extension to all of T. This supplies an extension of g to a neighborhood of t.]

2. If S is dense in Y, and a point $y \in Y - S$ is both a G_δ in $S \cup \{y\}$ and the limit of a sequence $(y_n)_{n \in \mathbf{N}}$ of distinct points of Y, then S is not C^*-embedded in $S \cup \{y\}$. [Note that y is a G_δ-point in the space

$$X = S \cup \{y\} \cup \{y_n : n \in \mathbf{N}\},$$

so that $\{y\} = \mathbf{Z}(f)$ for some $f \in C^*(X)$. Find $h \in C^*(\mathbf{R} - \{0\})$ such that $h \circ (f|S)$ has no continuous extension to y.]

3. Let X and Y be spaces in which every nonisolated point is both a G_δ and the limit of a sequence of distinct points. If $C^*(X)$ is isomorphic with $C^*(Y)$, then X is homeomorphic with Y. [Identify βX with βY. If $y \in Y - X$, then by *1*, $Y - \{y\}$ is C^*-embedded in $\beta Y - \{y\} \supset X$, and hence in $\beta X \supset Y$. This contradicts *2*.]

4. In *3*, the condition that every nonisolated point be a limit of a sequence of distinct points cannot be dropped.

5. In *3*, the condition that every point be a G_δ cannot be dropped. [8K.]

6. The analogue of *1* with C^* replaced by C is valid.

9O. THE CARDINAL OF $\beta\mathbf{N}$.

1. Let $X = E^E$, where E is the closed interval $[0,1]$, and let S be the set of all members of X that are polynomial functions with rational coefficients. Then $|X| = 2^{\mathfrak{c}}$, and S is a countable, dense subset of X.

2. Prove that $|\beta\mathbf{N}| \geqq 2^{\mathfrak{c}}$ by considering the Stone extension of a mapping from $\mathbf{N}$ onto S. Hence by 9A, $|\beta\mathbf{N}| = 2^{\mathfrak{c}}$.

Chapter 10

HOMOMORPHISMS AND CONTINUOUS MAPPINGS

10.1. We proved in Theorem 8.3 that two realcompact spaces X and Y are homeomorphic if and only if their respective function rings $C(X)$ and $C(Y)$ are isomorphic. The correspondence between the set of all homeomorphisms from X onto Y, and the set of all isomorphisms from $C(Y)$ onto $C(X)$, is one-one; this was not pointed out explicitly at the time, but the information is readily obtainable from an examination of the proof. We shall begin the present chapter by analyzing, in considerable detail, the duality relations expressed by this correspondence. More generally, we describe the relations between arbitrary continuous mappings from X into Y and homomorphisms from $C(Y)$ into $C(X)$. We shall find that, in a sense, every homomorphism from one function ring into another is induced by a continuous mapping.

10.2. *Induced mappings.* Let φ be a given mapping from a set A into a set B. For each mapping g from B into a set E, the composition $g \circ \varphi$ carries A into E. Thus, φ induces a mapping $\varphi': E^B \rightarrow E^A$; explicitly,

$$\varphi' g = g \circ \varphi.$$

There is a duality between the properties *one-one* and *onto* (provided that E has more than one element): φ' is one-one if and only if φ is onto, and φ' is onto if and only if φ is one-one. The verification of these facts is left to the reader.

In most applications, the object of interest is a restriction of the induced mapping to an appropriate subset. We shall employ the symbol φ' to denote the restricted mapping as well.

Here we are concerned with a continuous mapping τ from X into Y, with the role of E taken by $\mathbf{R}$. The appropriate subset of $\mathbf{R}^Y$ will be either $C(Y)$ or $C^*(Y)$. The induced mapping τ', defined by

$$\tau' g = g \circ \tau \qquad (g \in C(Y) \text{ [resp. } C^*(Y)]),$$

is evidently a homomorphism from $C(Y)$ into $C(X)$ [resp. $C^*(Y)$ into $C^*(X)$]. It carries the constant functions onto the constant functions, identically: for any $x \in X$ and $r \in \mathbf{R}$, we have $(\tau'\mathbf{r})(x) = \mathbf{r}(\tau x) = r$.

The homomorphism τ' determines the mapping τ uniquely: if $\sigma' = \tau'$, then for each $x \in X$, $g(\sigma x) = g(\tau x)$ for all g; by complete regularity, $\sigma x = \tau x$.

We now examine the duality relations between τ and τ'. Some complications are introduced by the requirement of continuity.

10.3. THEOREM. *Let τ be a continuous mapping from X into Y, and τ' the induced homomorphism $g \to g \circ \tau$ from $C(Y)$ into $C(X)$* [resp. *$C^*(Y)$ into $C^*(X)$*].

(a) *τ' is an isomorphism (into) if and only if $\tau[X]$ is dense in Y.*

(b) *τ' is onto if and only if τ is a homeomorphism whose image is C-embedded* [resp. *C^*-embedded*] *in Y.*

PROOF. (a). The following are evidently equivalent: τ' is an isomorphism; $\tau' g = \mathbf{0}$ implies $g = \mathbf{0}$; $g(\tau x) = 0$ for all $x \in X$ implies $g = \mathbf{0}$; $\tau[X]$ is dense in Y (since Y is completely regular).

(b). *Necessity.* The hypothesis asserts that for each $f \in C(X)$, there exists $g \in C(Y)$ such that $\tau' g = f$. To see that τ is one-one, we note that if $\tau x_1 = \tau x_2$, then $(\tau' g)(x_1) = (\tau' g)(x_2)$, i.e., $f(x_1) = f(x_2)$. As this holds for each f, $x_1 = x_2$. Thus, $\tau^{\leftarrow}$ is well defined as a mapping from $\tau[X]$ to X. Continuity of $\tau^{\leftarrow}$ follows from Theorem 3.8, since $f \circ \tau^{\leftarrow}$ is the continuous function $g|\tau[X]$; hence τ is a homeomorphism. Therefore every function in $C(\tau[X])$ has the form $g|\tau[X]$, i.e., $\tau[X]$ is C-embedded.

Sufficiency. By hypothesis, $\tau^{\leftarrow}$ is a continuous mapping from $\tau[X]$ onto X. Consider any $f \in C(X)$. The function $f \circ \tau^{\leftarrow}$ belongs to $C(\tau[X])$, and, by hypothesis, it has a continuous extension g to all of Y. Clearly, $f = g \circ \tau$, i.e., $f = \tau' g$.

The proof for C^* is similar.

Of course, if τ is a homeomorphism of X onto Y, then τ' is an isomorphism of $C(Y)$ onto $C(X)$.

10.4. COROLLARY. *If τ is a homeomorphism (into), and X is compact, then τ' is onto.*

The hypothesis that X is compact guarantees that $\tau[X]$ be C-embedded, under any homeomorphism τ into any space Y (3.11(c)).

For such an all-encompassing property, compactness is almost unavoidable; the precise condition is $|\beta X - X| \leqq 1$ (6J).

10.5. We examine, now, the inverse problem of determining when a given homomorphism of $C(Y)$ into $C(X)$ is induced by some continuous mapping of X into Y. In this section, we consider homomorphisms from $C(Y)$ into $\mathbf{R}$—in other words, the case in which X consists of just one point. The results depend upon the special property that the only nonzero homomorphism from $\mathbf{R}$ into itself is the identity (0.22).

(a) *Any nonzero homomorphism* $\mathfrak{u}$ *from* $C(Y)$ (*or* $C^*(Y)$) *into* $\mathbf{R}$ *is onto* $\mathbf{R}$; *in fact* $\mathfrak{u}\mathbf{r} = r$ *for all* $r \in \mathbf{R}$.

Since $\mathfrak{u}g = \mathfrak{u}g \cdot \mathfrak{u}\mathbf{1}$ for all g, and $\mathfrak{u}$ is not identically zero, we must have $\mathfrak{u}\mathbf{1} = 1$. Therefore the mapping $r \to \mathfrak{u}\mathbf{r}$ is a nonzero homomorphism of $\mathbf{R}$ into $\mathbf{R}$, and hence is the identity.

The kernel of a homomorphism of $C(Y)$ onto $\mathbf{R}$ is a maximal ideal in $C(Y)$—by definition, a real maximal ideal. On the other hand, each real maximal ideal is the kernel of such a homomorphism. Moreover, distinct homomorphisms onto $\mathbf{R}$ have distinct kernels (0.23). Similar remarks apply to $C^*(Y)$, where now the qualifier "real" may be dropped, since a maximal ideal in C^* is always real (Theorem 5.8(a)). We have proved:

(b) *The correspondence between the homomorphisms of* $C(Y)$ (*or* $C^*(Y)$) *onto* $\mathbf{R}$, *and the real maximal ideals, is one-one.*

For $y \in Y$, the fixed maximal ideal M_y (or M^*_y) is the kernel of the homomorphism $g \to g(y)$. Now, Y is *compact* if and only if every maximal ideal in $C^*(Y)$ is fixed (Theorem 4.11); and, by definition, Y is *realcompact* precisely when every real maximal ideal in $C(Y)$ is fixed. Hence we have:

(c) *Y is* [*real*]*compact if and only if, to each homomorphism* $\mathfrak{u}$ *from* $C^*(Y)$ [resp. $C(Y)$] *onto* $\mathbf{R}$—*i.e., each nonzero homomorphism into* $\mathbf{R}$—*there corresponds a point y of Y such that* $\mathfrak{u}g = g(y)$ *for all* g.

Briefly, the condition states that "every real homomorphism is fixed."

10.6. Our first result about homomorphisms from $C(Y)$ into $C(X)$, for arbitrary X, is a generalization of (c).

THEOREM. *Let* $\mathfrak{t}$ *be a homomorphism from* $C(Y)$ *into* $C(X)$ *with the property that* $\mathfrak{t}\mathbf{1} = \mathbf{1}$. *If Y is realcompact, then there exists a unique continuous mapping* τ *of X into Y such that* $\tau' = \mathfrak{t}$.

REMARKS. The condition $\mathfrak{t}\mathbf{1} = \mathbf{1}$ is plainly necessary. The hypothesis that Y be realcompact is also indispensable, for otherwise the conclusion will not hold even for a one-point space X. Likewise, the analogue of this theorem, with C replaced by C^*, will require that Y be compact. Since $C^*(Y) = C(Y)$ under this hypothesis, the analogue is merely a special case of the theorem itself.

PROOF. For each point x of X, the mapping $g \to (\mathfrak{t}g)(x)$ is a homomorphism from $C(Y)$ into $\mathbf{R}$; and, since $(\mathfrak{t}\mathbf{1})(x) = \mathbf{1}(x) \neq 0$, it is not the zero homomorphism. By 10.5(c), since Y is realcompact, there is a point τx of Y such that $(\mathfrak{t}g)(x) = g(\tau x)$ for all g in $C(Y)$. The mapping τ from X into Y, thus defined, evidently satisfies $\mathfrak{t}g = g \circ \tau$, for each g. Since $\mathfrak{t}g$ is a continuous function, Theorem 3.8 shows that τ is continuous. As observed in 10.2, τ is the *unique* continuous mapping for which $\tau' = \mathfrak{t}$.

10.7. As an application of this theorem, we present an alternative proof that (2) implies (1) in Theorems 6.4 and 8.6.

THEOREM. *Let X be dense in T. If X is C^*-* [resp. *C-*] *embedded in T, then any continuous mapping φ from X into a* [*real*]*compact space Y has an extension to a continuous mapping from all of T into Y.*

PROOF. It is enough to give the details for C. For any $g \in C(Y)$, the function $g \circ \varphi$ belongs to $C(X)$, and hence has an extension to a function $(g \circ \varphi)_0$ in $C(T)$. The mapping $g \to (g \circ \varphi)_0$ is a homomorphism $\mathfrak{t}$ of $C(Y)$ into $C(T)$, and it carries $\mathbf{1}$ to $\mathbf{1}$. Therefore, there is a continuous mapping τ from T into Y such that $\tau' = \mathfrak{t}$. For $x \in X$, we have

$$g(\tau x) = (\mathfrak{t}g)(x) = (g \circ \varphi)_0(x) = g(\varphi x)$$

for every $g \in C(Y)$, which implies that $\tau x = \varphi x$. Thus, τ is an extension of φ.

10.8. In spite of the remarks made in 10.6, every homomorphism is induced, in essence, by a continuous mapping.

THEOREM. *If $\mathfrak{s}$ is a homomorphism from $C(Y)$ into $C(X)$, then the set*

$$E = \{x \in X : (\mathfrak{s}\mathbf{1})(x) = 1\}$$

is open-and-closed in X. Furthermore, there exists a unique continuous mapping τ from E into vY, such that for all $g \in C(Y)$,

$$(\mathfrak{s}g)(x) = g^{v}(\tau x) \quad \textit{for all } x \in E,$$

and

$$(\mathfrak{s}g)(x) = 0 \quad \textit{for all } x \in X - E.$$

Similarly, for a homomorphism $\mathfrak{s}$ from $C^*(Y)$ into $C^*(X)$, τ will map E into βY, and satisfy $(\mathfrak{s}g)(x) = g^{\beta}(\tau x)$ for $x \in E$.

PROOF. As with any homomorphism, the element $e = \mathfrak{s}\mathbf{1}$ is an idempotent in $C(X)$. Hence it is the characteristic function of the set $E = e^{\leftarrow}(1)$. Since e is continuous, E is open-and-closed in X. Furthermore (again as with any homomorphism), e is the unity element of the image ring $\mathfrak{s}[C(Y)]$; it follows that $(\mathfrak{s}g)[X - E] = \{0\}$ for every $g \in C(Y)$.

Consider the homomorphism $\mathfrak{t}$ from $C(\upsilon Y)$ into $C(E)$ (assuming $E \neq \emptyset$) defined by: $\mathfrak{t}g^{\upsilon} = (\mathfrak{s}g)|E$. Evidently, $\mathfrak{t}$ sends $\mathbf{1}$ to $e|E$, i.e., to the function $\mathbf{1}$ in $C(E)$. By Theorem 10.6, there exists a unique continuous mapping τ of E into υY such that $\tau' = \mathfrak{t}$. For $x \in E$, we have

$$(\mathfrak{s}g)(x) = (\mathfrak{t}g^{\upsilon})(x) = g^{\upsilon}(\tau x);$$

and for $x \in X - E$, $(\mathfrak{s}g)(x) = 0$.

The proof for C^* is similar.

10.9. The fact that every (ring) homomorphism from $C(Y)$ into $C(X)$ is both an algebra homomorphism (1I) and a lattice homomorphism (Theorem 1.6) is exhibited clearly by its representation as given in the theorem.

Some of the results that have been obtained so far may be summarized in the following way.

(a) *The continuous mappings of X into υY are in one-one correspondence with the homomorphisms of $C(Y)$ into $C(X)$ whose images include the constants on X.*

For, if $\mathfrak{s}$ is a homomorphism whose image includes the constants, then the set E of the theorem is all of X. (See 10A.)

Incorporating the duality expressed in Theorem 10.3(a), we obtain:

(b) *υY contains a dense, continuous image of X if and only if $C(X)$ contains an isomorphic image of $C(Y)$ that includes the constants on X.*

In particular, *if X and Y are compact, then Y is a continuous image of X if and only if $C(X)$ contains an isomorphic image of $C(Y)$ that contains the constants on X.* (See 16E.*2*.)

The duality expressed in Theorem 10.3(b) yields:

(c) *υY contains a C-embedded copy of X if and only if $C(X)$ is a homomorphic image of $C(Y)$;*

(c*) *βY contains a C^*-embedded copy of X if and only if $C^*(X)$ is a homomorphic image of $C^*(Y)$.*

By 6.9(a), βY contains a C^*-embedded copy of X if and only if it contains a copy of βX; and a copy of βX, being compact, is necessarily C^*-embedded. Thus, (c*) is equivalent to the following corollary of (c): *if X and Y are compact, then Y contains a copy of X if and only if $C(X)$ is a homomorphic image of $C(Y)$.*

From (c), we have: *when Y is realcompact, $C^*(Y)$ is a homomorphic image of $C(Y)$ if and only if Y contains a copy of βY.* An example of a noncompact space with this property is the space $Y = N_1 \cup \beta N_2$ of 6K. (Since Y is the union of a compact space with a realcompact space, it is realcompact, by Theorem 8.16. The conclusion is also easy to reach directly.)

10.10. EXAMPLES. Let us examine C and C^* for the spaces $\mathbf{N}$, $\mathbf{Q}$, and $\mathbf{R}$. When is there a homomorphism of one of these rings onto another?

Since $\mathbf{N}$ is C-embedded in both $\mathbf{Q}$ and $\mathbf{R}$, $C(\mathbf{N})$ is a homomorphic image of both $C(\mathbf{Q})$ and $C(\mathbf{R})$, and $C^*(\mathbf{N})$ is a homomorphic image of both $C^*(\mathbf{Q})$ and $C^*(\mathbf{R})$.

There are no other homomorphisms. For most of the combinations, this follows from a straightforward application of the embedding criteria: $\mathbf{N}$ contains no copy of $\mathbf{Q}$, etc. Some of these cases can be handled even more directly by recalling that the image of every bounded function is bounded (Corollary 1.8).

A few combinations require some additional comment.

$C(\mathbf{Q})$ is not a homomorphic image of $C(\mathbf{R})$; in fact, $\mathbf{R}$ cannot contain even a C^*-embedded copy of $\mathbf{Q}$. For, by 1F.*4*, such a copy must be closed in $\mathbf{R}$. As a closed subspace of a locally compact space, then, it must be locally compact. But this is absurd.

$C^*(\mathbf{R})$ is not a homomorphic image of $C^*(\mathbf{N})$ or $C^*(\mathbf{Q})$. For, the totally disconnected spaces $\beta\mathbf{N}$ and $\beta\mathbf{Q}$ cannot contain copies of $\mathbf{R}$.

Finally, $C^*(\mathbf{Q})$ is not a homomorphic image of $C^*(\mathbf{R})$ or $C^*(\mathbf{N})$. Suppose that $\beta\mathbf{R}$ contains a C^*-embedded copy Q of $\mathbf{Q}$. If Q meets the open set $\mathbf{R}$, then $Q \cap \mathbf{R}$ is an open set in Q, and hence contains a closed interval of Q. As above, this is impossible. So $Q \subset \beta\mathbf{R} - \mathbf{R}$. But this, too, is impossible, for Q contains a countably infinite compact set, while by Corollary 9.12, every infinite closed set in $\beta\mathbf{R} - \mathbf{R}$ is uncountable. A similar argument shows that $\beta\mathbf{N}$ contains no copy of $\mathbf{Q}$.

QUOTIENT MAPPINGS

10.11. Let τ be a mapping from a space X onto a set Y. We recall that the quotient topology on Y is the largest topology such that τ is continuous. The set Y, endowed with the quotient topology, is called the quotient space of X, relative to the quotient mapping τ; we shall denote this space by Y_τ. The quotient topology is characterized as follows: a subset F of Y is closed in Y_τ if and only if $\tau^{\leftarrow}[F]$ is closed in X. Dually, the corresponding condition holds for open sets.

The next theorem tells us, in terms of the induced homomorphism, when a continuous mapping is a quotient mapping. The theorem is complicated by the fact that a quotient space of a completely regular space need not be completely regular (3J).

THEOREM. *Let τ be a continuous mapping from X onto a completely regular space Y, and suppose that the quotient space Y_τ is also completely regular. Then τ is a quotient mapping if and only if $\tau'[C(Y)]$ contains (hence is) the set A of all functions in $C(X)$ that are constant on all sets $\tau^{\leftarrow}(y)$, for $y \in Y$.*

PROOF. *Necessity.* If $f \in A$, then there exists $g \in \mathbf{R}^Y$ such that $f = g \circ \tau$. Let F be a closed set in $\mathbf{R}$. The set

$$\tau^{\leftarrow}[g^{\leftarrow}[F]], \quad \text{that is,} \quad f^{\leftarrow}[F],$$

is closed, by continuity of f. Hence $g^{\leftarrow}[F]$ is closed, since τ is a quotient mapping. Thus, g is continuous, so that $f = \tau' g \in \tau'[C(Y)]$.

Sufficiency. Since τ is onto Y, the mapping $g \to g \circ \tau$ from $\mathbf{R}^Y$ ($= \mathbf{R}^{Y_\tau}$) into $\mathbf{R}^X$ is one-one. By hypothesis, the image of $C(Y)$ under this induced mapping is A. Since the topology of Y_τ contains that of Y, so that $C(Y_\tau) \supset C(Y)$, the image of $C(Y_\tau)$ is also A. Thus, $C(Y)$ and $C(Y_\tau)$ have the same image under the one-one mapping, and therefore $C(Y) = C(Y_\tau)$. By Theorem 3.6, the completely regular spaces Y and Y_τ are homeomorphic.

The theorem holds with C replaced by C^*. The subrings of $C^*(X)$ that are images of homomorphisms are described in 10D.*4* and 16E.

10.12. We know that if X is compact, then every continuous mapping τ from X onto Y is a quotient mapping, in fact, a closed mapping. This suggests that for arbitrary X, the space Y might be realized as a quotient space relative to some extension of τ toward a compactification of X. The suggestion is bolstered by the observation that there will be fewer continuous functions on the enlarged space, and larger preimages of points under the extended mapping, making it more likely that the induced image of $C(Y)$ will include all functions that are constant on these preimages.

The result is stated in the next theorem. Because we are dropping the goal that τ itself be a quotient mapping, we are able to relax the requirement that it map X *onto* Y.

We begin with a lemma, which generalizes Lemma 6.11.

LEMMA. *Let φ be a continuous mapping from T into Y, and let S be a proper dense subset of T. If $\varphi^{\leftarrow}(y) \cap S$ is compact, for every $y \in Y$, then the restriction of φ to S is not a closed mapping.*

PROOF. Choose $p \in T - S$, and let $q = \varphi p$. The point p has an open neighborhood whose closure V (in T) does not meet the compact set $\varphi^{\leftarrow}(q) \cap S$. Then $q \notin \varphi[V \cap S]$. But $\mathrm{cl}_T (V \cap S) = V$, and $q \in \varphi[V]$. Hence

$$q \in \varphi[\mathrm{cl}_T (V \cap S)] \subset \mathrm{cl}_Y \varphi[V \cap S].$$

Therefore $\varphi|S$ takes the closed set $V \cap S$ in S to a set that is not closed in Y.

10.13. THEOREM. *Let τ be a continuous mapping from X onto a dense subspace of Y, $\bar{\tau}$ its Stone extension into βY, and $X_0 = \bar{\tau}^{\leftarrow}[Y]$. Then $\tau_0 = \bar{\tau}|X_0$ is a closed mapping from X_0 onto Y, and the preimage of every compact set is compact.*

The space X_0 is the largest subspace of βX to which τ has a continuous extension into Y. It is the only one for which the extension is a closed mapping such that the preimage of each point is compact.

PROOF. Since βX is compact, $\bar{\tau}$ is a closed mapping, and the inverse image of every compact set is compact. The mapping τ_0 has the same properties, because it is the restriction of $\bar{\tau}$ to a *total* preimage. Evidently, X_0 is the largest subspace of βX to which τ has a continuous extension. That it is the unique space for which the extension has the described properties is a consequence of the lemma.

It follows from 10.9(b) that υY is a quotient space of some space between X and βX if and only if $C(X)$ contains an isomorphic image of $C(Y)$ that includes the constants on X.

10.14. COROLLARY. *Given Y, let D be any discrete space equipotent with some dense subset of Y. Then Y is a quotient space of a suitable subspace of βD.*

For example, $\mathbf{R}$ is a quotient space of a subspace of $\beta\mathbf{N}$.

As another example, let D be the discrete subspace of $\mathbf{W}^*$ consisting of all isolated points, τ the identity map on D, $\bar{\tau}$ its Stone extension from βD onto $\mathbf{W}^*$, and D_0 the total preimage of $\mathbf{W}$. By Lemma 6.11, the total preimage of D is D. If a point p of $\beta D - D$ belongs to the closure of a countable subset A of D, then $\bar{\tau}$ carries p into $\mathrm{cl}_{\mathbf{W}^*} A$, and hence to a countable ordinal. Conversely, if α is a countable ordinal, then

$W(\alpha + 1)$ is a countable neighborhood of α in $\mathbf{W}^*$, so that $\tau^{\leftarrow}[W(\alpha + 1)]$ is a countable set in D whose closure in βD contains $\bar{\tau}^{\leftarrow}(\alpha)$. Thus, $D_0 - D$ consists precisely of those points of βD that are limit points of countable subsets of D. The closed set $\bar{\tau}^{\leftarrow}(\omega_1)$ consists of all points in βD that are not in the closure of any countable subset of D. By 9F, there are $2^{2^{\aleph_1}}$ such points.

10.15. In the general situation of Theorem 10.13, if τ is a one-one mapping, then all we can add to the conclusion is the trivial remark that τ_0 is one-one on X. As we have just seen, a more substantial improvement is obtainable from the requirement that τ be a homeomorphism, for then Lemma 6.11 can be applied.

COROLLARY. *If τ is a homeomorphism, then τ_0 carries $X_0 - X$ to $Y - \tau[X]$.*

In particular, every compactification Y of X is a continuous image of βX, under a mapping that leaves X pointwise fixed, and carries $\beta X - X$ to $Y - X$ (Theorem 6.12).

The corollary provides a description of all spaces in which X is dense. Each such space may be constructed from βX by deleting part of $\beta X - X$, and identifying the points in certain remaining compact subsets of $\beta X - X$. In this sense, βX is the largest (completely regular) space, and hence the largest compact space, in which X is dense. A compactification of X is constructed from βX by making identifications but no deletions.

Recall our description of compactifying X as the process of fixing the free z-ultrafilters (6.1). In an arbitrary compactification Y, distinct free z-ultrafilters on X may converge to the same point in Y; but in the largest compactification, βX, each z-ultrafilter has its own cluster point. The z-ultrafilters on X that converge to a given point y of Y are precisely those that converge in βX to the various points of $\tau_0^{\leftarrow}(y)$.

10.16. The space X_0 of Theorem 10.13 reflects certain properties of Y, as is shown in the next theorem.

THEOREM.

(a) *Y is locally compact if and only if X_0 is locally compact.*

(b) *Y is compact if and only if $X_0 = \beta X$.*

(c) *If Y is realcompact, then X_0 is realcompact.*

PROOF. (a). Since $\beta X_0 = \beta X$, the closed mapping $\bar{\tau}$ carries $\beta X_0 - X_0$ onto $\beta Y - Y$, as well as X_0 onto Y. The result now follows from the fact that a space S is locally compact if and only if it is open in βS (6.9(d)).

Conclusion (b) is obvious, while (c) was given in Theorem 8.13 and in 8B.*4*.

We do not know whether every closed, continuous image of a realcompact space is realcompact, or even whether the converse of (c) is true. It is known, however, that an *open*, continuous image need not be realcompact; see the example in 8I.

10.17. *Filters and z-filters.* The concept of a z-filter on a space X was introduced (in Chapter 2) as a tool for the study of ideals in $C(X)$. It turned out that z-filters can be used in much the same way as filters in the theory of convergence in completely regular spaces. In this section, we make some general comments on the relations between filters and z-filters on X, especially as they concern convergence. Additional details, along with some illuminating examples, are presented as problems at the end of the chapter.

Let D denote the discrete space whose points are those of X. The filters on D are the same as those on X, but the z-filters on the two spaces will be the same only when X is discrete. The space βX was constructed by supplying one cluster point for each z-ultrafilter on X, while βD, in general much larger than βX, requires a separate cluster point for each *ultrafilter.*

We have noted (2.2) that the zero-sets belonging to a given filter form a z-filter, and that every z-filter is a base for a filter. There is thus a natural (many-one) mapping from the set of all filters $\mathscr{F}$ onto the set of all z-filters. Let δ denote the identity map of D onto X. Then the mapping in question is precisely $\delta^{\#}$:

$$\delta^{\#}\mathscr{F} = \mathscr{F} \cap \mathbf{Z}(X).$$

Not only is $\delta^{\#}$ the natural mapping from the set-theoretic point of view, but it also preserves convergence (10J).

Let $\mathscr{U}$ be an ultrafilter. The z-filter $\delta^{\#}\mathscr{U}$ is prime (4.12), and so it is contained in a unique z-ultrafilter; we denote this z-ultrafilter by $\Delta\mathscr{U}$, thus defining a mapping Δ from the set of all ultrafilters on X onto the set of all z-ultrafilters. This mapping shares with $\delta^{\#}$ the property of preserving convergence; as a matter of fact, it is characterized by this property (10J.*3*).

Let $\bar{\delta}$ denote the Stone extension of δ into βX. Then $\bar{\delta}$ induces a mapping from the set of all ultrafilters on X onto the set of all z-ultrafilters—precisely, the mapping Δ (10J.*4*). The convergence-preserving property of Δ reflects, first of all, the fact that $\bar{\delta}$ is continuous, and, secondly, the fact that a z-ultrafilter converges in any compactification.

PROBLEMS

10A. INDUCED MAPPINGS.

Let Y be realcompact, let $\mathfrak{t}$ be a homomorphism from $C(Y)$ into $C(X)$ such that $\mathfrak{t}\mathbf{1} = \mathbf{1}$, and let $\mathfrak{t}'$ denote the mapping τ of X into Y described in Theorem 10.6.

1. If each point x of X is identified with the mapping $f \to f(x)$ of $C(X)$ into $\mathbf{R}$, and similarly for Y, then $\mathfrak{t}'$ may be interpreted as a mapping induced by $\mathfrak{t}$ in the sense of 10.2.

2. $\tau'' = \tau$ and $\mathfrak{t}'' = \mathfrak{t}$.

3. If τ is a homeomorphism of X onto Y, then τ' is an isomorphism of $C(Y)$ onto $C(X)$.

4. If $\mathfrak{t}$ is an isomorphism of $C(Y)$ onto $C(X)$, and if X as well as Y is realcompact, then $\mathfrak{t}'$ is a homeomorphism from X onto Y.

5. If X is connected, then every nonzero homomorphism from $C(Y)$ into $C(X)$ is induced by a continuous mapping from X into Y.

6. Let σ be a continuous mapping from X into Y, and σ': $C^*(Y) \to C^*(X)$ the induced homomorphism. Define a homomorphism $\mathfrak{s}$: $C(\beta Y) \to C(\beta X)$ by replacing the functions in $C^*(Y)$ and $C^*(X)$ by their extensions to βY and βX, respectively. Then $\mathfrak{s}' = \bar{\sigma}$. (This contains the essence of the proof of Theorem 10.7.)

7. Interpret $\varphi^{\#}$ (4.12) as a mapping induced by a mapping induced by φ.

10B. ISOMORPHISM OF RESIDUE CLASS FIELDS.

1. Let X be C-embedded in T; then $\beta X \subset \beta T$. For $p \in \beta X$, let $\boldsymbol{M}_X{}^p$ and $\boldsymbol{M}_T{}^p$ denote the ideals $\boldsymbol{M}^p$ in $C(X)$ and $C(T)$, respectively. Then $C(X)/\boldsymbol{M}_X{}^p$ is isomorphic with $C(T)/\boldsymbol{M}_T{}^p$. [The kernel of the homomorphism $g \to \boldsymbol{M}_X{}^p\,(g|X)$ is contained in $\boldsymbol{M}_T{}^p$.]

2. The above need not hold if X is only C^*-embedded in T. [Take $T = \beta X$.]

10C. THE SMALLEST COMPACTIFICATION.

1. If T is a compactification of X, and $T - X$ has more than one point, then X has a compactification T' that is smaller than T, i.e., there is a quotient mapping of T onto T' leaving X pointwise fixed.

2. X has a smallest compactification if and only if X is locally compact.

10D. KERNEL AND IMAGE OF A HOMOMORPHISM.

1. Let $\mathfrak{s}$ be a homomorphism from $C(Y)$ into a ring of continuous functions. If Y is realcompact, there exists a unique closed set F in Y such that the kernel of $\mathfrak{s}$ is the z-ideal of all functions that vanish on F. When Y is not realcompact, the kernel is still a z-ideal, but not necessarily of the stated form.

2. In *1*, if Y is both realcompact and normal, then $\mathfrak{s}[C(Y)]$ is isomorphic with $C(F)$. [3D.*1*.]

3. If Y is realcompact, but not normal (see 8.18), then there exists a homomorphism of $C(Y)$ into a function ring whose image is not isomorphic with $C(X)$ for any X. [If the family of restrictions to a closed set F is isomorphic with $C(X)$, then Theorem 10.3(b) leads to a proof that F is C-embedded.]

4. For any Y, *a homomorphic image of* $C^*(Y)$ *in a function ring is isomorphic with* $C^*(S)$ *for some* S.

10E. EXTENSION OF THE IDENTITY MAP.

1. Let Y_1 and Y_2 be compactifications of X, and suppose that whenever two z-ultrafilters on X converge to the same point in Y_1, they also converge to the same point in Y_2. Then there exists a continuous mapping of Y_1 onto Y_2 that leaves X pointwise fixed, and sends $Y_1 - X$ onto $Y_2 - X$. [6G.*4* and 6F.*3*.]

2. Let Y_1 and Y_2 be realcompactifications of X. The functions in $C(X)$ that are extendable to $C(Y_1)$ are the same as those extendable to $C(Y_2)$ when and only when there exists a homeomorphism of Y_1 onto Y_2 that leaves X pointwise fixed. [Theorem 10.6.] A similar result holds for C^*.

10F. TOPOLOGIES ON $C(X)$.

Let $\mathfrak{t}$ be a homomorphism from $C(Y)$ or $C^*(Y)$ into $C(X)$.

1. $\mathfrak{t}$ is continuous in the uniform norm topologies (2M). [By 1J.*4*, a homomorphism reduces the norm.]

2. $\mathfrak{t}$ need not be continuous in the m-topologies (2N). [Let Y be a one-point space, and X not pseudocompact.]

10G. LEMMA 10.12.

Derive Lemma 6.11 from Lemma 10.12.

10H. FILTERS.

1. A filter is contained in a unique ultrafilter only if it is an ultrafilter. Contrast with z-filters.

2. A filter on a discrete space D converges in βD if and only if it is an ultrafilter. [6F.*2*.]

3. *Every ultrafilter on* X *converges in any compactification of* X.

4. If $\mathscr{V}$ is an ultrafilter on a subset of X, then the filter on X with base $\mathscr{V}$ is an ultrafilter.

5. If X is a finite union of disjoint sets $X_1, \cdots, X_n$, and $\mathscr{U}$ is an ultrafilter on X, then some ultrafilter on one of the X_k is a base for $\mathscr{U}$.

10I. CONVERGENCE OF z-FILTERS.

1. A filter or z-filter on X converges to p in βX if and only if it contains $\boldsymbol{Z}[\boldsymbol{O}^p]$.

2. A z-filter on X converges in βX if and only if it is contained in a unique z-ultrafilter on X.

Let T be any space containing X.

3. State and prove an analogue of *1* with βX replaced by T.

4. If a z-filter $\mathscr{Z}$ is contained in a unique z-ultrafilter on X, and if the latter converges in T, then $\mathscr{Z}$ converges. The converse is false.

10J. THE MAPPINGS $\delta^{\#}$ AND Δ.

Let $X \subset T$.

1. A filter $\mathscr{F}$ on X converges to p in T if and only if $\delta^{\#}\mathscr{F}$ converges to p.

2. An ultrafilter $\mathscr{U}$ on X converges to p in T if and only if $\Delta\mathscr{U}$ converges to p.

3. If Δ_1 is a mapping from the set of all ultrafilters on X to the z-ultrafilters, and if Δ_1 satisfies *2* for $T = \beta X$, then $\Delta_1 = \Delta$.

4. Let $\mathscr{U}$ be the ultrafilter on X that converges to a point q in βD (where D is the discrete space with the points of X). Then $\mathscr{U}$ converges to p in βX if and only if $\delta q = p$, and if and only if $\Delta\mathscr{U} = \boldsymbol{A}^p$.

10K. ULTRAFILTERS CONTAINING A z-ULTRAFILTER.

1. A necessary and sufficient condition that $\boldsymbol{A}^p$ be contained in *every* ultrafilter on X that converges to p is that $\boldsymbol{M}^p = \boldsymbol{O}^p$. [See 10J.*4*. *Necessity.* If $Z \in \boldsymbol{A}^p - \boldsymbol{Z}[\boldsymbol{O}^p]$, then $p \in \operatorname{cl}_{\beta X}(X - Z)$.] Thus, this holds for a point p of X if and only if p is a P-point (4L); and it holds for *every* point p of βX if and only if X is a P-space [7L].

2. For $p \in X$, $\boldsymbol{A}_p$ is contained in a *unique* ultrafilter if and only if $\{p\}$ is a zero-set in X. [*Necessity.* If S meets every member of $\boldsymbol{A}_p$ in a point distinct from p, then $\boldsymbol{A}_p \cup \{S\}$ has the finite intersection property.]

3. For $p \in \beta X$, $\boldsymbol{A}^p$ is contained in a *unique* ultrafilter if and only if $\boldsymbol{A}^p$ generates an ultrafilter on X, i.e., there exists an ultrafilter each of whose members contains a member of $\boldsymbol{A}^p$. [10H.*1*.]

4. If $\boldsymbol{A}^p$ contains a discrete zero-set Z, and if Z is C^*-embedded in X, then $\boldsymbol{A}^p$ is contained in a *unique* ultrafilter. [The trace $\mathscr{F}$ of $\boldsymbol{A}^p$ on Z is an ultrafilter on Z and is contained in $\boldsymbol{A}^p$ [1F.*3*]; and $\mathscr{F}$ generates an ultrafilter on X [10H.*4*].]

10L. z-ULTRAFILTERS ON $\mathbf{R}$, Λ, AND $\mathbf{T}$.

1. In the space $\mathbf{R}$: If $p \in \beta\mathbf{N}$, then $\boldsymbol{A}^p$ is contained in a unique ultrafilter. (This is a special case of 10K.*4*.)

2. In the space $\mathbf{R}$: If p is a point in $\beta\mathbf{R}$ such that every member of $\boldsymbol{A}^p$ is of cardinal $\mathfrak{c}$ (see, e.g., 6U.*5*), then $\boldsymbol{A}^p$ is *not* contained in a unique ultrafilter. [As in 4G, find two disjoint sets in $\mathbf{R}$, each of which meets every member of $\boldsymbol{A}^p$.]

3. In the space $\Lambda = \beta\mathbf{R} - (\beta\mathbf{N} - \mathbf{N})$: If $p \in \beta\mathbf{N} - \mathbf{N}$, then $\boldsymbol{A}^p$ is *not* contained in a unique ultrafilter. [By 6P.*2* and 9E, every member of $\boldsymbol{A}^p$ is of cardinal $\geqq \mathfrak{c}$.]

4. In the space $\mathbf{T}$: An ultrafilter $\mathscr{U}$ contains $\boldsymbol{A}^t$ if and only if all tails of the top edge W of $\mathbf{T}$ belong to $\mathscr{U}$, i.e., some ultrafilter on W that includes all tails is a base for $\mathscr{U}$. Hence there exist $2^{2^{\aleph_1}}$ ultrafilters that contain A^t. [9F.]

5. In the space $\mathbf{T}$: There exist $2^{2^{\aleph_1}}$ ultrafilters converging to t that do *not* contain A^t. [For each ultrafilter $\mathscr{V}$ on $\mathbf{W}$, consider the filter base on $\mathbf{T}$ consisting of all sets $V \times \mathbf{N}$, for $V \in \mathscr{V}$.]

10M. REAL IDEALS AND REALCOMPACT SPACES.

1. If $\mathscr{U}$ is an ultrafilter on X, and p its limit in βX (10H.*3*), then $\delta^\# \mathscr{U}$ has the countable intersection property if and only if $\boldsymbol{M}^p$ is a real ideal. [7H.*3* and 10I.*1*.]

2. X is realcompact if and only if every ultrafilter $\mathscr{U}$, for which $\delta^\# \mathscr{U}$ has the countable intersection property, has a limit in X.

3. Use *2* to prove that every closed subspace of a realcompact space is realcompact. [10H.*4*.]

10N. COMPONENTS OF ZERO-SETS IN βR^+.

Let $f \in C(\beta R^+)$, with $\boldsymbol{Z}(f) - R^+ \neq \emptyset$.

1. There exist sequences (r_n) and (t_n) in R^+, with $r_n < t_n < r_{n+1}$, and $\lim_n r_n = \infty$, such that $|f(x)| \leqq 1/n$ whenever $r_n \leqq x \leqq t_n$.

2. Let $\mathscr{U}$ be a free ultrafilter on $\mathbf{N}$. There exist unique p, $q \in \beta R^+$, satisfying

$$p \in \bigcap_{U \in \mathscr{U}} \operatorname{cl}\{r_n : n \in U\}, \quad \text{and} \quad q \in \bigcap_{U \in \mathscr{U}} \operatorname{cl}\{t_n : n \in U\}.$$

Moreover, $p \neq q$, and $p, q \in \boldsymbol{Z}(f)$.

3. If V is any closed neighborhood of p not containing q, then the boundary of V meets $\boldsymbol{Z}(f)$. [Let $E \in \mathscr{U}$ be such that $\{r_n\}_{n \in E} \subset V$ and $\{t_n\}_{n \in E} \cap V = \emptyset$. For each $n \in E$, there exists s_n on the boundary of V, with $r_n < s_n < t_n$.]

4. p and q belong to the same component of $\boldsymbol{Z}(f)$. [The component of a point in a compact space is the intersection of all the open-and-closed sets containing it (Theorem 16.15).]

5. Every nonempty zero-set in $\beta R^+ - R^+$ contains an infinite connected set. Contrast with 6L.*4*.

Chapter 11

EMBEDDING IN PRODUCTS OF REAL LINES

11.1. The existence of a compactification of an arbitrary completely regular space X was first established, by Tychonoff, by embedding X in a product of a suitable number of copies of the interval $[0, 1]$. His process yields βX. The ideas involved will be presented in this chapter, along with the development of an analogous embedding that yields the realcompactification υX.

As a matter of mathematical interest, we wish to obtain βX and υX without recourse to our earlier constructions of these spaces. We shall not use material beyond Chapter 5, except for the portion of Chapter 10 through 10.7.

Imposition of this restriction raises the question of precisely what is to be meant by βX and υX. The answer is: βX is a compact space in which X is dense and C^*-embedded, and υX is a realcompact space in which X is dense and C-embedded. We must prove anew that these definitions determine βX and υX up to homeomorphisms that leave X pointwise fixed. From Theorem 10.7, we get Stone's theorem, which with 0.12(a) yields the uniqueness of βX as before. The proof of uniqueness of υX is similar.

We shall construct products P and P_* of real lines, and define homeomorphisms σ and σ_* of X into P and P_*, respectively, in such a way that $\sigma[X]$ will be C-embedded in P, and $\sigma_*[X]$, C^*-embedded in P_*. The closures of $\sigma[X]$ and $\sigma_*[X]$ will serve as models for υX and βX, respectively—as soon as it is proved that the former is realcompact and the latter compact. We shall present two approaches to these proofs, one algebraic, and the other topological. The algebraic method is better suited to the discussion of υX, while the topological method handles βX more efficiently.

11.2. Define

$$P = \mathbf{R}^{C(X)}, \quad \text{and} \quad P_* = \mathbf{R}^{C^*(X)}.$$

The points of P are, then, precisely the real-valued functions whose domain is $C(X)$: each point p of P has the form $p = (p_f)_{f \in C}$, where the real number p_f is the value of p at f.

For each $f \in C$, π_f will denote the projection of P onto $\mathbf{R}$ defined by

$$\pi_f(p) = p_f.$$

By definition, the topology of the product space P is the weak topology induced by the family of all functions π_f (see 3.10).

The points, the projections π_{*f}, and the topology of P_* are described in an analogous manner.

11.3. THEOREM.

(a) *The mapping σ, defined by*

$$\sigma x = (f(x))_{f \in C(X)} \qquad (x \in X),$$

is a homeomorphism of X into P, and $\sigma[X]$ is C-embedded in P.

(a*) *The mapping σ_*, defined by*

$$\sigma_* x = (f(x))_{f \in C^*(X)} \qquad (x \in X),$$

is a homeomorphism of X into P_*, and $\sigma_*[X]$ is C^*-embedded in P_*.

PROOF. (a). By definition of σ, $\pi_f \circ \sigma = f$, for each $f \in C(X)$. Therefore σ is continuous, by 3.10(c). Next, the induced homomorphism σ' (10.2) maps $C(P)$ *onto* $C(X)$; for,

$$\sigma' \pi_f = \pi_f \circ \sigma = f.$$

By Theorem 10.3(b), σ is a homeomorphism, and $\sigma[X]$ is C-embedded.

The proof of (a*) is similar.

11.4. Since σ maps X homeomorphically onto $\sigma[X]$, there is an induced isomorphism of $C(\sigma[X])$ onto $C(X)$, as follows. The image of a function $g \in C(\sigma[X])$ is the function $f \in C(X)$ for which $f(x) = g(\sigma x)$ $(x \in X)$. But this means simply that $g = \pi_f | \sigma[X]$; thus, $C(\sigma[X])$ consists precisely of the restrictions to $\sigma[X]$ of the projections π_f. That $\sigma[X]$ is C-embedded in P reflects the fact that π_f is a continuous extension of its restriction. Now, a function in $C(\sigma[X])$ may have many continuous extensions to all of P; but all of these extensions must agree on $\sigma[X]$—and hence also on its closure. Thus, the process of extension provides an isomorphism from $C(\sigma[X])$ onto $C(\operatorname{cl} \sigma[X])$. As a consequence, we have:

(a) *The mapping $f \to \pi_f | \operatorname{cl} \sigma[X]$ is an isomorphism of $C(X)$ onto $C(\operatorname{cl} \sigma[X])$.*

Similarly:

(a*) *The mapping* $f \to \pi_{*f}|\mathrm{cl}\ \sigma_*[X]$ *is an isomorphism of* $C^*(X)$ *onto* $C^*(\mathrm{cl}\ \sigma_*[X])$.

11.5. THEOREM. *The closure of* $\sigma_*[X]$ *in* P_* *is compact, and hence is* $\beta(\sigma_*[X])$.

PROOF. By Tychonoff's theorem (4.14), the subspace

$$\mathsf{X}_{f \in C^*} \mathrm{cl}_{\mathbf{R}}\ f[X]$$

of P_* is compact; and it evidently contains $\sigma_*[X]$. Therefore cl $\sigma_*[X]$ is a compactification in which $\sigma_*[X]$ is C^*-embedded.

Since $\sigma_*[X]$ is homeomorphic with X, $\beta(\sigma_*[X])$ is a model for βX.

11.6. In order to prove that $\mathrm{cl}_P\ \sigma[X]$ is $v(\sigma[X])$, we derive an algebraic characterization of $\mathrm{cl}_P\ \sigma[X]$.

A point p of P is a mapping of $C(X)$ into $\mathbf{R}$, the value of p at f being p_f. For each $x \in X$, the point σx of P is determined by: $(\sigma x)_f = f(x)$. Now, by the very definitions of the ring operations in $C(X)$,

$$(\sigma x)_{f+g} = (\sigma x)_f + (\sigma x)_g \quad \text{and} \quad (\sigma x)_{fg} = (\sigma x)_f \cdot (\sigma x)_g$$

for all f, $g \in C(X)$. Thus, σx is a *homomorphism* of $C(X)$ into $\mathbf{R}$. Moreover, σx maps $C(X)$ *onto* the field $\mathbf{R}$; in fact, its kernel is, evidently, the maximal ideal $\boldsymbol{M}_x$.

Let H denote the set of all elements of P that are homomorphisms of $C(X)$ *onto* $\mathbf{R}$.

(By 10.5(a), these are all the nonzero homomorphisms into $\mathbf{R}$.) Thus,

$$p_{f+g} = p_f + p_g \quad \text{and} \quad p_{fg} = p_f \cdot p_g \qquad (f, g \in C)$$

for p in H.

Since P is the set of *all* mappings of $C(X)$ into $\mathbf{R}$, H is the set of *all* homomorphisms of $C(X)$ onto $\mathbf{R}$. Therefore H is in one-one correspondence with the set of all real maximal ideals in $C(X)$—the kernels of these homomorphisms (10.5(b)). As we have seen, H contains $\sigma[X]$, and the correspondence just cited carries $\sigma[X]$ onto the set of all *fixed* maximal ideals.

11.7. THEOREM. $H = \mathrm{cl}_P\ \sigma[X]$.

PROOF. First, we prove that H is closed. The set

$$H' = \bigcap_{f,\, g \in C} \{p \in P: p_{f+g} - p_f - p_g = 0 = p_{fg} - p_f p_g\}$$

is, clearly, the set of *all* homomorphisms from C into $\mathbf{R}$ (the zero homomorphism included). Since $p \to p_f$ is a continuous function (namely, π_f), H' is an intersection of closed sets, and hence is closed. Now, if p is a homomorphism *onto* **R**, then $p_1 = 1$. Therefore,

$$H = H' \cap \{p: p_1 = 1\},$$

and so H, too, is closed.

Secondly, $\sigma[X]$ is dense in H. For, an arbitrary basic neighborhood of a point p of H is a set

$$\bigcap_{k=1}^{n} \{q \in H: |q_{f_k} - p_{f_k}| < \epsilon\} \qquad (f_k \in C, \epsilon > 0).$$

The kernel of p is a real maximal ideal M in $C(X)$, and, for each k, the real number $M(f_k)$ is equal to p_{f_k}. Since $\mathbf{Z}[M]$ has the finite intersection property, there is a point $x \in X$ such that $f_k(x) = M(f_k)$ for all $k = 1, \cdots, n$. Thus, $(\sigma x)_{f_k} = p_{f_k}$, so that σx is in the given neighborhood of p.

11.8. THEOREM. *The closure, H, of $\sigma[X]$ in P is realcompact, and hence $H = \upsilon(\sigma[X])$.*

PROOF. Let M be any real maximal ideal in $C(H)$. Now, as was noted in 11.4(a), $C(X)$ is isomorphic with $C(H)$, under the mapping $f \to \pi_f | H$. Therefore the set

$$K = \{f \in C(X): \pi_f | H \in M\}$$

is a real maximal ideal in $C(X)$. This implies that K is the kernel of some homomorphism $p \in H$, so that $f \in K$ if and only if $\pi_f(p) = 0$. Hence, M is the fixed ideal $\mathbf{M}_p$ in $C(H)$. Thus, H is realcompact.

Since $\sigma[X]$ is homeomorphic with X, $\upsilon(\sigma[X])$ is a model for υX.

11.9. It will be instructive to examine a proof of existence of βX that parallels the one just given for υX. Let H_* denote the set of all homomorphisms of $C^*(X)$ onto **R**. As before, H_* is in one-one correspondence with the set of all maximal ideals in $C^*(X)$, and this correspondence carries the subset $\sigma_*[X]$ of H_* onto the set of all fixed maximal ideals. The analogue of Theorem 11.7 is

(a) $$H_* = \mathrm{cl}_{P_*} \sigma_*[X].$$

The proof that H_* is closed in P_* is as before; but the proof that $\sigma_*[X]$ is dense in H_* is somewhat different. We consider a basic neighborhood of a point p of H_*:

$$\bigcap_{k=1}^{n} \{q \in H_*: |q_{f_k} - p_{f_k}| < \epsilon\} \qquad (f_k \in C^*, \epsilon > 0).$$

The function f, defined by

$$f(x) = \sum_{k=1}^{n} (f_k(x) - p_{f_k})^2 \qquad (x \in X)$$

belongs to $C^*(X)$; and, since p is a homomorphism, $p_f = 0$. Therefore, f belongs to an ideal in C^*, and hence it is not a unit. Thus, f is not bounded away from zero, and so there is a point x of X such that $f(x) < \epsilon^2$. Then for $k \leqq n$, we have

$$|(\sigma_* x)_{f_k} - p_{f_k}| = |f_k(x) - p_{f_k}| < \epsilon,$$

so that $\sigma_* x$ is in the given neighborhood of p.

The reason for the difference between this proof and the earlier one is that the zero-sets of an ideal in C^* need not have the finite intersection property. Essentially, what was done here was to work with e-filters (2L) rather than z-filters.

Finally:

(b) *Every maximal ideal in $C^*(H_*)$ is fixed.*

The proof is exactly like that of Theorem 11.8. In the case of H, there was no more to be done, because H is realcompact, by definition, when every real maximal ideal is fixed. But to conclude from (b) that H_* is compact involves Theorem 4.11, whose proof demands that an arbitrary ideal be embeddable in a maximal ideal. It is noteworthy that this last requires the intervention of the axiom of choice, as, indeed, do all proofs of the existence of βX. In contrast, the proof of existence of υX given in 11.2 to 11.8 does not seem to depend upon this axiom. See *Notes*.

11.10. Next we look at the proof of existence of υX modeled after the one given in 11.2 to 11.5 for βX. The problem is to parallel the proof of 11.5 in order to show that the closure of $\sigma[X]$ in P is realcompact.

The first step is to observe that P is a product of real lines (one factor for each member of $C(X)$), and hence is a product of realcompact spaces. Now, the product theorem for realcompact spaces does not require a prior construction of υX, as was pointed out in 8.12. Consequently, we may claim that P is realcompact.

The remaining step is to prove that the closed subspace cl $\sigma[X]$ of P is realcompact. We are not at liberty to invoke the closed subspace theorem (8.10), as its proof did require the existence of υX. However, we notice an additional fact here: cl $\sigma[X]$ is not only closed, but it is also C-embedded. It is enough, then, to provide an independent proof of the following result:

(a) *A closed, C-embedded subset S of a realcompact space Y is realcompact.*

Let $\mathfrak{u}$ be a homomorphism of $C(S)$ onto $\mathbf{R}$. Since S is C-embedded in Y, the mapping $g \to g|S$ is a homomorphism of $C(Y)$ *onto* $C(S)$. Therefore $g \to \mathfrak{u}(g|S)$ is a homomorphism of $C(Y)$ onto $\mathbf{R}$. Since Y is realcompact, there is, by 10.5(c), a point y of Y such that $\mathfrak{u}(g|S) = g(y)$ for every g in $C(Y)$. We assert that $y \in S$. Indeed, if this is not the case, then, because S is closed, there exists $g \in C(Y)$ such that $g(y) = 0$, while $g[S] = \{1\}$. But $\mathfrak{u}\mathbf{1} = 1$, and so

$$0 = g(y) = \mathfrak{u}(g|S) = \mathfrak{u}\mathbf{1} = 1,$$

which is absurd. Thus, $y \in S$; and $\mathfrak{u}f = f(y)$ for every f in $C(S)$. By 10.5(c) again, this implies that S is realcompact.

11.11. The constructions of βX and υX presented in this chapter fail to emphasize one essential property of these spaces, namely, that υX can be embedded in βX. To derive this result, we observe simply that $\beta(\upsilon X)$ is a compactification of X in which X is C^*-embedded, and hence is βX. Additional insight is provided by the following theorem, which shows how H, the model for υX, is embeddable naturally in H_*, the model for βX.

Let τ denote the restriction to H of the projection of P onto P_*. Clearly, τ carries H into H_*.

THEOREM. *The projection τ of H into H_* is a homeomorphism, and* $\tau \circ \sigma = \sigma_*$.

PROOF. Certainly, τ is continuous, and $\tau \circ \sigma = \sigma_*$. The latter states that the restriction of τ to $\sigma[X]$ is $\sigma_* \circ \sigma^{\leftarrow}$, and hence is a homeomorphism.

We now show that the homomorphism τ' induced by τ carries $C^*(H_*)$ *onto* $C^*(H)$. Consider any $f \in C^*(H)$; we are to find $g \in C^*(H_*)$ such that $\tau' g = f$. Since τ maps $\sigma[X]$ homeomorphically onto $\sigma_*[X]$, we may specify the values of g on $\sigma_*[X]$ as follows:

$$g(\tau p) = f(p) \qquad (p \in \sigma[X]).$$

Inasmuch as $\sigma_*[X]$ is C^*-embedded and dense in H_*, the values of g on H_* are determined; by continuity, the above identity holds for every $p \in H$. This means that $f = g \circ \tau$, that is, $\tau' g = f$. Thus, τ' is onto. By Theorem 10.3(b), τ is a homeomorphism.

11.12. If the space X is compact, then $\sigma_*[X] = H_*$. As we have seen, H_* is contained in the subspace

$$\times_{f \in C^*} \mathrm{cl}_{\mathbf{R}} f[X]$$

of P_*, so that the compact space X is embedded as a closed subspace of a product of compact subsets of $\mathbf{R}$.

If X is realcompact, then $\sigma[X] = H$, and σ embeds X as a closed subspace of the product P. Conversely, any closed subspace of a product of real lines is realcompact, by Theorems 8.11 and 8.10 or by 8.12 and 11E.*1*. Hence we have:

THEOREM. *A space is realcompact if and only if it is homeomorphic with a closed subspace of a product of real lines.*

PROBLEMS

11A. THE PRODUCT P.

1. Describe P and $\sigma[X]$ for a one-point space X, and for a two-point space.

2. The subspace $\times_{f \in C} f[X]$ of P is realcompact and contains H.

11B. RESTRICTION OF PROJECTIONS.

1. Let $S \subset H$. In order that the functions $\pi_f | S$ be distinct for distinct $f \in C(X)$, and constitute all of $C(S)$, it is necessary and sufficient that S be dense and C-embedded in H. The analogue holds for H_*.

2. Find a subset E of P, disjoint from H, such that the functions $\pi_f | E$ are distinct for distinct f, and constitute all of $C(E)$. [Let $E = \{p' : p \in H\}$, where $p'_f = 2p_f$ for all f.]

11C. EMBEDDING OF AN ARBITRARY COMPACTIFICATION.

Let T be any [real]compactification of X. There exists a homeomorphism τ of X into a suitable product of real lines such that $\operatorname{cl} \tau[X]$ is homeomorphic with T.

11D. ANALOGUE OF THEOREM 11.7.

Prove that $H_* = \operatorname{cl} \sigma_*[X]$ by using 11.4(a*) and the fact that $\operatorname{cl} \sigma_*[X]$ is compact.

11E. CLOSED SUBSPACES OF A REALCOMPACT SPACE.

1. Give an alternative proof that a closed subspace S of a realcompact space Y is realcompact, as follows: let φ denote the identity map of S into Y; given a real z-ultrafilter $\mathscr{A}$ on S, consider $\varphi^\#\mathscr{A}$, and take note of Lemma 8.12.

2. In 11.10(a), it was assumed in addition that S is C-embedded in Y. What special property does $\tau^\#\mathscr{Z}$ have when $\mathscr{Z}$ is a z-ultrafilter on a C-embedded subspace and τ is the identity map into the whole space? [Theorem 1.18.]

Chapter 12

DISCRETE SPACES. NONMEASURABLE CARDINALS

12.1. We have seen many times that the discrete space $\mathbf{N}$ is realcompact. More generally, according to 8.18, every discrete space whose cardinal is $\leqq \mathfrak{c}$ is realcompact. The question arises whether *all* discrete spaces are realcompact. Since, among discrete spaces, the cardinal is the only significant variable, this is, in fact, a question about cardinal numbers.

By a $\{0, 1\}$-*valued measure* on a set X, we mean a countably additive function defined on the family of all subsets of X, and assuming only the values 0 or 1. Every set X admits certain trivial measures. One is the zero measure: the function $\mathbf{0}$. Another type is obtained by selecting a point $x \in X$, and defining $\mu(A) = 1$ if $x \in A$, and $\mu(A) = 0$ if $x \notin A$. We call a cardinal $\mathfrak{m}$ *measurable* if a set X of cardinal $\mathfrak{m}$ admits a $\{0, 1\}$-valued measure μ that is not one of these trivial types, i.e., such that $\mu(X) = 1$, and $\mu(\{x\}) = 0$ for every $x \in X$. As we shall prove in a moment, a discrete space is realcompact if and only if its cardinal is *nonmeasurable*, i.e., every nonzero, $\{0, 1\}$-valued measure assigns measure 1 to some one-element set. Our problem, then, is the purely set-theoretic one of finding the nonmeasurable cardinals. What will be proved is that the class of all nonmeasurable cardinals is very extensive; in fact, it is closed under all the standard operations of cardinal arithmetic. Whether *every* cardinal is nonmeasurable is a celebrated unsolved problem.

The chapter closes with some theorems concerning cardinal numbers of residue class fields of $C(X)$ for discrete X.

12.2. The realcompact discrete spaces are just those for which every ultrafilter with the countable intersection property is fixed (Theorem 5.14).

An ultrafilter $\mathscr{U}$ on a set X may be specified by its characteristic function $\chi_{\mathscr{U}}$ (defined on the set of all subsets of X): for $A \subset X$,

$$\chi_{\mathscr{U}}(A) = 1 \quad \text{if} \quad A \in \mathscr{U}, \qquad \chi_{\mathscr{U}}(A) = 0 \quad \text{if} \quad A \notin \mathscr{U}.$$

Evidently, $\chi_{\mathscr{U}}(X) = 1$. Next, if $A \cap B = \emptyset$, then $A \cup B \in \mathscr{U}$ if and only if exactly one of A, B belongs to $\mathscr{U}$. Therefore,

$$\chi_{\mathscr{U}}(A \cup B) = \chi_{\mathscr{U}}(A) + \chi_{\mathscr{U}}(B) \qquad (A \cap B = \emptyset).$$

Thus, $\chi_{\mathscr{U}}$ is a nonzero, *finitely additive* function defined on the family of all subsets of X.

Conversely, let μ be any nonzero, finitely additive function from the family of all subsets of X into $\{0, 1\}$. Define $\mathscr{U}$ to be the collection of all sets A for which $\mu(A) = 1$. Like any nonnegative, additive set-function, μ is monotone: if $A \subset B$, then $\mu(A) \leqq \mu(A) + \mu(B - A) = \mu(B)$; hence $A \in \mathscr{U}$ implies $B \in \mathscr{U}$. Also, $\mu(X) = 1$ (since $\mu \neq \mathbf{0}$). In addition, $\mu(A \cup B) = 0$ whenever $\mu(A) = \mu(B) = 0$—whether A and B are disjoint or not; therefore $\mathscr{U}$ is closed under finite intersection. Since $\mu(A) + \mu(X - A) = 1$, one of A, $X - A$ belongs to $\mathscr{U}$. Finally, $\mu(\emptyset) = \mu(\emptyset) + \mu(\emptyset)$, whence $\mu(\emptyset) = 0$, so that $\emptyset \notin \mathscr{U}$.

Consequently, $\mathscr{U}$ is an ultrafilter. Clearly, $\chi_{\mathscr{U}} = \mu$. Thus, the correspondence $\mathscr{U} \to \chi_{\mathscr{U}}$ is one-one from the set of all ultrafilters on X onto the set of all nonzero, finitely additive, $\{0, 1\}$-valued set-functions defined on X.

If $\mathscr{U}$ has the countable intersection property, then it is closed under countable intersection; for if $\bigcap_n A_n = A \notin \mathscr{U}$, then $X - A \in \mathscr{U}$, and $\bigcap_n (A_n - A) = \emptyset$ (this is a special case of 5.14). We aver that this condition is equivalent to countable additivity, that is, to the stipulation that $\mu = \chi_{\mathscr{U}}$ be a *measure*. By definition, μ is countably additive provided that

$$\mu(\textstyle\bigcup_n A_n) = \sum_n \mu(A_n)$$

whenever (A_n) is a sequence of pairwise disjoint sets. No difficult question of convergence of the infinite sum can arise here, because we must have $\mu(A_n) = 0$ for all but at most one A_n (since two disjoint sets cannot both belong to $\mathscr{U}$). If $\mu(A_n) = 1$ for some (hence just one) n, the equation in question holds trivially; accordingly, the problem reduces to the case in which $\mu(A_n) = 0$ for all n. And here, as in the finite case, we can drop the requirement that the sets be disjoint. Hence μ is a measure if and only if

$$\mu(A_n) = 0 \ (n \in \mathbf{N}) \quad \text{implies} \quad \mu(\textstyle\bigcup_n A_n) = 0.$$

But this is simply the dual of the statement that $\mathscr{U}$ is closed under countable intersection.

Let us call $\chi_{\mathscr{U}}$ *fixed* or *free* according as the ultrafilter $\mathscr{U}$ is fixed or free. Then we have: X is realcompact if and only if every measure $\chi_{\mathscr{U}}$ is fixed. Now, by definition, $|X|$ is nonmeasurable if every measure $\chi_{\mathscr{U}}$ is fixed. Therefore we have, finally:

THEOREM. *A discrete space is realcompact if and only if its cardinal is nonmeasurable.*

12.3. All measures referred to henceforth are understood to be $\{0, 1\}$-valued. We shall say that a measure μ is $\mathfrak{m}$-*additive* if $\mu(\bigcup_{s \in S} A_s) = 0$ whenever $(A_s)_{s \in S}$ is a family of disjoint sets of measure zero, with $|S| = \mathfrak{m}$. Evidently, an $\mathfrak{m}$-additive measure is automatically $\mathfrak{n}$-additive for every $\mathfrak{n} \leqq \mathfrak{m}$.

Just as in the countable case, we may drop the requirement that the sets be disjoint. (To reduce an arbitrary family of sets A_s of measure zero to an equivalent family of disjoint sets A'_s, well-order the index set, and define $A'_s = A_s - \bigcup_{t<s} A_t$.) We find, then, that μ is $\mathfrak{m}$-additive if and only if the intersection of any $\mathfrak{m}$ sets of measure 1 is of measure 1.

(a) *Each measure is* $\mathfrak{m}$-*additive for every nonmeasurable cardinal* $\mathfrak{m}$.

If μ is not $\mathfrak{m}$-additive, there exists a family $(A_s)_{s \in S}$ of $\mathfrak{m}$ disjoint sets of measure 0 whose union is of measure 1. Define a set-function λ on the index set S, as follows:

$$\lambda(E) = \mu(\textstyle\bigcup_{s \in E} A_s) \qquad (E \subset S).$$

Evidently, λ is a measure on S, with $\lambda(S) = 1$. Since $\lambda(\{s\}) = \mu(A_s) = 0$ for each s, λ is free. Therefore the cardinal $\mathfrak{m} = |S|$ is measurable.

(b) *If μ is a free measure, and $|A|$ is nonmeasurable, then* $\mu(A) = 0$.

For, the restriction of μ to the set of all subsets of A is a free measure on A. The result also follows from (a). The restatement of (b) in terms of topology of discrete spaces is of some interest: no member of a free ultrafilter with the countable intersection property is realcompact.

12.4. We shall say that a class of cardinal numbers is *closed* if it is closed under all the standard processes for forming cardinals from given ones: addition, multiplication, the formation of suprema, exponentiation, and the passage from a given cardinal to its immediate successor or to any smaller cardinal. In the case of the first three of these processes, it is understood that the cardinal number of the index set is itself a member of the class in question.

Plainly, the class of *all* cardinals is closed. Likewise, the class of all *finite* cardinals is closed; in fact, it is the smallest (nonempty) closed class. Of course, our interest will be in closed classes that also contain the number $\aleph_0$.

LEMMA. *A nonempty class $\mathfrak{C}$ is closed if and only if, whenever* $\mathfrak{m} \in \mathfrak{C}$,

(i) $\mathfrak{n} \in \mathfrak{C}$ *for all* $\mathfrak{n} < \mathfrak{m}$;

(ii) *the sum of any* $\mathfrak{m}$ *members of* $\mathfrak{C}$ *is in* $\mathfrak{C}$;

(iii) $2^{\mathfrak{m}} \in \mathfrak{C}$.

PROOF. The necessity of the conditions is obvious. Conversely, suppose that $\mathfrak{C}$ satisfies these conditions; we are to prove that $\mathfrak{C}$ contains successors, suprema, products, and exponentials, as described. Let $\mathfrak{m} \in \mathfrak{C}$. The successor of $\mathfrak{m}$ is less than or equal to $2^{\mathfrak{m}}$, and hence belongs to $\mathfrak{C}$, by (iii) and (i). Next, given $\mathfrak{m}$ cardinals $\mathfrak{n}_s$ belonging to $\mathfrak{C}$, define $\mathfrak{n} = \sum_s \mathfrak{n}_s$. Since $\sup_s \mathfrak{n}_s \leqq \mathfrak{n}$, and since $\mathfrak{n} \in \mathfrak{C}$, by (ii), we have $\sup_s \mathfrak{n}_s \in \mathfrak{C}$, by (i). Furthermore,

$$\prod_s \mathfrak{n}_s \leqq \prod_s 2^{\mathfrak{n}_s} = 2^{\mathfrak{n}} \in \mathfrak{C},$$

so that $\prod_s \mathfrak{n}_s \in \mathfrak{C}$. Finally, this last implies that if $\mathfrak{l} \in \mathfrak{C}$, then $\mathfrak{l}^{\mathfrak{m}}$—which is the product of $\mathfrak{l}$ by itself, $\mathfrak{m}$ times—is also in $\mathfrak{C}$.

12.5. THEOREM. *The class of all nonmeasurable cardinals is a closed class containing* $\aleph_0$.

PROOF. $\aleph_0$ is already known to be nonmeasurable. We proceed to verify the conditions of the lemma.

(i) *Every cardinal smaller than a nonmeasurable cardinal is nonmeasurable*, in other words, every subspace X of a realcompact, discrete space Y is realcompact. For, if μ is a nonzero, free measure on X, then ν, defined by:

$$\nu(B) = \mu(B \cap X),$$

is a nonzero, free measure on Y. (Alternatively, the result is a special case of Theorem 8.10 or Corollary 8.14.)

(ii) *Every nonmeasurable sum of nonmeasurable cardinals is nonmeasurable.* Let $|X|$ be a sum of $\mathfrak{m}$ nonmeasurable cardinals, where $\mathfrak{m}$ is nonmeasurable. Then X is expressible as a union of $\mathfrak{m}$ disjoint subsets A_s, each of nonmeasurable cardinal. Let μ be any free measure on X. By 12.3(b), $\mu(A_s) = 0$ for each s. By 12.3(a), μ is $\mathfrak{m}$-additive, so that $\mu(X) = \mu(\bigcup_s A_s) = 0$. Thus, $\mu = \mathbf{0}$. Therefore $|X|$ is nonmeasurable.

(iii) *If* $\mathfrak{m}$ *is nonmeasurable, then* $2^{\mathfrak{m}}$ *is nonmeasurable.* Let X be any set of power $\mathfrak{m}$, and $\mathfrak{X}$ the family of all subsets of X. Then $|\mathfrak{X}| = 2^{\mathfrak{m}}$. Let μ be any nonzero measure on $\mathfrak{X}$. By 12.3(a), μ is $\mathfrak{m}$-additive. For each $x \in X$, consider the family $\mathfrak{S}_x$ of all subsets of X that contain x, and its complementary family $-\mathfrak{S}_x$:

$$\mathfrak{S}_x = \{A \subset X: x \in A\}, \quad -\mathfrak{S}_x = \{B \subset X: x \notin B\}.$$

Next, define

$$S = \{x \in X: \mu(\mathfrak{S}_x) = 1\}.$$

Then the family

$$\mathfrak{S} = \bigcap_{x \in S} \mathfrak{S}_x \cap \bigcap_{x \notin S} (-\mathfrak{S}_x)$$

is the intersection of $\mathfrak{m}$ families of measure 1, and hence is itself of measure 1 (since μ is $\mathfrak{m}$-additive). But $\mathfrak{S}$ is a one-element subfamily of $\mathfrak{X}$:

$$\mathfrak{S} = \{A \in \mathfrak{X}: A \supset S\} \cap \{B \in \mathfrak{X}: B \subset S\} = \{S\}.$$

Therefore the measure μ is not free. Thus, $|\mathfrak{X}|$ is nonmeasurable.

12.6. REMARKS. The smallest closed class containing $\aleph_0$ comprises a vast collection of cardinals; and we know from Theorem 12.5 that all of them are nonmeasurable. Not only is $\aleph_0$ nonmeasurable, but so are $\aleph_1, \cdots, \aleph_\omega, \cdots, \aleph_{\omega_1}, \cdots, \aleph_{\omega_\omega}, \cdots$, as well as $\mathfrak{c}$, $2^{\mathfrak{c}}$, $2^{2^{\mathfrak{c}}}$, $\cdots$; more generally, so are all cardinals that can be defined in terms of given nonmeasurable cardinals by the standard processes of cardinal arithmetic. For all these cardinals, the corresponding discrete spaces are realcompact.

A cardinal is said to be *strongly inaccessible* if the set of all smaller cardinals is a closed class containing $\aleph_0$. Thus, Theorem 12.5 states that the smallest measurable cardinal (if any exist) is strongly inaccessible. We have indicated that it must be huge; in Hausdorff's words, if such numbers exist, "so ist die kleinste unter ihnen von einer so exorbitanten Grösse, dass sie für die üblichen Zwecke der Mengenlehre kaum jemals in Betracht kommen wird."

Whether strongly inaccessible cardinals exist at all is still an unsettled matter. Conceivably, one could prove that no such cardinal exists—whence no measurable cardinal exists. In any event, it cannot be proved (in standard axiom systems for set theory) that these cardinals do exist. For, if we have a model for set theory that includes strongly inaccessible cardinals, there will, of course, be a smallest such cardinal $\mathfrak{a}$. Let us then reinterpret "set" to apply only to those sets whose power is less than $\mathfrak{a}$. (This requirement applies as well to the *members* of these sets.) The corresponding reduced collection of cardinals will constitute a closed class containing $\aleph_0$. Now, the arithmetical processes for constructing cardinals are simply the counterparts of the standard processes for the construction of sets. Therefore the new model will again satisfy all the axioms. (A similar situation occurs if one tries to prove the existence of infinite cardinals, in the absence of the axiom of infinity. The class of all finite sets satisfies the remaining axioms, and therefore the existence of an infinite set cannot be derived from these axioms.)

If strongly inaccessible cardinals do not exist, then, of course, every cardinal is nonmeasurable. On the other hand, even if such cardinals do exist, it may still be true that every cardinal is nonmeasurable. As an example of an intermediate possibility, it is conceivable that the first measurable cardinal is equal to the second strongly inaccessible cardinal.

The hypothesis that every cardinal is nonmeasurable differs in an important respect from, say, the continuum hypothesis (which is also known to be consistent with the usual axiom systems): measurable cardinals are remote from most mathematical work, whereas the continuum problem is encountered frequently, in a natural way.

CARDINALS OF RESIDUE CLASS FIELDS

12.7. We close this chapter with some results having to do with cardinals generally, but not with measurability. The problem is to get some information about the cardinal of $C(X)/M$, where M is a maximal ideal. Since C/M contains the real field, we always have $|C/M| \geqq \mathfrak{c}$. And we know that equality can hold, even when M is hyper-real—for example, in case $X = \mathbf{N}$, so that $C(X)$ itself is of power $\mathfrak{c}$.

We now prove that there exist residue class fields of arbitrarily large cardinal.

THEOREM. *Let X be the discrete space of power $\mathfrak{m}$, where $\mathfrak{m}$ is an infinite cardinal. Then there exists a maximal ideal M in $C(X)$ such that $|C/M| > \mathfrak{m}$.*

PROOF. Since X is infinite, it is equipotent with the set of all its finite subsets. Let $(F_x)_{x \in X}$ be a one-one indexing of the nonempty finite subsets of X, and, for each point y of X, define

$$Z_y = \{x \in X : y \in F_x\}.$$

The family of all sets Z_y has the finite intersection property: $Z_{y_1} \cap \cdots \cap Z_{y_n}$ contains the point x for which $\{y_1, \cdots, y_n\} = F_x$. Therefore this family is embeddable in an ultrafilter $\mathbf{Z}[M]$ (where M denotes a maximal ideal in $C(X)$).

To show that $|C/M| > \mathfrak{m}$, we consider any subset B of C of power $\leqq \mathfrak{m}$, and exhibit f such that $M(f) > M(g)$ for every $g \in B$. Let B be indexed as $(g_y)_{y \in X}$ (with duplicate indexing in case $|B| < |X|$). Define

$$f(x) = 1 + \max_{y \in F_x} g_y(x) \qquad (x \in X).$$

Since F_x is finite and not empty, this is meaningful, and defines f on all of X. Now consider any function $g_y \in B$. For every $x \in Z_y$, we have $y \in F_x$, and therefore

$$g_y(x) \leqq \max_{z \in F_x} g_z(x) < f(x).$$

Thus, $f - g_y$ is positive on the zero-set Z_y of M. Hence $M(f) > M(g_y)$ (5.4(b)).

REMARK. We have actually proved the stronger result that no set of power $\leqq \mathfrak{m}$ is cofinal in the totally ordered field C/M.

Notice that $|C/M| \leqq 2^{\mathfrak{m}}$ (5J). Hence if there is no cardinal between $\mathfrak{m}$ and $2^{\mathfrak{m}}$, then $|C/M| = 2^{\mathfrak{m}}$.

12.8. For the special case $\mathfrak{m} = \mathfrak{c}$, this result can be sharpened. We need the following general lemma.

LEMMA. *Let X and Y be sets of power $\mathfrak{m}$, where $\mathfrak{m}$ is infinite. There exists a family E of mappings from X into Y, with $|E| > \mathfrak{m}$, such that any two members of E agree only on a set of power $< \mathfrak{m}$.*

PROOF. By the maximal principle, there exists a maximal subfamily E of Y^X, no two members of which agree on a set of power $\mathfrak{m}$. We suppose that $|E| \leqq \mathfrak{m}$ and obtain a contradiction.

If $|E| \leqq \mathfrak{m}$, we can index E by a suitable subset of X: $E = (f_s)$, where s ranges over the subset. Now we define a new mapping f. Well-order X according to the least ordinal of cardinal $\mathfrak{m}$. Then, for each $x \in X$, we have $|\{s\colon s < x\}| < \mathfrak{m}$. Hence we can define f by choosing $f(x) \in Y$ so as to differ from $f_s(x)$ for every $s < x$. Thus, for a given $f_s \in E$, $f(x) = f_s(x)$ only if $x \leqq s$; hence f agrees with f_s only on a set of power $< \mathfrak{m}$. Accordingly, the family $E \cup \{f\}$ satisfies the stated requirement, and so E is not maximal.

12.9. THEOREM. *Let X be the discrete space of power $\mathfrak{c}$. If M is any maximal ideal in $C(X)$ such that $|Z| = \mathfrak{c}$ for every $Z \in \mathbf{Z}[M]$, then $|C/M| > \mathfrak{c}$.*

PROOF. By Lemma 12.8, $C(X)$ $(= \mathbf{R}^X)$ contains a family of more than $\mathfrak{c}$ functions, no two of which agree on a set of power $\mathfrak{c}$. Hence no two agree on any zero-set of M, i.e., no two are congruent modulo M.

Of course, such maximal ideals exist. In fact, let $\mathfrak{m} \geqq \mathfrak{n} \geqq \aleph_0$, and let X be a set of cardinal $\mathfrak{m}$. The family of all subsets A for which $|X - A| < \mathfrak{n}$ has the finite intersection property, and hence is contained in an ultrafilter $\mathscr{U}$. Certainly, $|Z| \geqq \mathfrak{n}$ for every $Z \in \mathscr{U}$.

Notice, incidentally, that the point in βX to which $\mathscr{U}$ converges is not in the closure of any subset of X whose cardinal is less than $\mathfrak{n}$. Cf. the second example in 10.14.

There also exists a maximal ideal M in $C(\mathbf{R})$ such that $|Z| = \mathfrak{c}$ for every $Z \in \mathbf{Z}[M]$ (see, e.g., 4F). But $|C(\mathbf{R})/M| = \mathfrak{c}$, nonetheless.

PROBLEMS

12A. CLOSED DISCRETE SUBSPACES.

Let $X = W(\alpha) \times \mathbf{N}$, where $|\alpha|$ is a (hypothetical) measurable cardinal.

1. Every closed discrete subspace of X is realcompact.

2. There exists a space Y, in which X is dense, such that not every closed discrete subspace of Y is realcompact.

12B. ALMOST DISJOINT SUBSETS.

Any set of cardinal $\mathfrak{m} \geqq \aleph_0$ has a family of more than $\mathfrak{m}$ subsets, each of power $\mathfrak{m}$, such that the intersection of any two is of power $< \mathfrak{m}$. [In Lemma 12.8, $X \times Y$ has such a family.] (For the special case $\mathfrak{m} = \aleph_0$, a stronger result is given in 6Q.*1*.)

12C. ZERO-SETS OF LARGE CARDINAL.

Let $\mathfrak{m} \geqq \mathfrak{n} \geqq \aleph_0$, and let X be the discrete space of power $\mathfrak{m}$.

1. There exists an ultrafilter $\mathscr{U}$ on X such that $|Z| \geqq \mathfrak{n}$ for every $Z \in \mathscr{U}$, and $|Z| = \mathfrak{n}$ for at least one $Z \in \mathscr{U}$.

2. If M is the maximal ideal in $C(X)$ for which $\mathscr{U} = \mathbf{Z}[M]$, then $|C/M| \leqq 2^{\mathfrak{n}}$, by 5J. In case $\mathfrak{n} = \mathfrak{c}$, we have $|C/M| > \mathfrak{c}$. [10H.*4*.]

12D. FREE REAL IDEALS.

Let X be a discrete space, let M be a free maximal ideal in $C(X)$, and let $\mathfrak{m}$ denote the smallest measurable cardinal. If M is real, then $|Z| \geqq \mathfrak{m}$ for every $Z \in \mathbf{Z}[M]$, but not conversely. [12.3(b).] Thus, Theorem 12.9 does not go through with $\mathfrak{m}$ in place of $\mathfrak{c}$ in the hypothesis.

12E. RESIDUE CLASS FIELDS OF LARGE CARDINAL.

1. Let X be a discrete space of power $\mathfrak{m} \geqq \aleph_0$. There exist $2^{\mathfrak{m}}$ maximal ideals M in $C(X)$ such that $|C/M| > \mathfrak{m}$. [Modify the proof of Theorem 12.7, taking account of the following:

(i) X has $2^{\mathfrak{m}}$ subsets S of power $\mathfrak{m}$;

(ii) for each such S, the family

$$\{Z_y: y \in S\} \cup \{X - Z_y: y \notin S\}$$

has the finite intersection property, and so is contained in an ultrafilter $\mathscr{U}_S$;

(iii) if $S \neq S'$, then $\mathscr{U}_S \neq \mathscr{U}_{S'}$.]

2. Let X be the discrete space of cardinal $\mathfrak{c}$. There exist $2^{2^{\mathfrak{c}}}$ maximal ideals M in $C(X)$ for which $|C/M| > \mathfrak{c}$. [9F or 12I.]

12F. CARDINALS OF RESIDUE CLASS RINGS.

Let $\mathfrak{n}$ be a given cardinal, and let M^p be a hyper-real ideal in $C(X)$ such that no set of cardinal $\leqq \mathfrak{n}$ is cofinal in C/M^p.

1. There exists a set A of $\mathfrak{n}$ elements in C/M^p such that the ratio of the larger to the smaller of any two is infinitely large. [Construct A inductively.]

2. Let T denote the subspace $X \cup \{p\}$ of βX. There exists $g \in C(T)$ such that $|C(T)/(g)| \geqq \mathfrak{n}$. [Consider f^{-1}, where $M^p(f)$ exceeds every member of A.]

3. For an ideal I in C that is not a z-ideal, $|C/I|$ may be greater than $\mathfrak{c}^{\mathfrak{m}}$, where $\mathfrak{m}$ is the smallest of the cardinal numbers of the members of $\mathbf{Z}[I]$. Compare 5J.

12G. NONMEASURABLE SUMS OF SPACES.

If X is the topological sum of spaces X_α, where α ranges over an index set A of nonmeasurable cardinal, then υX is the topological sum of the spaces υX_α. [Let φ denote the obvious mapping of X onto A. Then $\varphi^\circ: \upsilon X \to A$ satisfies $\varphi^{\circ\leftarrow}(\alpha) = \mathrm{cl}_{\upsilon X} X_\alpha$.] Hence if each X_α is realcompact, then so is X.

12H. EXTREMALLY DISCONNECTED P-SPACES.

1. Let p be a nonisolated point of an extremally disconnected space X (1H). There exists a family $\mathscr{S}$ of disjoint open-and-closed sets whose union is dense in X but does not contain p. [Consider a maximal family.]

2. Let $\mathfrak{U}$ be the collection of all subfamilies $\mathscr{T}$ of $\mathscr{S}$ such that $p \in \mathrm{cl}_X \bigcup \mathscr{T}$. Then $\mathfrak{U}$ is a free ultrafilter on $\mathscr{S}$.

3. If $|\mathscr{S}|$ is nonmeasurable, then there exists a bounded function φ on $\mathscr{S}$ that is positive everywhere, and converges to zero on the ultrafilter $\mathfrak{U}$ (i.e., for each $n \in \mathbf{N}$, there exists $\mathscr{T} \in \mathfrak{U}$ such that $\varphi(T) < 1/n$ for all $T \in \mathscr{T}$). [Consider $\mathscr{S}$ as a discrete space.]

4. There exists $f \in C(X)$ such that $f(s) = \varphi(S)$ for $s \in S \in \mathscr{S}$. [1H.*6* or 6M.*2*.]

5. If $|\mathscr{S}|$ is nonmeasurable, then p is not a P-point (4L) of X.

6. Every extremally disconnected P-space (4J) *of nonmeasurable cardinal is discrete.*

7. An extremally disconnected P-space of measurable cardinal need not be discrete. [Consider υX for discrete X.]

12I. βX FOR DISCRETE X.

Let X be any infinite discrete space, and let S denote the set of all points $p \in \beta X$ such that every neighborhood of p meets X in a set of cardinal $|X|$. Then S is compact, and S contains a copy of βX. [Cf. 6.10(a).] Hence $|S| = 2^{2^{|X|}}$.

Chapter 13

HYPER-REAL RESIDUE CLASS FIELDS

13.1. In this chapter, we continue the study, initiated in Chapter 5, of the algebraic structure and order structure of hyper-real residue class fields of $C(X)$. Although none of the material developed after Chapter 5 will be called upon, we shall need quite a bit more of the abstract theory of fields than heretofore. We begin with a summary of these algebraic prerequisites.

Let E be a subfield of an arbitrary field F. An element of F is said to be *algebraic* over E if it is a root of a polynomial equation with coefficients in E; otherwise, it is said to be *transcendental* over E.

The extension of E to the smallest subfield of F that also contains subsets $S, \cdots$, and elements $x, \cdots$, is denoted by $E(S, \cdots, x, \cdots)$. In case x is a transcendental element, $E(x)$ consists of all rational functions in x with coefficients in E.

A set T is said to be (*algebraically*) *independent* over E if $\mathfrak{p}(x_1, \cdots, x_n) \neq 0$ whenever $n \in \mathbf{N}$, $\mathfrak{p}$ is a nonzero polynomial over E in n indeterminates, and $x_1, \cdots, x_n$ are distinct elements of T. Thus, T is a set of independent transcendentals if and only if each x in T is transcendental over the subfield $E(T - \{x\})$.

A *transcendence base* for F over E is a maximal set of independent transcendentals. Any set of independent transcendentals is extendable to a transcendence base. (This follows easily from the maximal principle.) The cardinal of a transcendence base is an invariant, called the *transcendence degree* of F over E.

When every member of S is algebraic over E, $E(S)$ is called an algebraic extension of E. We state the following well-known facts without proof: if E' is an algebraic extension of E, then every element of E' is algebraic over E; if E'' is an algebraic extension of E', then E'' is an algebraic extension of E; the set of *all* elements of F that are algebraic over E is a field.

Every algebraic extension of an infinite field E is equipotent with E: algebraic elements are zeros of polynomials, and each polynomial has only a finite number of coefficients (so that the set of all polynomials in one indeterminate is equipotent with E), and a finite number of zeros.

Throughout this chapter, *ordered*, unmodified, is to mean *totally ordered.*

The intersection of all the subfields of an *ordered* field E is a replica of the rational field **Q**; for simplicity of notation, we shall identify it with **Q**. From the above, if E is uncountable, then its cardinal is the same as its transcendence degree over **Q**.

An ordered field F is said to be *real-closed* if it satisfies the following conditions—which are known to be equivalent:

(1) every positive element is a square, and every polynomial over F (in one indeterminate) of odd degree has a zero in F;

(2) $K = F(\sqrt{-1})$ is algebraically closed (i.e., every polynomial over K has a zero in K);

(3) F has no proper algebraic extension to an ordered field.

It follows from (2) that every polynomial over a real-closed field, of degree greater than two, is reducible.

The field **R** of real numbers is real-closed, of course. Any real-closed field has a unique ordering: an element is nonnegative if and only if it is a square. Hence every isomorphism of a real-closed field (to an ordered field) is order-preserving. The fundamental result on existence of real-closed fields is the following theorem, which we state without proof.

THEOREM (ARTIN-SCHREIER). *Every ordered field E has an algebraic extension to a real-closed field $\mathscr{R}E$ whose order is an extension of the order on E. Furthermore, $\mathscr{R}E$ is unique up to an isomorphism that leaves the elements of E fixed.*

$\mathscr{R}E$ may be called the *real-closure* of E. Since it is an algebraic extension, it is equipotent with E.

The uniqueness of $\mathscr{R}E$ means that every order-preserving isomorphism from E onto an ordered field E' can be extended to an isomorphism from $\mathscr{R}E$ onto $\mathscr{R}E'$. On the other hand, an isomorphism from E to E' that does not preserve order *cannot* be so extended, because the order in a real-closed field is determined by the algebraic structure. In this connection, see 13C.

COROLLARY. *If E is a subfield of a real-closed field F, then E has a unique real-closed, algebraic extension—a copy of $\mathscr{R}E$—in F.*

PROOF. Let A denote the subfield of all elements of F that are algebraic over E. A routine check shows that A is real-closed. Now, obviously, every algebraic extension of E in F is contained in A; and, by (3), no proper subfield of A containing E is real-closed. Therefore, A is the only real-closed, algebraic extension of E in F. By the theorem, $A = \mathscr{R}E$.

As an example: $\mathscr{R}\mathbf{Q}$ is the set of all algebraic real numbers.

13.2. Let $f \in C(X)$, with $f \geqq \mathbf{0}$, and let r be any positive real number. The function f^r, defined by

$$f^r(x) = (f(x))^r \qquad (x \in X),$$

is continuous, since it is the composition of two continuous functions. This is an extension of the definition of exponentiation for positive real numbers (appearing here as constant functions).

Indeed, it is possible to define exponentiation in any residue class field C/M. Given $r > 0$, and a positive element u of C/M, let f and g be any two nonnegative preimages of u in C. Since $f - g \equiv \mathbf{0}$ (mod M), the set $\boldsymbol{Z}(f^r - g^r)$, which is the same as $\boldsymbol{Z}(f - g)$, belongs to $\boldsymbol{Z}[M]$. Therefore $f^r \equiv g^r$ modulo the z-ideal M. Accordingly, the definition $u^r = M(f^r)$ depends only upon u (and r), not upon the particular representative f.

The laws of exponents, $u^r u^s = u^{r+s}$ and $(u^r)^s = u^{rs}$, are clearly valid. Furthermore, if $u < v$, then $u^r < v^r$. It follows that if u is infinitely large, then so is u^r for every r (> 0).

THEOREM. *The transcendence degree over* $\mathbf{R}$ *of a hyper-real field is at least* $\mathfrak{c}$.

PROOF. Given a hyper-real ideal M in $C(X)$, let u be any infinitely large element of C/M. We shall show that the powers u^r of u in C/M are algebraically independent over $\mathbf{R}$, whenever r ranges over a set of positive real numbers that are linearly independent over $\mathbf{Q}$. Since there exist $\mathfrak{c}$ linearly independent, positive reals, this will establish the theorem.

Let

$$\mathfrak{p}(\lambda_1, \cdots, \lambda_q) = \sum_{k=1}^{m} a_k \lambda_1^{n_{k1}} \cdots \lambda_q^{n_{kq}}$$

be any polynomial in q indeterminates λ_j, with nonzero coefficients $a_k \in \mathbf{R}$, where it is assumed that like monomials have been collected in a single term. We are to show that the element

$$v = \mathfrak{p}(u^{r_1}, \cdots, u^{r_q})$$

of C/M (where the r_j are positive and linearly independent) is different from zero. We shall prove, in fact, that $|v|$ is infinitely large. We have

$$v = \sum_{k=1}^{m} a_k u^{s_k}, \quad \text{where} \quad s_k = n_{k1}r_1 + \cdots + n_{kq}r_q.$$

Since the numbers r_j are linearly independent, the exponents s_k are all distinct. Obviously, they are all positive; therefore, since u is infinitely large, so is each power u^{s_k}.

For $m = 1$, it is trivial that $|v|$ is infinitely large. In case $m > 1$, we pick the largest of the numbers s_k—say it is s_l—and assume, as we may, that $a_l = m - 1$. Then we have

$$v = \sum_{k \neq l} u^{s_k}(u^{s_l - s_k} + a_k).$$

In each term, both factors are infinitely large, and therefore v is infinitely large.

It is clear from the proof that the conclusion of the theorem is valid in any nonarchimedean field containing **R** in which exponentiation can be suitably defined.

13.3. Our next objective is to show that every residue class field of C is real-closed.

Fix $n \in \mathbf{N}$. For each point

$$a = (a_1, \cdots, a_n) \in \mathbf{R}^n,$$

let $\rho_1 a, \cdots, \rho_n a$ denote the real parts of the (complex) zeros of the polynomial

$$\mathfrak{p}_a(\lambda) = \lambda^n + a_1\lambda^{n-1} + \cdots + a_n$$

(listing each according to its multiplicity), indexed so that

$$\rho_1 a \leqq \cdots \leqq \rho_n a.$$

This defines n functions $\rho_1, \cdots, \rho_n$ from $\mathbf{R}^n$ into $\mathbf{R}$. We prove:

(a) *Each of the functions ρ_k from $\mathbf{R}^n$ to $\mathbf{R}$ is continuous.*

The n functions can be handled simultaneously. Consider any point $a \in \mathbf{R}^n$, and let $r_1, \cdots, r_q$ denote the distinct values among $\rho_1 a, \cdots, \rho_n a$. Given neighborhoods W_j of r_j, we are to find a neighborhood U of a such that for every $b \in U$, if $\rho_k a = r_j$, then $\rho_k b \in W_j$. We may assume that the W_j are disjoint intervals of $\mathbf{R}$. For each j, let Γ_j be the boundary of a rectangle in the complex plane that projects into W_j, and that encloses all of the zeros of $\mathfrak{p}_a(\lambda)$ whose real part is r_j. The continuous function

$$z \rightarrow |\mathfrak{p}_a(z)|$$

from the complex plane to $\mathbf{R}$ has a positive lower bound ϵ on the compact set $\bigcup_j \Gamma_j$. The mapping

$$(b, z) \to \mathfrak{p}_b(z)$$

is continuous; therefore, for each complex number w, there exist neighborhoods U_w of a and V_w of w such that

$$|\mathfrak{p}_b(z) - \mathfrak{p}_a(w)| < \epsilon/2$$

whenever $(b, z) \in U_w \times V_w$. A finite number of the V_w cover the compact set $\bigcup_j \Gamma_j$, and the intersection of the corresponding sets U_w is a neighborhood U of a. For each $b \in U$ and $z \in \bigcup_j \Gamma_j$, there exists w such that

$$|\mathfrak{p}_b(z) - \mathfrak{p}_a(z)| \leqq |\mathfrak{p}_b(z) - \mathfrak{p}_a(w)| + |\mathfrak{p}_a(w) - \mathfrak{p}_a(z)| < \epsilon.$$

Thus, for any $b \in U$, and for each j,

$$|\mathfrak{p}_a(z)| > |\mathfrak{p}_b(z) - \mathfrak{p}_a(z)|$$

for all $z \in \Gamma_j$. By Rouché's theorem, Γ_j encloses exactly as many zeros of $\mathfrak{p}_b(\lambda)$ as of $\mathfrak{p}_a(\lambda)$ (counting multiplicities). It follows at once that if $\rho_k a = r_j$, then $\rho_k b \in W_j$.

13.4. THEOREM. *Every residue class field $C(X)/M$ is real-closed.*

PROOF. We are to prove that every positive element has a square root, and that every polynomial of odd degree has a zero.

Given $u > 0$ in C/M, let f be a nonnegative preimage of u in C; then $M(f^{1/2}) = u^{1/2}$.

In showing that a given polynomial of odd degree, with coefficients in $C(X)/M$, has a zero in $C(X)/M$, we may assume, of course, that the leading coefficient is 1. Rephrased in terms of preimages in $C(X)$, the problem becomes: given a polynomial

$$\mathfrak{P}(\lambda) = \lambda^n + f_1\lambda^{n-1} + \cdots + f_n,$$

with n odd, and with coefficients $f_k \in C(X)$, to find $g \in C(X)$ such that $\mathfrak{P}(g) \in M$.

Define a mapping φ from X into $\mathbf{R}^n$ as follows:

$$\varphi x = (f_1(x), \cdots, f_n(x)).$$

Certainly, φ is continuous. It follows from 13.3(a) that:

(a) *Each of the functions $g_k = \rho_k \circ \varphi$ from X to $\mathbf{R}$ is continuous.*

For each $x \in X$, the numbers $g_1(x), \cdots, g_n(x)$ are the real parts of the zeros of the polynomial

$$\mathfrak{p}_{\varphi x}(\lambda) = \lambda^n + f_1(x)\lambda^{n-1} + \cdots + f_n(x).$$

Since this is a polynomial of odd degree, with real coefficients, it has at least one real zero. Thus:

(b) *For each* $x \in X$, *there exists an index* k *such that* $g_k(x)$ *is a zero of* $\mathfrak{p}_{\varphi x}(\lambda)$.

Now, $g_k \in C(X)$, as was noted in (a). Therefore $\mathfrak{P}(g_k) \in C(X)$. By definition of the ring operations in $C(X)$, $\mathfrak{P}(g_k)$ satisfies

$$\mathfrak{P}(g_k)(x) = \mathfrak{p}_{\varphi x}(g_k(x))$$

for every $x \in X$. Hence by (b), for each $x \in X$, there exists an index k such that $\mathfrak{P}(g_k)(x) = 0$. Therefore,

$$\mathfrak{P}(g_1)\mathfrak{P}(g_2)\cdots\mathfrak{P}(g_n) = \mathbf{0}.$$

Since $\mathbf{0} \in M$, and M is a prime ideal, we conclude that there exists k for which $\mathfrak{P}(g_k) \in M$.

ORDER STRUCTURE OF HYPER-REAL FIELDS

13.5. LEMMA. *Let* I *be an absolutely convex ideal in a lattice-ordered ring* A. *Every countable family* E *in the partially ordered ring* A/I *has a family* F *of preimages in* A—*that is,* $E = \{I(a): a \in F\}$—*such that, for all* $a, b \in F$,

$$I(a) \leqq I(b) \text{ (in } E) \text{ implies } a \leqq b \text{ (in } A).$$

Furthermore, F *can be chosen so as to include a specified preimage of any one element of* E.

PROOF. Let $(c_n)_{n \in \mathbf{N}}$ be an indexed family of preimages of the elements of E, with c_1 the specified preimage. Consider any $n \in \mathbf{N}$, and suppose that for all $k < n$, a_k has been defined so that $a_k \equiv c_k$ (mod I), and $I(a_k) \leqq I(a_l)$ implies $a_k \leqq a_l$. Let

$$a' = \sup\{a_k : I(a_k) \leqq I(c_n), k < n\},$$

and

$$a'' = \inf\{a_l : I(a_l) \geqq I(c_n), l < n\}.$$

From the induction hypothesis, $a' \leqq a''$. Define

$$a_n = (a' \vee c_n) \wedge a'',$$

with the convention that a' or a'' is simply omitted in case the set defining it is empty. Then, clearly, $a_1 = c_1$, and $I(a_k) \leqq I(c_n) \leqq I(a_l)$ implies $a_k \leqq a' \leqq a_n \leqq a'' \leqq a_l$. Since the mapping $a \to I(a)$ is a lattice homomorphism (Theorem 5.3), we have $I(a_n) = I(c_n)$.

We recall (Theorem 5.5) that any prime ideal P in a ring $C(X)$ is absolutely convex, and so the lemma applies to this case. Moreover, C/P, and hence both E and F, are (totally) ordered.

13.6. *η_1-sets.* Let E be a (totally) ordered set. For brevity, we write $A < x$ to mean that $a < x$ for every $a \in A$; and correspondingly, $A \leqq x$, $A < B$, etc. These conditions are satisfied vacuously for the empty set: $\emptyset < x < \emptyset$ for every $x \in E$.

(a) *Every countably infinite, ordered set E without last element contains a cofinal, increasing sequence.*

Let $(x_n)_{n \in \mathbf{N}}$ be an enumeration of the elements of E. Define

$$S = \{x_n \colon x_k < x_n \quad \text{for all} \quad k < n\}.$$

Clearly, the elements of S form an increasing sequence. To see that S is cofinal, consider any $x \in E$. Let x_n be the element of E of smallest index n such that $x \leqq x_n$; then $x_k < x \leqq x_n$ for all $k < n$. Hence, $x_n \in S$.

By applying this result to the set E with its ordering reversed (or by a similar proof), we find that if E has no first element, then it contains a coinitial, decreasing sequence.

An ordered set E is called an *η_1-set* if for any countable subsets A and B, with $A < B$, there exists $v \in E$ satisfying $A < v < B$.

With A or B, respectively, taken to be the empty set, this definition implies that in an η_1-set, no countable subset is coinitial or cofinal. The definition may also be put into the following convenient form: E has more than one element, and whenever A and B are nonempty, countable subsets, with $A < B$, there exist elements u, v, and w satisfying

$$u < A < v < B < w.$$

Observe that the sets $\mathbf{Q}$ and $\mathbf{R}$ satisfy the analogous condition for finite subsets (see 13B). The existence of η_1-sets, which is not at all obvious, will follow from Theorem 13.8 or Theorem 13.20. It is plain that the cardinal of an η_1-set must be at least $\aleph_1$: no countable subset is cofinal. In the absence of the continuum hypothesis, the following result is stronger.

(b) *The cardinal of any η_1-set E is at least* $\mathfrak{c}$.

We prove that $\mathbf{R}$ can be embedded in E. First of all, it is easy to embed $\mathbf{Q}$: enumerate its elements in a sequence, $(r_n)_{n \in \mathbf{N}}$, and, inductively, define r'_n in E so that $r'_k < r'_n < r'_l$ whenever $r_k < r_n < r_l$. (More generally, see the remark at the end of 13.9.) The η_1-property then ensures that each irrational, being determined by a Dedekind cut in **Q**, can be accommodated as well.

13.7. LEMMA. *Let P be a prime ideal contained in a hyper-real maximal ideal M in $C(X)$. Given countable sets A and B in C/P, with $A < B$, there exists an element v such that*

$$A \leqq v \leqq B.$$

REMARK. Our present application of this lemma is to the maximal ideal M itself: the result for arbitrary $P \subset M$ will not be needed until Chapter 14. The portion of the proof for $P \neq M$ that refers to material beyond Chapter 5 is enclosed in brackets.

PROOF. We assume that A has no greatest element, and B no least, since otherwise there is nothing left to prove. The proof will be set up so that the case $B = \emptyset$ can be handled merely by omitting all statements that refer to B. The case $A = \emptyset$ will then follow from symmetry.

There exist a cofinal increasing sequence (a_n) in A, and a coinitial decreasing sequence (b_n) in B. By Lemma 13.5, there exist preimages f_n of a_n, and g_n of b_n, such that

$$f_k \leqq f_n \leqq g_n \leqq g_k \quad \text{whenever} \quad k < n.$$

We shall define v in terms of the functions f_n alone.

There exists $s \in C(X)$, with $s \geqq \mathbf{1}$, such that $M(s)$ is infinitely large. This function will be employed to obtain a weighted average of the f_k. First we introduce auxiliary functions $\varphi_k \in C(\mathbf{R})$, for each $k \in \mathbf{N}$, as follows:

$$\varphi_k(r) = \begin{cases} r - k + 1 & \text{for} \quad k - 1 \leqq r \leqq k, \\ k + 1 - r & \text{for} \quad k \leqq r \leqq k + 1, \\ 0 & \text{otherwise.} \end{cases}$$

Then $\mathbf{0} \leqq \varphi_k \leqq \mathbf{1}$, for each k. Next, $\varphi_k(r) = 0$ if $r \leqq k - 1$ or if $r \geqq k + 1$. Therefore, if $n - 1 < r < n + 1$ (where $n \in \mathbf{N}$), then $\varphi_k(r) = 0$ unless $k = n - 1$, n, or $n + 1$. Finally,

$$\sum_{k \in \mathbf{N}} \varphi_k(r) = 1 \quad \text{for all} \quad r \geqq 1.$$

We now define the weighted average, h; its image modulo M [modulo P] will be the required element v. Put

$$h(x) = \sum_{k \in \mathbf{N}} \varphi_k(s(x))\, f_k(x) \qquad (x \in X)$$

Then h is defined on all of X. On each open set

$$\{x \in X : n - 1 < s(x) < n + 1\} \qquad (n \in \mathbf{N}),$$

we have

$$h(x) = \sum_{n-1}^{n+1} \varphi_k(s(x))\, f_k(x);$$

it follows that h is continuous on X.

If $r \geqq n$, then $\varphi_k(r) = 0$ for all $k < n$; hence

$$\text{if} \quad s(x) \geqq n, \quad \text{then} \quad h(x) = \sum_{k \geqq n} \varphi_k(s(x))\, f_k(x) \qquad (n \in \mathbf{N}).$$

Since the sequence (f_k) is increasing, this shows that $h(x) \geqq f_n(x)$ on the zero-set

$$Z_n = \{x \in X: s(x) \geqq n\}.$$

Now, $Z_n \in \mathbf{Z}[M]$, by 5.7(a). Thus, $h - f_n$ is nonnegative on a zero-set of M. By 5.4(a), $M(h) \geqq M(f_n)$. Putting $v = M(h)$, then, we have $v \geqq a_n$, for every $n \in \mathbf{N}$, and therefore $v \geqq A$.

[Since $s^*(p) = \infty$, where $M = \mathbf{M}^p$, we have $Z_n \in \mathbf{Z}[\mathbf{O}^p]$. By 7.15(b), $P(h) \geqq P(f_n)$. Putting $v = P(h)$, then, we have $v \geqq A$.]

Finally, we recall that $f_k \leqq g_n$ for all k and n. Since h is an average of the f_k, this yields $h \leqq g_n$. Therefore $v \leqq b_n$, for all n, and so $v \leqq B$.

13.8. An ordered field that is an η_1-set will be called an *η_1-field*. Obviously, an η_1-field is nonarchimedean.

THEOREM. *Every hyper-real residue class field of $C(X)$ is an η_1-field.*

PROOF. The problem not handled in the lemma is that of interpolating an element properly between A and B. If B, say, is empty, then by the lemma, $A < v + 1 < B$. If A has no last element and B no first, then obviously $A < v < B$. Finally, suppose that A has a last element a. The elements $y - a$, for $y \in B$, are all positive. From what we have already proved, there exists v exceeding every element $1/(y - a)$; then $A < a + 1/v < B$.

The cardinal of every residue class field of $C(X)$ is at least $\mathfrak{c}$. There exist hyper-real fields whose cardinal is exactly $\mathfrak{c}$. For example, the rings $C(\mathbf{N})$, $C(\mathbf{Q})$, and $C(\mathbf{R})$ are all of power $\mathfrak{c}$, and so the hyper-real residue class fields of these rings *are η_1-fields of power* $\mathfrak{c}$. Since $\mathfrak{c}$ is the smallest cardinal of an η_1-set, it follows that the existence of an η_1-set of cardinal $\aleph_1$ is equivalent to the continuum hypothesis ($\mathfrak{c} = \aleph_1$).

ISOMORPHISM OF η_1-FIELDS

13.9. We are going to prove that all η_1-fields of power $\aleph_1$ are isomorphic. Because the proof of this result involves some intricate interactions between the algebraic structure and the order structure, we present first an analogous theorem for ordered sets without algebraic structure. Its proof will provide an unobstructed view of the purely order-theoretic aspects of the later one.

Two ordered sets are said to be *similar* if there exists a one-one, order-preserving mapping of one onto the other.

THEOREM. *All η_1-sets of cardinal $\aleph_1$ are similar.*

REMARK. As was pointed out in 13.8, this result is vacuous without the continuum hypothesis.

PROOF. Let S and T be η_1-sets of power $\aleph_1$. We well-order these sets:

$$S = \{s_\alpha\}_{\alpha<\omega_1}, \quad T = \{t_\alpha\}_{\alpha<\omega_1},$$

and define a suitable mapping φ, from S onto T, by transfinite induction.

Consider any ordinal $\alpha < \omega_1$. Our induction assumption is that a subset $\{s'_\xi\}_{\xi<\alpha}$ of S, and a subset $\{t'_\xi\}_{\xi<\alpha}$ of T, have been defined, and that φ has been defined (so far) as a one-one, order-preserving mapping from the countable set

$$S_\alpha = \{s_\xi\}_{\xi<\alpha} \cup \{s'_\xi\}_{\xi<\alpha}$$

to the set

$$T_\alpha = \{t_\xi\}_{\xi<\alpha} \cup \{t'_\xi\}_{\xi<\alpha},$$

such that $\varphi s_\xi = t'_\xi$, and $\varphi s'_\xi = t_\xi$, for each $\xi < \alpha$.

We now define t'_α. In case $s_\alpha \in S_\alpha$, take t'_α to be the element φs_α, already defined. In case $s_\alpha \notin S_\alpha$, we decompose S_α into the complementary subsets A_S and B_S determined by the condition

$$A_S < s_\alpha < B_S.$$

We have $\varphi[A_S] < \varphi[B_S]$ in T, and, clearly, these subsets are countable. Since T is an η_1-set, some element t of T lies between them. We put $t'_\alpha = t$, and $\varphi s_\alpha = t'_\alpha$.

Next, we define s'_α, in a similar way. If

$$t_\alpha \in T_\alpha \cup \{t'_\alpha\},$$

we take for s'_α the element $\varphi^{\leftarrow}(t_\alpha)$, already defined. If

$$t_\alpha \notin T_\alpha \cup \{t'_\alpha\},$$

we decompose this set into the complementary subsets A_T and B_T determined by the condition

$$A_T < t_\alpha < B_T.$$

Since S is an η_1-set, some element s of S lies between $\varphi^{\leftarrow}[A_T]$ and $\varphi^{\leftarrow}[B_T]$. We put $s'_\alpha = s$, and $\varphi s'_\alpha = t_\alpha$.

It is clear that the mapping φ, defined inductively in this way, is one-one and order-preserving from S onto T.

REMARK. In the foregoing proof, if we had defined only φs_α, inductively, without supplying a value for $\varphi^{\leftarrow}(t_\alpha)$, we could conclude only that φ maps S *into* T. To arrive at this conclusion, however, not all the hypotheses are needed. In fact, it is easy to see that such a procedure leads to the following result: *An η_1-set (of any cardinal) contains a copy of each ordered set of cardinal* $\leqq \aleph_1$.

13.10. Let F be a nonempty, ordered set having no pair of consecutive elements. A subset E is *dense* in F if between any two elements of F, there lies an element of E. (Because F has no consecutive elements, this is the same as saying that E is dense as a subspace of F in the interval topology (3O).). We remark that an ordered *field* never contains consecutive elements: $(a + b)/2$, for example, lies between a and b $(\neq a)$.

(a) *If S is dense in an η_1-set F, then whenever A and B are countable subsets of F, with $A < B$, there exists $v \in S$ satisfying $A < v < B$.*

For, there exist $v_1 \in F$ such that $A < v_1 < B$, and, in turn, $v_2 \in F$ such that $v_1 < v_2 < B$. Since S is dense, there exists $v \in S$ satisfying $v_1 < v < v_2$.

(b) *If a subfield E of an ordered field F contains an interval $a \leqq x < b$ of F, then $E = F$.*

Subtraction of a shows that the field E contains every positive element less than $b - a$, and taking reciprocals yields the set of all elements greater than $1/(b - a)$. Every element of F is a difference of two elements in the latter set.

13.11. LEMMA. *Every uncountable ordered field F has a dense transcendence base over* $\mathbf{Q}$.

PROOF. Since F is infinite, it is equipotent with the set of all intervals of the form $a < x < b$, where $a, b \in F$. Let $\prec$ be a well-ordering, according to the smallest ordinal of cardinal $|F|$, of the set of all these intervals. Inductively, let J be any such interval, and suppose that a family

$$S_J = (s_I)_{I \prec J}$$

of independent transcendentals has been defined, with $s_I \in I$ for each $I \prec J$. Let F_J denote the field consisting of all elements of F that are algebraic over $\mathbf{Q}(S_J)$. Since $|S_J| < |F|$ and F is uncountable, we have $|F_J| < |F|$, so that F_J is a proper subfield of F. Therefore F_J covers no interval of F, and we may choose an element

$$s_J \in J - F_J;$$

it will automatically be independent of S_J. In this way, we construct a dense set of independent transcendentals; we then extend it (if necessary) to a transcendence base.

13.12. LEMMA. *If $E(x)$ is an ordered field, where E is real-closed and x is transcendental over E, then the ordering in $E(x)$ is determined by the ordering in $E \cup \{x\}$.*

PROOF. We may confine our attention to the polynomial ring $E[x]$, as its ordering determines that of its field of quotients $E(x)$: $\mathfrak{p}(x)/\mathfrak{q}(x) > 0$ if and only if $\mathfrak{p}(x)\mathfrak{q}(x) > 0$. It suffices to show that if $\mathfrak{p}$ is any nonconstant polynomial in $E[x]$ with leading coefficient unity, then the condition $\mathfrak{p} > 0$ depends only upon the location of x relative to E. Since E is real-closed, $\mathfrak{p}$ is expressible as a product of linear and quadratic factors; and $\mathfrak{p} > 0$ if and only if the number of negative factors is even. So it is enough to handle the cases in which $\mathfrak{p}$ is of degree 1 or 2.

Case 1. $\mathfrak{p} = x - c$ (where $c \in E$). Then $\mathfrak{p} > 0$ if and only if $x > c$.

Case 2. $\mathfrak{p} = (x - h)^2 + k$ (where $h, k \in E$). If $k \geqq 0$, then $\mathfrak{p} > 0$, independently of x. If $k < 0$, then $-k$ has a positive square root a in the real-closed field E. Then $\mathfrak{p} > 0$ if and only if $|x - h| > a$, hence if and only if either $x > h + a$ or $x < h - a$.

13.13. THEOREM. *All real-closed η_1-fields of cardinal $\aleph_1$ are isomorphic.*

REMARK. This result has far-reaching consequences if the continuum hypothesis ($\mathfrak{c} = \aleph_1$) is assumed, but is vacuous otherwise.

PROOF. Let F and G be real-closed η_1-fields of power $\aleph_1$. If E is any subfield of F, then, by the corollary to the Artin-Schreier theorem, F contains a unique copy of $\mathscr{R}E$ containing E—the unique real-closed, algebraic extension of E in F. Likewise, any subfield of G has a unique extension to its real-closure in G.

By Lemma 13.11, F has a dense transcendence base S over $\mathbf{Q}$, and G has a dense transcendence base T over $\mathbf{Q}$. Their cardinal is $\aleph_1$. We begin by well-ordering them:

$$S = \{s_\alpha\}_{\alpha<\omega_1}, \quad T = \{t_\alpha\}_{\alpha<\omega_1}.$$

An isomorphism $\mathfrak{u}$ of F onto G will be defined, inductively, by mapping the base S onto the base T, and extending the mapping algebraically. We shall denote by t'_α the element $\mathfrak{u}s_\alpha$ of T, and by s'_α the preimage $\mathfrak{u}^{\leftarrow}(t_\alpha)$ in S. For each $\alpha \leqq \omega_1$, we shall let

$$S_\alpha = \{s_\xi\}_{\xi<\alpha} \cup \{s'_\xi\}_{\xi<\alpha}$$

and

$$T_\alpha = \{t_\xi\}_{\xi<\alpha} \cup \{t'_\xi\}_{\xi<\alpha}.$$

Identifying the copies of $\mathbf{Q}$ in both F and G with $\mathbf{Q}$ itself, we note that $\mathbf{Q}(S_\alpha)$ and $\mathbf{Q}(T_\alpha)$ are the smallest subfields of F and G containing S_α and T_α, respectively. Their real-closures are also present; we set

$$F_\alpha = \mathscr{R}(\mathbf{Q}(S_\alpha)), \quad \text{and} \quad G_\alpha = \mathscr{R}(\mathbf{Q}(T_\alpha)).$$

Consider, now, any ordinal $\alpha \leqq \omega_1$. Our induction assumption is that for each $\delta < \alpha$, $\mathfrak{u}$ has been defined (so far) as an isomorphism from F_δ onto G_δ, such that

$$\mathfrak{u}s_\xi = t'_\xi, \quad \text{and} \quad \mathfrak{u}s'_\xi = t_\xi, \quad \text{for every} \quad \xi < \delta.$$

As an isomorphism of a real-closed field, $\mathfrak{u}$ will necessarily preserve order. In extending $\mathfrak{u}$ to an isomorphism from F_α onto G_α, we distinguish three cases.

Case I: $\alpha = 0$. Noting that $S_0 = T_0 = \emptyset$, we stipulate that $\mathfrak{u}$ be the identity from $\mathbf{Q} = \mathbf{Q}(S_0)$ (in F) to $\mathbf{Q} = \mathbf{Q}(T_0)$ (in G). By the Artin-Schreier theorem, this mapping has an extension to an isomorphism between their real-closures, i.e., from F_0 onto G_0.

Case II: $\alpha = \lambda + 1$. By our induction assumption, $\mathfrak{u}$ maps F_λ isomorphically onto G_λ. First we shall define t'_λ; then, later on, s'_λ.

In case $s_\lambda \in S_\lambda$, we take t'_λ to be the element $\mathfrak{u}s_\lambda$, already defined.

Assume, now, that $s_\lambda \notin S_\lambda$. Since F_λ is an *algebraic* extension of $\mathbf{Q}(S_\lambda)$, it does not contain the transcendental s_λ. In addition, it is countable, since S_λ is countable. Decompose F_λ into the complementary subsets A_F and B_F, determined by:

$$A_F < s_\lambda < B_F.$$

Since $\mathfrak{u}$ preserves order, $\mathfrak{u}[A_F] < \mathfrak{u}[B_F]$; and each of these subsets of the η_1-set G is countable. By 13.10(a), there exists an element t of the dense set T such that

$$\mathfrak{u}[A_F] < t < \mathfrak{u}[B_F].$$

We put $t'_\lambda = t$.

Now define $\mathfrak{u}s_\lambda = t'_\lambda$. Next, extend $\mathfrak{u}$ to the field extension $F_\lambda(s_\lambda)$ so as to preserve sums and products; since s_λ and t'_λ are transcendentals, this mapping will be an isomorphism. By Lemma 13.12, it preserves order. By the Artin-Schreier theorem, then, it can be extended to an isomorphism from the real-closure $\mathscr{R}(F_\lambda(s_\lambda))$ of $F_\lambda(s_\lambda)$ into G. It follows from the corollary to that theorem that

$$\mathscr{R}(F_\lambda(s_\lambda)) = \mathscr{R}(\mathbf{Q}(S_\lambda, s_\lambda)).$$

We are now ready to define s'_λ. If

$$t_\lambda \in T_\lambda \cup \{t'_\lambda\},$$

we take for s'_λ the element $\mathfrak{u}^{\leftarrow}(t_\lambda)$, already defined. Assume, now, that

$$t_\lambda \notin T_\lambda \cup \{t'_\lambda\}.$$

Since $\mathfrak{u}$ maps S_λ onto T_λ, and s_λ to t'_λ, the transcendental t_λ does not belong to

$$\mathfrak{u}\,[\mathscr{R}(F_\lambda(s_\lambda))].$$

We decompose this last into the complementary subsets A_G and B_G, determined by:

$$A_G < t_\lambda < B_G.$$

Arguing as above, we find an element s of S satisfying

$$\mathfrak{u}^{\leftarrow}[A_G] < s < \mathfrak{u}^{\leftarrow}[B_G],$$

and define $s'_\lambda = s$. Continuing as before, we extend $\mathfrak{u}$ to an isomorphism from F_α onto G_α.

Case III: *α is a limit ordinal.* In this case, we have, by definition,

$$S_\alpha = \bigcup_{\delta<\alpha} S_\delta, \quad \text{and} \quad T_\alpha = \bigcup_{\delta<\alpha} T_\delta.$$

Now, a union of a chain of real-closed fields is real-closed, as a straightforward check shows. Therefore $\bigcup_{\delta<\alpha} F_\delta$ is real-closed. Clearly, it is an algebraic extension of $\bigcup_{\delta<\alpha} \mathbf{Q}(S_\delta)$, i.e., of $\mathbf{Q}(S_\alpha)$. Hence we have

$$F_\alpha = \bigcup_{\delta<\alpha} F_\delta; \quad \text{and, similarly,} \quad G_\alpha = \bigcup_{\delta<\alpha} G_\delta.$$

It follows that $\mathfrak{u}$, as already defined, is an isomorphism of F_α onto G_α.

This completes the induction. Since $F_{\omega_1} = F$, and $G_{\omega_1} = G$, it also completes the proof of the theorem.

13.14. Every hyper-real residue class field of $C(X)$ is a real-closed η_1-field (Theorems 13.4 and 13.8). Applying the preceding theorem, we have:

THEOREM. *Under the continuum hypothesis* ($\mathfrak{c} = \aleph_1$), *all hyper-real fields of cardinal* $\mathfrak{c}$ *are isomorphic.*

This result applies to any hyper-real field $C(X)/M$ of power $\mathfrak{c}$, for arbitrary X. An important special case is that in which $C(X)$ itself is of power $\mathfrak{c}$. This case includes all X having a countable, dense subset—for example, all subspaces of euclidean spaces, and hence, in particular, **R**, **Q**, and **N**. Incidentally, subspaces of euclidean spaces are all realcompact (Theorem 8.2), so that, first of all, topologically distinct spaces among them have algebraically distinct function rings (Theorem 8.3), and, secondly, the fields in question are the residue class fields modulo arbitrary free maximal ideals.

It can also happen, for suitable X, that $|C(X)/M| = \mathfrak{c}$ for some, but not all, hyper-real M (12C.*2*).

13.15. The proof of Theorem 13.13 can be modified, like that of Theorem 13.9 on similarity of η_1-sets, to show that a real-closed η_1-field contains a copy of any real-closed field of cardinal $\leqq \aleph_1$. Since every ordered field E can be embedded in its real-closure, and $|E| = |\mathscr{R}E|$, this yields: *A real-closed η_1-field (of any cardinal) contains a copy of each ordered field of cardinal $\leqq \aleph_1$.* Moreover, it contains a copy of **R** as well; see 13M.

DEDEKIND COMPLETION OF η_1-SETS

13.16. The remainder of this chapter is purely set-theoretic. For application in Chapter 14, we wish to obtain more information about the structure of η_1-sets, and to ascertain the minimum cardinal of a Dedekind-complete set containing an η_1-set. Our discussion will illustrate a standard procedure for constructing η_1-sets and their generalizations to higher cardinals. Additional details are outlined in the problems at the end of the chapter.

Let $\mathfrak{s}$ denote the set of all $\{0, 1\}$-valued, transfinite sequences

$$x = (x_\xi)_{\xi < \omega_1}.$$

Obviously, $|\mathfrak{s}| = 2^{\aleph_1}$. If x and y are distinct members of $\mathfrak{s}$, then there exists an ordinal τ such that $x_\xi = y_\xi$ for all $\xi < \tau$, while $x_\tau \neq y_\tau$. Hence, the lexicographic order may be imposed on $\mathfrak{s}$; specifically, $x < y$ if $x_\tau = 0$ (and $y_\tau = 1$).

We point out that $\mathfrak{s}$ contains pairs of consecutive elements, and, in fact, we can easily characterize these pairs. Suppose that x has an immediate successor x^+. Let τ denote the first index at which $x_\tau \neq x^+{}_\tau$; then $x_\tau = 0$ and $x^+{}_\tau = 1$. Now, $x_\sigma = 1$ for every $\sigma > \tau$: for if there exists $\sigma > \tau$ such that $x_\sigma = 0$, then the element u, defined by

$$u_\xi = x_\xi \quad \text{for} \quad \xi < \sigma, \quad \text{and} \quad u_\xi = 1 \quad \text{for} \quad \xi \geqq \sigma,$$

satisfies $x < u < x^+$. Similarly, $x^+{}_\sigma = 0$ for all $\sigma > \tau$. Thus we have:

$$\begin{aligned} x_\xi &= x^+{}_\xi \quad \text{for} \quad \xi < \tau, \\ x_\tau &= 0 \quad \text{and} \quad x^+{}_\tau = 1, \quad \text{and} \\ x_\xi &= 1 \quad \text{and} \quad x^+{}_\xi = 0 \quad \text{for} \quad \xi > \tau. \end{aligned}$$

Conversely, if x and x^+ are as just given, then it is obvious that no element of S lies between them.

When x has an immediate successor x^+, we shall refer to x as a *lower* element of S, and to x^+ as an *upper* element. It is evident from the above description that no upper element is a lower element.

13.17. We prove next that S is Dedekind-complete. Since S has both a first element and a last element (the constant sequences **0** and **1**), it is, then, lattice-complete; indeed, this will follow from the proof.

LEMMA. *Every set in* S *has a supremum and an infimum.*

PROOF. Given a set A, we construct its supremum $s = (s_\xi)$ inductively, as follows. Given $\sigma < \omega_1$, assume that s_ξ has been defined for every $\xi < \sigma$. If there exists $a \in A$ such that $a_\xi = s_\xi$ for all $\xi < \sigma$, and if, in addition, $a_\sigma = 1$, then we define $s_\sigma = 1$; in the contrary case, we put $s_\sigma = 0$. It is a straightforward matter to verify that $s = \sup A$.

13.18.

The set of all upper elements of S *is denoted by* Q.

Thus, Q is the set of all $x \in$ S such that

$$\{\xi \colon x_\xi = 1\}$$

has a largest member.

For each α, we let Q_α denote the set of all elements of Q for which this largest member is less than α. Obviously, $\sigma < \alpha$ implies $Q_\sigma \subset Q_\alpha$.

By definition, $x \in Q_\alpha$ if and only if there exists $\tau < \alpha$ such that $x_\tau = 1$, while $x_\xi = 0$ for all $\xi > \tau$. In this case, $x \in Q_{\tau+1}$. Now, if α is a limit ordinal, then $\tau + 1 < \alpha$. Hence,

$$Q_\alpha = \bigcup\nolimits_{\sigma<\alpha} Q_\sigma \quad \text{for } \alpha \text{ a limit ordinal.}$$

In particular,

$$Q = Q_{\omega_1} = \bigcup\nolimits_{\alpha<\omega_1} Q_\alpha.$$

If $x \in Q_\alpha$, then, of course, $x_\xi = 0$ for all $\xi \geqq \alpha$. The converse is false, however, whenever $\alpha \geqq \omega$, even for a nonconstant sequence.

For, let σ be any limit ordinal $\leqq \alpha$. Define x so that $x_\xi = 1$ for arbitrarily large $\xi < \sigma$, while $x_\xi = 0$ for all $\xi \geqq \sigma$ (and hence for all $\xi \geqq \alpha$). Then $x \notin \mathrm{Q}$.

Every set in Q_α has a supremum and an infimum in S, though not necessarily in Q. When the extremum does not belong to the set itself, it takes on a special form.

(a) *Let $x = \sup A$, where $A \subset \mathrm{Q}_\alpha$. If $x \notin A$, then $x_\xi = 0$ for all $\xi \geqq \alpha$, but $x \notin \mathrm{Q}$.*

For, if $x_\sigma = 1$ for some $\sigma \geqq \alpha$, then the element u, given by $u_\xi = x_\xi$ for $\xi \neq \sigma$, and $u_\sigma = 0$, satisfies $A < u < x$. And if $x \in \mathrm{Q}$, then its immediate predecessor in S is also an upper bound of A.

(b) *Let $y = \inf B$, where $B \subset \mathrm{Q}_\alpha$. If $y \notin B$, then $y_\xi = 1$ for all $\xi \geqq \alpha$, but y is not a lower element of* S.

The proof is similar to that of (a).

13.19. LEMMA. *Let $x, y \in \mathrm{S}$, with $x < y$.*

(a) *If $y \neq x^+$, then there exists $u \in \mathrm{Q}$ such that $x < u < y$.*

(b) *If $x, y \in \mathrm{Q}_{\sigma+1}$, then there exists $u \in \mathrm{Q}_\sigma$ such that $x \leqq u \leqq y$.*

PROOF. Let τ denote the smallest index for which $x_\tau < y_\tau$; then $x_\tau = 0$ and $y_\tau = 1$. Define y' as follows:

$$y'_\xi = y_\xi \quad \text{for} \quad \xi \leqq \tau, \quad \text{and} \quad y'_\xi = 0 \quad \text{for} \quad \xi > \tau.$$

Then $y' \in \mathrm{Q}_{\tau+1}$, and $x < y' \leqq y$.

(a) If $y \notin \mathrm{Q}$, we define $u = y'$. If $y \in \mathrm{Q}$, then y is an upper element z^+, and the hypothesis in (a) implies that $z > x$; we then take $u = z'$, defined analogously.

(b) Since $y_\tau = 1$, while $y \in \mathrm{Q}_{\sigma+1}$, we must have $\tau \leqq \sigma$. If $\tau = \sigma$, then $x \in \mathrm{Q}_\sigma$, and we put $u = x$. If $\tau < \sigma$, then $\tau + 1 \leqq \sigma$, so that $y' \in \mathrm{Q}_{\tau+1} \subset \mathrm{Q}_\sigma$; we take $u = y'$.

13.20. THEOREM. *Q is an η_1-set of cardinal $\mathfrak{c}$.*

PROOF. Let A and B be countable sets in Q, with $A < B$. Each member x of the countable set $A \cup B$ belongs to Q_{α_x}, for some $\alpha_x < \omega_1$. Define $\alpha = \sup_x \alpha_x$. Then $\alpha < \omega_1$ (5.12(a)), and $A \cup B \subset \mathrm{Q}_\alpha$.

Define $x = \sup A$ and $y = \inf B$. Then $x \leqq y$. Since $A \cup B \subset \mathrm{Q}_\alpha$, 13.18(a,b) imply that if $x = y$, then $x \in A \cap B = \emptyset$. Thus, $x \neq y$, and therefore $x < y$.

Now, if $x \in A$ ($\subset \mathrm{Q}_\alpha$), then x is not a lower element; and if $x \notin A$, then again, by 13.18(a), x is not a lower element. So $y \neq x^+$.

By Lemma 13.19(a), there exists $u \in \mathrm{Q}$ such that $x < u < y$. This proves that Q is an η_1-set.

It is evident that $|\mathrm{Q}_\sigma| \leqq 2^{|\sigma|}$, so that $|\mathrm{Q}_\sigma| \leqq \mathfrak{c}$ for $\sigma < \omega_1$. On the other hand, it is clear that $|\mathrm{Q}_{\omega+1}| = \mathfrak{c}$. Since

$$\mathrm{Q}_{\omega+1} \subset \mathrm{Q} = \bigcup_{\sigma<\omega_1} \mathrm{Q}_\sigma,$$

we have

$$\mathfrak{c} \leqq |\mathrm{Q}| \leqq \sum_{\sigma<\omega_1} |\mathrm{Q}_\sigma| \leqq \aleph_1 \cdot \mathfrak{c} = \mathfrak{c},$$

i.e., $|\mathrm{Q}| = \mathfrak{c}$. (We also know that $|\mathrm{Q}| \geqq \mathfrak{c}$ from 13.6(b).)

13.21. Lemma. *For $\alpha < \omega_1$, every set in Q_α has a countable cofinal and coinitial subset.*

Proof. Given $A \subset \mathrm{Q}_\alpha$, let $x = \sup A$ (in S). If $x \in A$, then $\{x\}$ is cofinal. Assume, now, that $x \notin A$. By 13.18(a), there exists an ordinal $\leqq \alpha$—and hence a smallest such ordinal τ—such that $x_\xi = 0$ for all $\xi \geqq \tau$; also, $x \notin \mathrm{Q}$, so that τ must be a limit ordinal.

By definition of τ, there exist arbitrarily large indices $\sigma < \tau$ for which $x_\sigma = 1$. To each such σ, associate an element s, as follows:

$$s_\xi = x_\xi \quad \text{for} \quad \xi \leqq \sigma, \quad \text{and} \quad s_\xi = 0 \quad \text{for} \quad \xi > \sigma.$$

Since τ is countable, there are only countably many such elements s; and, clearly, their supremum is x. Hence, if for each such s, we pick an element a of A for which $a \geqq s$, then the set of all such elements a will be a countable, cofinal subset of A.

The construction of a countable, coinitial subset is similar.

13.22. We prove next that Q is a *minimal* η_1-set.

Theorem. *Every η_1-set contains a copy of* Q.

Proof. Let E be a given η_1-set. Inductively, we shall define a mapping φ that embeds Q_α in E, for each $\alpha \leqq \omega_1$. Consider any such ordinal α. Our induction assumption is that for every $\sigma < \alpha$, φ has been defined (so far) as a one-one, order-preserving mapping of Q_σ into E. We now extend φ to Q_α.

Assume, first, that $\alpha = \sigma + 1$. Consider any element x of $\mathrm{Q}_\alpha - \mathrm{Q}_\sigma$. Let A denote the set of all predecessors of x in Q_σ, and B the set of all its successors. By Lemma 13.19(b), x is the unique element of Q_α for which $A < x < B$. By Lemma 13.21, A has a countable, cofinal subset A', and B has a countable, coinitial subset B'. Hence $\varphi[A']$ and $\varphi[B']$ are countable subsets of the η_1-set E, with $\varphi[A'] < \varphi[B']$; consequently, some element y of E lies between them, and hence also lies between $\varphi[A]$ and $\varphi[B]$. We define $\varphi x = y$.

Assume, finally, that α is a limit ordinal or 0. Then $\mathrm{Q}_\alpha = \bigcup_{\sigma<\alpha} \mathrm{Q}_\sigma$. (Note that $\mathrm{Q}_0 = \emptyset$.) Hence φ, as already defined, embeds Q_α in E. This completes the induction. Since $Q_{\omega_1} = Q$, it also completes the proof of the theorem.

13.23. Let R denote the subset of S obtained by deleting all lower elements, as well as the end elements **0** and **1**.

THEOREM. *R is the Dedekind completion of* Q, *and* $|R| = 2^{\aleph_1}$.

PROOF. Obviously, $R \supset Q$. To show that R is Dedekind-complete, consider any bounded subset A, and let $s = \sup A$ in S. Then, either $s \in R$, or s^+ is the supremum of A in R. Next, since R contains no lower elements, it follows from Lemma 13.19(a) that between any two elements of R, there lies an element of Q. Clearly, then, no proper subset of R containing Q is Dedekind-complete. Therefore R is the Dedekind completion of Q.

We know that $|S| = 2^{\aleph_1}$. Since the set $S - R - \{\mathbf{0}, \mathbf{1}\}$ of all lower elements is in one-one correspondence with the set Q of all upper elements, $|S - R| = |Q| \leqq |R|$. With the equation

$$S = (S - R) \cup R,$$

this yields

$$|S| \leqq |R| + |R| = |R|.$$

Therefore, $|R| = |S| = 2^{\aleph_1}$.

13.24. COROLLARY. *A Dedekind-complete set containing an* η_1*-set contains a copy of* R, *and hence its cardinal is at least* $2^{\aleph_1}$.

PROOF. Since every η_1-set contains a copy of Q (Theorem 13.22), the given complete set must contain a copy of R.

PROBLEMS

13A. CONTINUITY OF ZEROS OF POLYNOMIALS.

1. Let f be a real-valued function on $\mathbf{R}^3$. If, for each $a = (a_1, a_2, a_3)$, $f(a)$ is a zero of the polynomial

$$\lambda^3 + a_1\lambda^2 + a_2\lambda + a_3,$$

then f is not continuous. [Vary a_3 only.]

2. Define $\rho_1, \cdots, \rho_n$ as in 13.3, but where now $a_1, \cdots, a_n$ are complex. The functions ρ_k are still continuous.

13B. η_0-SETS.

A nonempty ordered set without first or last element, and containing no pair of consecutive elements, is called an *η_0-set.*

1. All countable η_0-sets are similar to **Q**.

2. Every countable ordered set is embeddable in **Q**.
3. All ordered fields are η_0-sets.
4. A countable ordered field need not be isomorphic with **Q**.

13C. COUNTABLE SUBFIELDS OF **R**.

Let $s, t \in \mathbf{R}$ be transcendentals over **Q**. Then $\mathbf{Q}(s)$ and $\mathbf{Q}(t)$ are isomorphic.

1. The ordered fields **Q**, $\mathbf{Q}(s)$, and $\mathbf{Q}(t)$, and their real-closures, are all similar. [13B.]

2. There exists no isomorphism from $\mathbf{Q}(s)$ into $\mathscr{R}\mathbf{Q}$.

3. The only isomorphism from $\mathscr{R}(\mathbf{Q}(s))$ into **R** is the identity. [An isomorphism of a real-closed field preserves order.]

4. If s and t are independent transcendentals, there exists no isomorphism from $\mathscr{R}(\mathbf{Q}(s))$ into $\mathscr{R}(\mathbf{Q}(t))$. Thus, the analogue of Theorem 13.13 for η_0-sets does not hold, even when the fields have the same transcendence degree over **Q**.

5. Let s and t be independent transcendentals. Although $\mathbf{Q}(s)$ and $\mathbf{Q}(t)$ are both similar and isomorphic, there exists no order-preserving isomorphism from one onto the other.

13D. ORDERS ON FIELD EXTENSIONS.

Let F be an ordered field.

1. If $F(\xi)$ is an ordered extension of F, with $\xi > F$, then

$$\xi^n > \sum_{k=0}^{n-1} c_k \xi^k$$

for all $c_k \in F$ and $n \in \mathbf{N}$.

2. If $F(\alpha)$ is an algebraic extension of F, then it is not possible to order $F(\alpha)$ so that $\alpha > F$.

3. If $F(\tau)$ is a transcendental extension of F, then the condition $\tau > F$ determines an ordering of $F(\tau)$; in this ordering, if $c_n \neq 0$, then

$$c_n\tau^n + \cdots + c_0 > 0$$

if and only if $c_n > 0$. [$F(\tau)$ is the set of all rational functions in τ.]

13E. SIMPLE EXTENSIONS OF **Q**.

1. Let $\mathbf{Q}(\tau)$ be an ordered extension of **Q**, with τ infinitely large. The element τ is necessarily transcendental. If $\xi \in \mathbf{Q}(\tau)$, and $|\xi|$ is not infinitely large, then there exists a unique number r in **Q** such that $|\xi - r|$ is infinitely small or zero. [13D.]

2. An order in a transcendental extension $\mathbf{Q}(\sigma)$ of **Q** is determined by the conditions: $\sigma > 0$, and $\sigma^2 - 2$ is infinitely small. There exists *no* $r \in \mathbf{Q}$ for which $|\sigma - r|$ is infinitely small or zero. [Every polynomial $\mathfrak{p}(\sigma)$ is of the form

$$\mathfrak{p}(\sigma) = (\sigma^2 - 2)^k \big((\sigma^2 - 2)\, \mathfrak{q}(\sigma) + a\sigma + b\big),$$

where a and b are not both 0.]

13F. TRANSCENDENTAL EXTENSIONS.

Let $F(\tau)$ be an ordered transcendental extension of F, with $\tau > F$ (13D.*3*).

1. τ has no square root in $F(\tau)$.

2. $F(\tau)$ is not dense in $\mathscr{R}(F(\tau))$. [Consider the elements greater than F but less than $\tau^{1/2}$.]

3. A transcendental extension $F(\tau)(x)$ can be ordered so that $F < x^n < \tau$ for all $n \in \mathbf{N}$.

4. $F(\tau)(x)$ can also be ordered so that $F < x < c\tau < x^2$, for all $c > 0$ in F. [Let

$$\sum_{k,m} c_{k,m} x^k \tau^m$$

be positive if $c_{l,n} > 0$, where n is the largest exponent of τ occurring among the terms of highest degree.]

5. The order on $F(\tau) \cup \{x\}$ is the same in *3* and *4*. Contrast with Lemma 13.12.

13G. ISOMORPHISM OF REAL-CLOSED FIELDS.

Let F be an ordered field.

1. The only automorphism of $\mathscr{R}F$ that leaves F pointwise fixed is the identity. [An automorphism preserves the order in $\mathscr{R}F$ and permutes the zeros of each polynomial over F.]

2. Let $F(\tau)$ be an ordered extension of F, with $\tau > F$. The isomorphism $\mathfrak{u}$ determined by the conditions: $\mathfrak{u}c = c$ for all $c \in F$, and $\mathfrak{u}\tau = \tau^2$, is order-preserving. [13D.*3*.]

3. The mapping $\mathfrak{u}$ has an extension to an isomorphism from $\mathscr{R}(F(\tau))$ *onto* itself. Contrast with *1*.

4. Let $G = F(\tau_1, \tau_2, \cdots)$ be an ordered extension of F, with

$$\tau_n > F(\tau_1, \cdots, \tau_{n-1})$$

for all $n \in \mathbf{N}$. Then $\mathscr{R}G$ is isomorphic with a proper subfield of itself.

13H. LEMMA 13.7.

1. Reconcile the following facts:

(i) The construction of h in the proof of the lemma depends upon the f_n alone, not upon the g_n.

(ii) C/M is an η_1-set (Theorem 13.8), so that $M(h) \neq \sup_n M(f_n)$.

2. Verify that for all $n \in \mathbf{N}$,

$$h(x) = f_n(x) + \big(s(x) - n\big)\big(f_{n+1}(x) - f_n(x)\big)$$

whenever $n \leqq s(x) \leqq n + 1$, and then establish continuity of h by means of 1A.*3*.

13I. COFINAL WELL-ORDERED SUBSETS.

1. Every ordered set of cardinal $\aleph_\alpha$ has a cofinal, well-ordered subset of type $\leqq \omega_\alpha$. [Argue as in 13.6(a).]

2. If an ordered set has a cofinal, well-ordered subset of type ω_1, then it has no countable cofinal subset. [5.12(a).]

3. Let E be an ordered set. Every set in E has a countable cofinal subset if and only if every well-ordered set in E is countable.

13J. COFINALITY IN SUBSETS OF S.

The following information will lead to an alternative proof that Q is an η_1-set.

1. Let $x \in$ S, and suppose that (x_ξ) has a cofinal (transfinite) subsequence of 1's [resp. 0's]. Exhibit an [inversely] well-ordered set of type ω_1, contained in Q, having x as supremum [infimum].

2. Suppose that (x_ξ) has a tail of 0's [resp. 1's]. Exhibit a countable, cofinal [coinitial] subset of the set of all predecessors [successors] of x.

3. Every set in S has a cofinal and coinitial subset of power $\leqq \aleph_1$. [Apply *1* and *2*. Alternatively, reason as in 13.21.]

4. If an element of R is the supremum of a countable subset of Q, then it is not the infimum of a countable subset of Q, unless it belongs to both sets. [*1* and 13I.*2*.]

5. Use *4* and *1* to prove that Q is an η_1-set.

13K. SIMILARITY OF η_1-SETS.

1. $W(\omega_2) \times$ Q, in the lexicographic order, is an η_1-set.

2. $W(\omega_2) \times$ Q is not similar to any subset of Q. [9K.*1* and 13J.*3*.]

3. All η_1-sets of power $\mathfrak{c}$ are similar if and only if the continuum hypothesis is true. [Consider the cardinal of $W(\omega_2) \times$ Q.]

13L. SUBSETS OF η_1-SETS.

Let H be an ordered set.

1. The following are equivalent.

(1) H is embeddable in every η_1-set.

(2) H is embeddable in Q (whence $|H| \leqq \mathfrak{c}$).

(3) H is a union of $\aleph_1$ sets, each of which satisfies:

(a) *Every subset has a countable cofinal and coinitial subset.*

(4) H is a union of an increasing family of sets, each of which satisfies (a). [(4) *implies* (1). Well-order H so that the set of all predecessors of any given element satisfies (a). Make use of (a) as in 13.22.]

2. If every set in H of power $< |H|$ satisfies (a), then H satisfies the conditions in *1*. The converse holds only under the continuum hypothesis.

3. R contains a copy of S. [The set of all triadic ω_1-sequences with final nonzero term can be embedded in Q.]

13M. EMBEDDING OF **R**.

Every real-closed η_1-field contains $2^{\mathfrak{c}}$ copies of the field **R**. [Combine the idea in 13L with the proof described in 13.15. At each stage in the induction, there are many possible choices for the image of the transcendental.]

13N. DEDEKIND-COMPLETE SETS.

1. No η_1-set is Dedekind-complete.

2. Q_α is Dedekind-complete if and only if α is not a limit ordinal.

13O. η_α-SETS.

An ordered set E is called an η_α*-set* if for any subsets A and B of power $< \aleph_\alpha$, with $A < B$, there exists $v \in E$ satisfying $A < v < B$. For $\alpha = 0$ or 1, this definition agrees with those given previously. (When α is not a limit ordinal, a construction analogous to that of Q yields an η_α-set.)

An ordered field that is an η_α-set is called an η_α-field. (We do not know any examples for $\alpha > 1$.)

The proofs of the following results (among others) are the same as for the case $\alpha = 1$, except for notation.

1. All η_α-sets of cardinal $\aleph_\alpha$ are similar.

2. An η_α-set contains a copy of every ordered set of cardinal $\leqq \aleph_\alpha$.

3. For $\alpha > 0$, all real-closed η_α-fields of cardinal $\aleph_\alpha$ are isomorphic. The condition $\alpha > 0$ is essential, as is shown by 13C.*4*.

4. For $\alpha > 0$, any real-closed η_α-field contains a copy of every ordered field of cardinal $\leqq \aleph_\alpha$. Again, the condition $\alpha > 0$ is essential.

13P. ORDERED P-SPACES.

1. Every η_1-set, under the interval topology (3O), is a P-space (4J) without isolated points. [5O.*1*.]

2. If E is any η_1-set, then $\mathbf{N} \times E$, ordered lexicographically, is a P-space without isolated points (compare 4K.*6*), but is not an η_1-set.

Chapter 14

PRIME IDEALS

14.1. We saw, in Chapter 7, that every prime ideal P in a ring $C(X)$ lies between O^p and M^p for some unique p in βX. We also know that P is absolutely convex (Theorem 5.5), and hence that the canonical homomorphism $f \to P(f)$ of C onto C/P is a lattice homomorphism as well. Moreover, the integral domain C/P is totally ordered. The set of images of the constant functions is a copy of $\mathbf{R}$, and we identify this copy with $\mathbf{R}$ itself.

The present chapter is devoted largely to a study of the order structure of C/P, and of the distribution of prime ideals in C. The ring C/P turns out to have properties akin to those of η_1-sets; and, in some cases, it *is* an η_1-set.

As in any residue class ring, the correspondence $I \to I/P$ is one-one from the ideals I in C that contain P onto the ideals in C/P. Furthermore, I/P is prime if and only if I is prime. Thus, to investigate the distribution of prime ideals in C containing P, we may examine the prime ideals in C/P. The advantage in doing this stems from the presence of a total order on C/P.

We shall find that the prime ideals containing P form a chain. If P is not maximal, then the chain is a Dedekind-complete set containing an η_1-set, and hence its cardinal is at least $2^{\aleph_1}$.

The chapter closes with some theorems about prime *z-ideals*, including some special results about O^p.

14.2. We begin with a purely algebraic result.

(a) *The union and intersection of any chain of prime ideals are prime.*

It is clear that the union of a chain $\mathfrak{A}$ of prime ideals is prime. To see that $\bigcap \mathfrak{A}$ is also prime, suppose that $a \notin \bigcap \mathfrak{A}$ and $b \notin \bigcap \mathfrak{A}$. There exist $P, Q \in \mathfrak{A}$ such that $a \notin P$ and $b \notin Q$. Say $P \subset Q$; then $b \notin P$. Since P is prime, $ab \notin P$. Hence $ab \notin \bigcap \mathfrak{A}$.

We shall call a prime ideal an *upper ideal* if the set of all its predecessors (in the partial order under set inclusion) has a maximal element. Briefly, an upper ideal is one that has an immediate predecessor. Similarly, a *lower ideal* is defined as a prime ideal that has an immediate successor.

PRIME IDEALS IN C/P

14.3. We now apply these ideas to the ring $C(X)/P$, where P is any prime ideal in $C(X)$.

(a) *Every prime ideal in the totally ordered ring C/P is convex.*

We prove the following more general proposition:

(b) *If I is a convex ideal in a partially ordered ring A, and J is any convex ideal containing I, then J/I is a convex ideal in the partially ordered ring A/I.*

Given $0 \leqq I(a) \leqq I(b)$, where $I(b) \in J/I$ (i.e., $b \in J$), there exist $u \geqq 0$ and $v \geqq 0$ in A, such that $a \equiv u \pmod{I}$ and $b - u \equiv b - a \equiv v \pmod{I}$. Then $b - (u + v) \in I \subset J$, and so, $u + v \in J$. But also, $0 \leqq u \leqq u + v$. Since J is convex, $u \in J$, whence $I(a) = I(u) \in J/I$. Therefore J/I is convex.

Every ideal in a ring is symmetric, i.e., it contains $-a$ whenever it contains a. A *convex* ideal in a totally ordered ring, therefore, is a symmetric *interval*: with every two of its elements, it also contains all elements that lie between them. But obviously, of any two symmetric intervals, one contains the other. We have proved:

(c) *The prime ideals in C/P form a chain.*

It follows, incidentally, that C/P contains at most one—and hence exactly one—maximal ideal. In turn, this yields a third proof that the prime ideal P in C is contained in a unique maximal ideal.

Because of (c), it is especially easy to describe the upper and lower ideals in C/P. Let a be a positive non-unit of C/P; such an element will exist whenever P is not maximal, i.e., C/P is not a field. As in any commutative ring with unity, a belongs to at least one prime ideal. By 14.2(a), the intersection of all the prime ideals containing a is prime. Denote it by P^a:

$$P^a = \text{the smallest prime ideal containing } a \qquad (a > 0).$$

By 0.18, $b \in P^a$ if and only if some power of b is a multiple of a.

Since a does not belong to the prime ideal (0), the chain of all prime ideals not containing a is nonempty. The union of this chain is a prime ideal; denote it by P_a:

$$P_a = \text{the largest prime ideal not containing } a \qquad (a > 0).$$

Plainly, $a \in P^a - P_a$, and P^a is the immediate successor of P_a in the chain of all prime ideals in C/P. Thus, P_a is a lower ideal, and P^a is an upper ideal.

Conversely, if I and J are successive prime ideals, with $I \subset J$, then for each positive $a \in J - I$, we have

$$I \subset P_a \neq P^a \subset J.$$

Hence, the lower ideal I is P_a, and the upper ideal J is P^a.

14.4. We have seen that if the prime ideal P is not maximal, then C/P contains at least one upper ideal and one lower ideal. Later on (14.13), we shall find that it actually contains infinitely many.

THEOREM. *Every nonzero prime ideal in C/P is a union of upper ideals. Every nonmaximal prime ideal in C/P is an intersection of lower ideals.*

PROOF. Let J be a nonzero prime ideal in C/P. If a is a positive element of J, then $P^a \subset J$. Therefore

$$J = \bigcup_{0 < a \in J} P^a.$$

This proves the first statement. The proof of the second is similar.

It is clear that when a nonzero prime ideal is not itself an upper ideal, then it is not only a union of upper ideals, but of the corresponding lower ideals as well. Similarly, any nonmaximal prime ideal that is not a lower ideal is an intersection of upper ideals.

14.5. Every non-unit in a commutative ring with unity belongs to some maximal ideal. In C/P, there is just one maximal ideal, namely M^p/P, where M^p is the unique maximal ideal in C that contains P. Hence M^p/P is just the set of all non-units of C/P. Since M^p/P, as a prime ideal in C/P, is a symmetric interval, every element greater than a positive unit is a unit. Thus:

(a) *If a is a non-unit, then $sa < 1$ for every $s \in C/P$.*

It follows that M^p/P contains only infinitely small elements, their negatives, and 0. In case $p \in \upsilon X$, M^p/P includes all such elements; but if $p \notin \upsilon X$, it does not (7.16).

Every positive element a of C/P has a unique positive n^{th} root $a^{1/n}$ (where $n \in \mathbf{N}$). For, if f is any preimage of a in C, with $f \geqq \mathbf{0}$, then $P(f^{1/n})$ is a positive n^{th} root of a; and, as in any totally ordered integral domain, such an element is necessarily unique. If $0 < a < 1$, then, of course,

$$\cdots < a^3 < a^2 < a < a^{1/2} < a^{1/3} < \cdots.$$

14.6. THEOREM.

$$P^a = \bigcup_{n \in \mathbf{N}} \{b \in C/P : |b| < a^{1/n}\},$$

and

$$P_a = \bigcap_{n \in \mathbf{N}} \{b \in C/P : |b| < a^n\}.$$

PROOF. If $b \in P^a$, then $b^n = sa$ for some $s \in C/P$ and $n \in \mathbf{N}$; then $b^{2n} = (s^2a)a < a$ by 14.5(a). Conversely, $|b| < a^{1/n}$ implies $b \in P^a$, since P^a is prime and convex.

Next, if $b \in P_a$, then $|b| < a^n$ for all n, since P_a is convex and contains no power of a. Finally, if $|b| < a^n$ for every n, then $a \notin P^{|b|}$, whence $b \in P^{|b|} \subset P_a$.

PRIME IDEALS IN C

14.7. THEOREM. *If I is a z-ideal in $C(X)$, and Q is minimal in the class of prime ideals containing I, then Q is a z-ideal.*

PROOF. Let P be a prime ideal containing I, and suppose that P is not a z-ideal. There exist $f \in P$, and $g \notin P$, such that $\mathbf{Z}(g) = \mathbf{Z}(f)$. Consider the set

$$S = (C - P) \cup \{hf^n : h \notin P, n \in \mathbf{N}\}.$$

Since P is prime, S is closed under multiplication. Furthermore, S does not meet I; for, if $hf^n \in I$, then hg belongs to the z-ideal I, and hence to the prime ideal P, whence $h \in P$. By Theorem 0.16, there is a prime ideal containing I and disjoint from S, and hence contained properly in P. So P is not minimal.

Since every ideal contains the z-ideal $(\mathbf{0})$, we have, as a corollary, that *every minimal prime ideal in $C(X)$ is a z-ideal.*

14.8. From 14.3(c), we have:

(a) *The prime ideals in C containing a given prime ideal form a chain.*

Let $\mathfrak{A}$ and $\mathfrak{B}$ be intersecting chains of prime ideals in C. Then $\mathfrak{A} \cap \mathfrak{B}$ is a chain, and, by 14.2(a), the intersection of all its members is a prime ideal, P.

In case both $\mathfrak{A}$ and $\mathfrak{B}$ are *maximal* chains, then $P \in \mathfrak{A} \cap \mathfrak{B}$, and, by (a), $\mathfrak{A} \cap \mathfrak{B}$ consists of all the prime ideals that contain P. By Theorem 14.7, the minimal prime ideals $\bigcap \mathfrak{A}$ and $\bigcap \mathfrak{B}$ are z-ideals. We shall show that P is a z-ideal as well. Here, for the first time in some while, we utilize properties of βX.

LEMMA. *If I and J are z-ideals, then (I, J) is a z-ideal (or all of C).*

PROOF. We are to prove that if $Z(h) \supset Z(g_1 + g_2)$, where $g_1 \in I$ and $g_2 \in J$, then $h \in (I, J)$. Because h is a multiple of $h \cdot (1 + h^2)^{-1}$, there is no loss of generality in assuming that h is bounded.

Let $Z_1 = Z(g_1)$, and $Z_2 = Z(g_2)$. Then $Z(h) \supset Z_1 \cap Z_2$. By (IV) of the compactification theorem (6.5),

$$\mathrm{cl}_{\beta X} Z(h) \supset \mathrm{cl}\, Z_1 \cap \mathrm{cl}\, Z_2.$$

Hence, we may define a function φ as follows:

$$\begin{aligned} \varphi(p) &= 0 && \text{for } p \in \mathrm{cl}\, Z_1, \\ \varphi(p) &= h^\beta(p) && \text{for } p \in \mathrm{cl}\, Z_2, \end{aligned}$$

and φ will be continuous on the compact set $\mathrm{cl}\, Z_1 \cup \mathrm{cl}\, Z_2$. Then φ has an extension to a function in $C(\beta X)$; this has the form f^β, for suitable $f \in C^*(X)$. Evidently, $Z(f) \supset Z_1 \in Z[I]$; so f belongs to the z-ideal I. Likewise, $Z(h - f) \supset Z_2$, so that $h - f \in J$. Therefore, $h = f + (h - f) \in (I, J)$.

14.9. THEOREM. *If $\mathfrak{A}$ and $\mathfrak{B}$ are intersecting maximal chains of prime ideals in C, then $\bigcap (\mathfrak{A} \cap \mathfrak{B})$ is a prime z-ideal, and is equal to (I, J), where I and J are the minimal members of $\mathfrak{A}$ and $\mathfrak{B}$, respectively.*

PROOF. The minimal prime ideals I and J are z-ideals. By the lemma, (I, J) is a z-ideal. Since (I, J) contains a prime ideal, it is prime, by Theorem 2.9. Since the chains $\mathfrak{A}$ and $\mathfrak{B}$ consist of all prime ideals containing I and J, respectively, the ideal (I, J) belongs to both, and is the smallest member of $\mathfrak{A} \cap \mathfrak{B}$.

14.10. The set of all prime ideals in C containing a given prime ideal is a chain (14.8(a)). Therefore each lower ideal, i.e., each prime ideal with an immediate successor, has a *unique* immediate successor. We prove next that each upper ideal has a unique immediate predecessor.

LEMMA. *A z-ideal in C is never an upper ideal.*

PROOF. Let Q be an upper ideal, and P any immediate predecessor of Q. Then Q/P is an upper ideal, say P^a, in C/P. Choose a preimage f of a in C, satisfying $\mathbf{0} \leq f \leq \mathbf{1}$, and define

$$g(x) = \sum_{n \in \mathbf{N}} 2^{-n} f^{1/n}(x) \qquad (x \in X).$$

Then $g \in C(X)$, $Z(g) = Z(f)$, and, for each $n \in \mathbf{N}$,

$$g \geq \mathbf{2}^{-2n} f^{1/2n}.$$

Since $a^{1/2n}$ is infinitely small,

$$P(g) \geq 2^{-2n} a^{1/2n} \geq a^{1/2n} a^{1/2n} = a^{1/n}.$$

By Theorem 14.6, $P(g)$ does not belong to the upper ideal $P^a = Q/P$. Thus, $g \notin Q$. Since $f \in Q$, this shows that Q is not a z-ideal.

It follows from the lemma that a maximal ideal is never an upper ideal.

A z-ideal cannot be a lower ideal, either; see 14D.*4*.

14.11. THEOREM. *The set of all successors of a lower ideal in C has a smallest member. The set of all predecessors of an upper ideal in C has a largest member.*

PROOF. The assertion about the lower ideal was noted above.

Let I be an immediate predecessor of an upper ideal J; we shall prove that I contains every predecessor H of J. Let $\mathfrak{I}$ be a maximal chain of prime ideals containing I (and J), and $\mathfrak{H}$ a maximal chain containing H (and J). Then I is the immediate predecessor of J in $\mathfrak{I}$.

By Theorem 14.9, $\mathfrak{I} \cap \mathfrak{H}$ has a smallest member, P, and P is a z-ideal. Obviously, $P \subset J$. By the lemma, P is not the upper ideal J. Therefore, P, as a member of $\mathfrak{I}$, is contained properly in J. Consequently, $P \subset I$.

Now, in the chain $\mathfrak{H}$, either $H \subset P$ or $P \subset H$. In the first case, we have $H \subset I$, at once. On the other hand, if $P \subset H$, then H belongs to the maximal chain $\mathfrak{I}$; since $H \neq J$, this implies again that $H \subset I$.

14.12. Let us summarize what we know about the incidence of prime ideals in $C(X)$. Every prime ideal lies between $\boldsymbol{O}^p$ and $\boldsymbol{M}^p$, for some unique p in βX (Theorem 7.15). For any given p, consider a maximal chain in the family of prime ideals contained in $\boldsymbol{M}^p$. At the top of the chain is $\boldsymbol{M}^p$, a z-ideal. At the bottom of the chain is a prime ideal which is minimal, and hence also a z-ideal. Whenever two maximal chains intersect, their intersection has a least member, which is again a z-ideal; and the intersection consists of all prime ideals containing this least member. In special cases, this least member is $\boldsymbol{M}^p$ itself; see 14G.*5*.

The z-ideal $\boldsymbol{O}^p$ is an intersection of prime ideals. Since each prime ideal contains a minimal prime ideal, $\boldsymbol{O}^p$ is, in fact, an intersection of minimal prime ideals. Hence, if $\boldsymbol{O}^p \neq \boldsymbol{M}^p$, then $\boldsymbol{M}^p$ contains prime z-ideals different from $\boldsymbol{M}^p$. The ideal $\boldsymbol{O}^p$ may itself be prime; then the prime ideals between $\boldsymbol{O}^p$ and $\boldsymbol{M}^p$ form a single chain. This happens, for example, for the space Σ of 4M: the z-ideal $\boldsymbol{O}_\sigma$ is prime, and there are actually no other z-ideals between $\boldsymbol{O}_\sigma$ and $\boldsymbol{M}_\sigma$. In general, however, $\boldsymbol{O}^p$ is not a prime ideal.

14.13. Having emphasized the presence of prime ideals that are also z-ideals, we take note now of some that are not. It follows from

Theorem 14.11 that if P is any prime ideal contained in an ideal I, then *I/P is a lower ideal in C/P if and only if I is a lower ideal in C.* Likewise, if P is contained properly in J, then *J/P is an upper ideal in C/P if and only if J is an upper ideal in C.*

We can now state the counterpart of Theorem 14.4 for C. Let J be a nonminimal prime ideal in C, and let $\mathfrak{A}$ be any maximal chain of prime ideals, with $J \in \mathfrak{A}$. Then J is a union of upper ideals belonging to $\mathfrak{A}$; and if J itself is not an upper ideal, then it is also a union of lower ideals belonging to $\mathfrak{A}$. Also, if I is a nonmaximal prime ideal in C, then it is an intersection of lower ideals (necessarily forming a chain); and if I is not a lower ideal, it is also an intersection of upper ideals.

It follows that *there exist infinitely many upper ideals containing any given nonmaximal prime ideal P*; equivalently, there exist infinitely many upper ideals in C/P. For, the maximal ideal containing P is a union of a chain of upper ideals that contain P, but, as a z-ideal, is not itself an upper ideal.

Thus, whenever $\boldsymbol{O}^p \neq \boldsymbol{M}^p$—i.e., whenever $\boldsymbol{M}^p$ contains a nonmaximal prime ideal—then $\boldsymbol{M}^p$ contains prime ideals that are not z-ideals. An extreme case is that of $\boldsymbol{M}_\sigma$, cited above. The considerations that follow will enable us to conclude that $\boldsymbol{M}_\sigma$ contains $\mathfrak{c} = 2^{\aleph_0}$ upper ideals, and $2^{\aleph_1}$ prime ideals altogether; none of these, excepting $\boldsymbol{M}_\sigma$ and $\boldsymbol{O}_\sigma$, is a z-ideal.

η_1-SETS

14.14. We recall (13.6) that an *η_1-set* is a totally ordered set with the following property: given countable subsets A and B, with $A < B$, there exists an element c satisfying $A < c < B$. The possibility that A or B be empty is included here, so that an η_1-set is necessarily nonempty, and, moreover, no countable subset is either coinitial or cofinal.

Theorem 13.8 states that every residue class field modulo a hyper-real maximal ideal is an η_1-set. Our next goal is to ascertain what η_1-properties are possessed by C/P for arbitrary prime P. The problem here is complicated by a certain lack of symmetry in C/P (when it is not a field): infinitely small elements always exist, but infinitely large elements need not. Moreover, even when infinitely large elements do exist, there will still remain some infinitely small elements that have no inverse. Let $\boldsymbol{M}^p$ denote the unique maximal ideal containing P. As we know, $\boldsymbol{M}^p/P$ consists of all non-units of C/P, and every positive non-unit is infinitely small. If $p \in \upsilon X$, then, conversely, no infinitely small element is a unit; but when $p \notin \upsilon X$, infinitely small units always exist.

14.15. LEMMA. *Let P be a prime ideal in C, with $P \subset M^p$.*

(a) *Given nonempty, countable sets A and B in C/P, with $A < B$, there exists $c \in C/P$ satisfying*

$$A \leqq c \leqq B.$$

(b) *C/P has a countable, cofinal subset if and only if M^p is real, i.e., $p \in \upsilon X$.*

PROOF. Lemma 13.7 yields (a) in case M^p is hyper-real, i.e., $p \notin \upsilon X$. The same lemma (with $B = \emptyset$) shows that no countable set is cofinal in this case. The converse in (b) is obvious: if $p \in \upsilon X$, then C/P has no infinitely large elements (7.16), so that the elements $P(\mathbf{n})$, for $n \in \mathbf{N}$, form a cofinal set.

There remains the proof of (a) for $p \in \upsilon X$. Because $C(X)$ and $C(\upsilon X)$ are isomorphic, there will be no loss of generality in assuming that $X = \upsilon X$, so that $p \in X$.

The proof will proceed along lines similar to those of Lemma 13.7. We may assume that A has no greatest element and B no least. By 13.6(a) and Lemma 13.5, there exist sequences (f_n) and (g_n), satisfying

$$f_k \leqq f_n \leqq g_n \leqq g_k \quad \text{whenever} \quad k < n,$$

and with $(P(f_n))$ cofinal in A and $(P(g_n))$ coinitial in B. If there exists a real number r such that $\sup_n f_n(p) < r < \inf_n g_n(p)$, then $P(f_n) < P(\mathbf{r}) = r < P(g_n)$ for all n, i.e., $A < r < B$. We assume, then, that $\sup_n f_n(p) = \inf_n g_n(p)$; and there is no harm in supposing this number to be 0. If $P(f_n) \leqq 0 \leqq P(g_n)$ for all n, then $A \leqq 0 \leqq B$, and so we may restrict our attention to the case where $P(f_n) > 0$ for some n. And by discarding finitely many f_n, if necessary, we secure the condition $P(f_n) > 0$ for all n. It follows, of course, that $P(g_n) > 0$ for all n. Also, $0 \leqq f_n(p) \leqq \sup_k f_k(p) = 0$, so that $f_n(p) = 0$ for every n.

Since $g_n \equiv g_n \vee \mathbf{0}$ and $f_n \equiv f_n \vee \mathbf{0}$, modulo P, we may replace the former by the latter in both cases, without destroying either our congruence or our order relations. Therefore we may suppose that $f_n \geqq \mathbf{0}$. Similarly, we may suppose that $g_n \leqq \mathbf{1}$.

In summary:

$$\mathbf{0} \leqq f_n \leqq g_n \leqq \mathbf{1},$$

and

$$f_n(p) = 0 = \inf_k g_k(p),$$

for all $n \in \mathbf{N}$.

Now we introduce a function s:

$$s(x) = \sum_{n \in \mathbf{N}} 4^{-n}(f_n(x) + |g_n(x) - g_n(p)|) \qquad (x \in X).$$

Because the convergence is uniform, s is continuous on X. Evidently, $\mathbf{0} \leqq s \leqq \mathbf{1}$, and $s(p) = 0$. Hence, for each n, the zero-set

$$Z_n = \{x \in X: s(x) \leqq 1/n\}$$

is a neighborhood of p.

Next, we define auxiliary functions $\varphi_k \in C(\mathbf{R})$, for each $k \in \mathbf{N}$, as follows:

$$\varphi_k(r) = \begin{cases} k(k+1)r - k & \text{for} \quad 1/(k+1) \leqq r \leqq 1/k, \\ k - k(k-1)r & \text{for} \quad 1/k \leqq r \leqq 1/(k-1), \\ 0 & \text{otherwise} \end{cases}$$

(with the obvious interpretation in the case $k = 1$). Then $\mathbf{0} \leqq \varphi_k \leqq \mathbf{1}$, for each k. Next, $\varphi_k(r) = 0$ if $r \leqq 1/(k+1)$ or if $r \geqq 1/(k-1)$. Therefore, if $1/(n+1) < r < 1/(n-1)$ (where $n \in \mathbf{N}$), then $\varphi_k(r) = 0$ unless $k = n-1$, n, or $n+1$. Finally,

$$\textstyle\sum_{k \in \mathbf{N}} \varphi_k(r) = 1 \quad \text{for} \quad r > 0,$$

and $\varphi_k(0) = 0$ for each k.

Now we define the function h whose image modulo P will be the required element c. Put

$$h(x) = \textstyle\sum_{k \in \mathbf{N}} \varphi_k(s(x))g_k(x) \qquad (x \in X).$$

Then h is defined on all of X, $h \geqq \mathbf{0}$, and $h(y) = 0$ whenever $s(y) = 0$. On each open set

$$\{x \in X: 1/(n+1) < s(x) < 1/(n-1)\} \qquad (n \in \mathbf{N}),$$

we have

$$h(x) = \textstyle\sum_{n-1}^{n+1} \varphi_k(s(x))g_k(x);$$

it follows that h is continuous at every point x at which s does not vanish.

We still have to prove that h is continuous at every point of $\mathbf{Z}(s)$ as well. First, recall that if $r \leqq 1/n$, then $\varphi_k(r) = 0$ for all $k < n$; hence

$$\text{if} \quad s(x) \leqq 1/n, \quad \text{then} \quad h(x) = \textstyle\sum_{k \geqq n} \varphi_k(s(x))g_k(x) \qquad (n \in \mathbf{N}).$$

Since the sequence (g_k) is decreasing, $h(x) \leqq g_n(x)$ whenever $s(x) \leqq 1/n$, i.e., on the neighborhood Z_n of p.

Now consider any point $y \in \mathbf{Z}(s)$, and let $\epsilon > 0$ be given. Since $\inf_k g_k(p) = 0$, there exists m for which $g_m(p) < \epsilon$. Let

$$U = \{x: s(x) < 4^{-m}\epsilon\}.$$

From the definition of s, we have, for $x \in U$,

$$4^{-m} |g_m(x) - g_m(p)| \leqq s(x) < 4^{-m}\epsilon,$$

so that $|g_m(x) - g_m(p)| < \epsilon$, whence $g_m(x) < 2\epsilon$. Thus, for all x belonging to the neighborhood $U \cap Z_m$ of y, we have $h(x) \leqq g_m(x) < 2\epsilon$. This establishes continuity of h at y.

We have seen that, for each n, $h - g_n$ is nonpositive on a neighborhood of p; therefore $P(h) \leqq P(g_n)$ (7.15(b)). Next, if $f_n(x) \neq 0$, then $s(x) \neq 0$, so that $h(x)$ is an average of the $g_k(x)$. Since $f_n \leqq g_k$ for all k, we have $f_n \leqq h$. Hence $P(f_n) \leqq P(h)$.

14.16. THEOREM. *Let P be a prime ideal in C.*

(a) *C/P is an η_1-set if and only if the set of positive elements has no countable coinitial or cofinal subset.*

(b) *The set of infinitely small elements in C/P never has a countable cofinal subset; and it is an η_1-set if and only if it has no countable coinitial subset.*

REMARK. By (b) of the preceding lemma, no countable set of positive elements is cofinal when and only when the maximal ideal containing P is hyper-real.

PROOF. We prove first that no countable subset D of the set E of infinitely small elements is cofinal. By (a) of the lemma, there exists $c \in C/P$ such that

$$D \leqq c \leqq \{1/n : n \in \mathbf{N}\}.$$

The second inequality implies that $c \in E$. Therefore $2c \in E$, and we have $D \leqq c < 2c$. So D is not cofinal.

Next, the necessity in (a) and (b) is plain. For the converses, we assume that no countable subset is coinitial in the set of positive elements, and consider any two nonempty, countable sets A and B in C/P, with $A < B$. By (a) of the lemma, there exists c satisfying $A \leqq c \leqq B$. Say $A < c$. By hypothesis, the countable set $\{c - a : a \in A\}$ has a positive lower bound d, and so we have $A < c - d/2 < c \leqq B$.

If, in addition, there is no countable cofinal set of positive elements, then C/P has no countable coinitial set either, and hence is an η_1-set. This proves (a).

Finally, if both A and B are subsets of E, then $c - d/2 \in E$. It follows that E is an η_1-set. This completes the proof of (b).

Whenever $\boldsymbol{M}^p$ is different from $\boldsymbol{O}^p$, it contains prime ideals P for which the set of positive elements in C/P does have a countable, coinitial subset, as well as others for which it does not; see 14J.

14.17. We examine, next, the η_1-properties of the totally ordered family of prime ideals in C/P—or of chains of prime ideals in C. If $P^a \subset P_b$, then by Lemma 14.15(a), there exists an element c satisfying $a^{1/n} < c < b^n$ for all n. By Theorem 14.6, $c \in P_b - P^a$. This shows that a prime ideal in C/P (or in C) cannot be simultaneously an upper ideal and a lower ideal. More generally, we have:

THEOREM. *Let P be a prime ideal in C, and let (a_i) and (b_j) be sequences of positive non-units of C/P. If*

$$P^{a_i} \subset P_{b_j}$$

for all i and j, then there exists $c \in C/P$ such that

$$P^{a_i} \subset P_c \subset P^c \subset P_{b_j} \text{ for all } i \text{ and } j;$$

furthermore, all the inclusions are necessarily proper.

PROOF. The hypothesis implies that $a_i^{1/m} < b_j^n$ for all i, j, m, and n. By Lemma 14.15(a), there exists c such that

$$a_i^{1/m} < c < b_j^n$$

for all i, j, m, and n. Hence

$$a_i < c^m \leqq c < b_j^n$$

for all i, j, m, and n, and the result follows from Theorem 14.6.

14.18. COROLLARY. *No countable union of upper ideals in C/P is a countable intersection of lower ideals.*

Hence no countable union of a chain of upper ideals in C is a countable intersection of lower ideals.

14.19. We pointed out in 14.13 that if P is a nonmaximal prime ideal, then C/P contains infinitely many upper ideals.

THEOREM. *Let P be a nonmaximal prime ideal in C. In C/P:*

(a) *The chain of all prime ideals is Dedekind-complete.*

(b) *The set of all upper ideals between any two given ones is an η_1-set, and hence its cardinal is at least $\mathfrak{c} = 2^{\aleph_0}$.*

(c) *There exist at least $2^{\aleph_1}$ prime ideals between any two upper ideals.*

PROOF. (a). The union of a chain of prime ideals is prime, and hence is the supremum of the chain.

(b). This follows directly from Theorem 14.17, as soon as one observes that if P^a is contained properly in P^b, then $P^a \subset P_b$.

(c). According to Corollary 13.24, the cardinal of a Dedekind-complete set containing an η_1-set is at least $2^{\aleph_1}$.

The corresponding results hold for lower ideals, of course; and both sets of results have their counterparts in C. In particular, *if* M^p *is different from* O^p, *then* M^p contains nonmaximal prime ideals, and hence *contains at least* $2^{\aleph_1}$ *prime ideals.*

14.20. Finally, we examine a special case.

THEOREM. *Suppose that there are just* $\mathfrak{c}$ *upper ideals between* P^a *and* P^b, *where* $P^a \subset P^b$, *and assume the continuum hypothesis. Then the set of all prime ideals in the closed interval* $[P^a, P_b]$ *is similar to the lexicographically ordered set* S *of all* $\{0, 1\}$*-valued, transfinite sequences* $(x_\xi)_{\xi < \omega_1}$.

PROOF. We work first with the open interval between P^a and P_b. The set of upper ideals in this interval is an η_1-set, as was just proved; under the continuum hypothesis, its cardinal is $\aleph_1$. The set Q of 13.18 is also an η_1-set of cardinal $\aleph_1$ (Theorem 13.20). By Theorem 13.9, these two sets are similar. Next, by Theorem 14.4, every prime ideal in the interval in question is a union of upper ideals. Consequently, given any two nonlower ideals, there is an upper ideal lying strictly between them. As in 13.23, it follows that the set of all nonlower ideals in this interval is the Dedekind completion of the set of upper ideals, and therefore is similar to the Dedekind completion R of Q. Next, the set of lower ideals maps naturally onto the set of lower elements of S (see 13.16). Finally, P^a corresponds to the constant sequence **0**, and P_b to **1**.

A noteworthy application of this theorem is to any space X for which $|C(X)| = \mathfrak{c}$—for example, to **R** or **Q**. For then, $|C/P| \leqq |C| = \mathfrak{c}$; and, since every upper ideal is determined by some element of C/P, there are at most $\mathfrak{c}$ upper ideals. Therefore, if P is not maximal, there are exactly $\mathfrak{c}$ upper ideals. Thus, let I be any upper ideal in $C(X)$, and J any lower ideal, with $I \subset J$; and suppose that $|C(X)| = \mathfrak{c} = \aleph_1$. Then the set of all prime ideals between I and J, inclusive, is similar to the set S (of cardinal $2^{\aleph_1}$).

PRIME z-IDEALS

14.21. Thus far, we have been concerned for the most part with arbitrary prime ideals in C. Now, we know that each prime ideal contains an ideal O^p; thus, the z-ideals O^p are lower bounds for the prime ideals. A natural undertaking in the study of prime ideals is, therefore, to examine conditions under which O^p itself can be prime.

Before specializing to this extent, we gather some additional information about prime z-ideals in general.

LEMMA. *Let I be an ideal in C, and let $f \in C$. If $(I(f), I(|f|))$ is a principal ideal (perhaps improper) in C/I, then there exists a zero-set $Z \in \mathbf{Z}[I]$ such that*

$$Z \cap \operatorname{pos} f \quad \textit{and} \quad Z \cap \operatorname{neg} f$$

are completely separated.

PROOF. By hypothesis, there exists $d \in C$ for which

$$(I(f), I(|f|)) = (I(d)).$$

Hence there exist g, h, s, $t \in C$ such that

$$f \equiv gd \pmod{I}, \qquad |f| \equiv hd \pmod{I},$$

and

$$sf + t|f| \equiv d \pmod{I}.$$

Now, congruence modulo I implies equality on some zero-set of I. We can therefore find a zero-set $Z \in \mathbf{Z}[I]$ on which all three of the above congruences reduce to equalities:

(a) $$f(x) = g(x)d(x), \quad |f(x)| = h(x)d(x), \quad \textit{and}$$
$$s(x)f(x) + t(x)|f(x)| = d(x), \quad \textit{for} \quad x \in Z.$$

On combining these equations, we get

(b) $$\big(s(x)g(x) + t(x)h(x)\big)d(x) = d(x) \qquad (x \in Z).$$

Now, by (a), d has no zeros on $Z - \mathbf{Z}(f)$; with (b), this yields

$$s(x)g(x) + t(x)h(x) = 1 \qquad (x \in Z - \mathbf{Z}(f)).$$

Note that $Z - \mathbf{Z}(f) = (Z \cap \operatorname{pos} f) \cup (Z \cap \operatorname{neg} f)$. Next, (a) implies that

$$g(x) = h(x) \qquad (x \in Z \cap \operatorname{pos} f),$$

and

$$g(x) = -h(x) \qquad (x \in Z \cap \operatorname{neg} f).$$

Consequently,

$$Z \cap \operatorname{pos} f \subset \mathbf{Z}(g - h) \cap \mathbf{Z}(sg + th - \mathbf{1}),$$

and

$$Z \cap \operatorname{neg} f \subset \mathbf{Z}(g + h) \cap \mathbf{Z}(sg + th - \mathbf{1}).$$

Since the second members of these inequalities are disjoint zero-sets, the first members are completely separated.

14.22. COROLLARY. *The following are equivalent for any* $f \in C$.

(1) $\operatorname{pos} f$ *and* $\operatorname{neg} f$ *are completely separated.*

(2) *There exists* $k \in C$ *such that* $f = k|f|$.

(3) $(f, |f|)$ *is a principal ideal in* C *(or all of* C*).*

PROOF. Both (1) and (2) are simply reformulations of the statement that there exists $k \in C$ that is equal to 1 on $\operatorname{pos} f$ and to -1 on $\operatorname{neg} f$. Trivially, (2) implies (3). Finally, if (3) holds, then the lemma, with $I = (\mathbf{0})$ (whence $Z = X$), yields (1).

Notice that if $f = k|f|$, then $|f| = kf$—for the same k. Since the validity of these equations is not affected by the values of k on $\mathbf{Z}(f)$, it may always be assumed that $|k| \leqq \mathbf{1}$.

14.23. LEMMA. *Let* I *be a z-ideal containing* $\mathbf{O}^p$ *(for some* p*). If the (perhaps improper) ideal* $(I(f), I(|f|))$ *in* C/I *is principal, for every* $f \in C$, *then* I *is prime.*

PROOF. By the preceding lemma, there exists $Z \in \mathbf{Z}[I]$ such that $Z \cap \operatorname{pos} f$ and $Z \cap \operatorname{neg} f$ are completely separated. These sets, therefore, have disjoint closures in βX. Hence there exists a zero-set $Z' \in \mathbf{Z}[\mathbf{O}^p]$ disjoint from, say, $Z \cap \operatorname{neg} f$. Evidently, f is nonnegative on the zero-set $Z \cap Z'$—which belongs to $\mathbf{Z}[I]$, because $Z' \in \mathbf{Z}[\mathbf{O}^p] \subset \mathbf{Z}[I]$. So $I(f) \geqq 0$ (5.4(a)). This shows that C/I is a totally ordered ring; and since I is a z-ideal, it is prime (5.4(c)).

The hypothesis that $I \supset \mathbf{O}^p$ is essential, as is shown by the example following 5.4(c).

14.24. THEOREM. *Let* I *be a z-ideal containing* $\mathbf{O}^p$ *(for some* p*). The following conditions on the residue class ring* C/I *are equivalent.*

(1) *Every ideal is convex.*

(2) *The ideals form a chain.*

(3) *The principal ideals form a chain.*

(4) *Every finitely generated ideal is principal.*

In addition, these conditions on C/I *imply that* I *is a prime ideal in* C.

REMARK. It is clear that every ideal will be convex when and only when every *principal* ideal is convex.

PROOF. (1) *implies* (2). Since $-I(|f|) \leqq I(f) \leqq I(|f|)$, and $(I(|f|))$ is convex, we have $I(f) \in (I(|f|))$; therefore,

$$(I(f), I(|f|)) = (I(|f|)).$$

By the lemma, C/I is totally ordered. Therefore its ideals, being convex, are symmetric intervals, and so they form a chain.

(2) *implies* (3). Trivial.

(3) *implies* (4). This holds in any ring; for if $(a) \subset (b)$, then $(a, b) = (b)$.

(4) *implies* (1). By the lemma, C/I is a totally ordered integral domain. Given $a, b \in C/I$, with $0 < a < b$, we are to show that a is a multiple of b. By hypothesis, $(a, b) = (d)$ for some d, and we may assume that $d > 0$. There exist $a_1, b_1, s, t \in C/I$ such that $a = a_1 d$, $b = b_1 d$, and $sa + tb = d$. Then $(sa_1 + tb_1)d = d$, whence $sa_1 + tb_1 = 1$, since C/I has no zero-divisors. Now, $0 < a_1 < b_1$. It follows that b_1 is a unit, since otherwise we would have

$$sa_1 + tb_1 \leqq (|s| + |t|)b_1 < 1,$$

by 14.5(a). Thus, b_1^{-1} exists, and $a = a_1 b_1^{-1} b$.

It has already been noted that condition (4) implies that I is a prime ideal. (See also 14B.*3*.) This completes the proof of the theorem.

A commutative ring with unity in which the principal ideals form a chain is known as a *valuation* ring. The condition can be expressed in this form: of any two elements, one is a multiple of the other. The ring C/I of the theorem satisfies a stronger condition: the smaller of two positive elements is always a multiple of the larger.

14.25. We now examine the situations in which, for *every* p, *all* the z-ideals containing $\boldsymbol{O}^p$ possess the listed properties. As it turns out, we need only postulate these properties for the ideals $\boldsymbol{O}^p$ themselves. Indeed, the hypothesis that each ideal $\boldsymbol{O}^p$ be prime is sufficient. Moreover, it leads to the conclusion that conditions (1) and (4) are satisfied by the entire ring C.

When every finitely generated ideal in $C(X)$ is principal, we call X an *F-space.*

THEOREM. *For any X, the following are equivalent.*

(1) *Every ideal $\boldsymbol{O}^p$ is prime.*

(2) *The prime ideals contained in any given maximal ideal form a chain.*

(3) *Given $p \in \beta X$ and $f \in C$, there is a zero-set of $\boldsymbol{O}^p$ on which f does not change sign.*

(4) *For each $f \in C$,* pos f *and* neg f *are completely separated.*

(5) *Given $f \in C$, there exists $k \in C$ such that $f = k|f|$.*

(6) *Every cozero-set in X is C^*-embedded.*

(7) *Every ideal in C is convex.*

(8) *For all $f, g \in C$, $(f, g) = (|f| + |g|)$.*

(9) *X is an F-space (i.e., every finitely generated ideal in $C(X)$ is principal).*

(10) *βX is an F-space (i.e., every finitely generated ideal in $C^*(X)$ is principal).*

PROOF. The prime ideals contained in $\boldsymbol{M}^p$ are just those containing $\boldsymbol{O}^p$. Hence if $\boldsymbol{O}^p$ is prime, the prime ideals contained in $\boldsymbol{M}^p$ form a chain (14.8(a)). Conversely, if they form a chain, then their intersection, which is $\boldsymbol{O}^p$, is prime (14.2(a)). Thus, (1) is equivalent to (2).

The equivalence of (1) with (3) is a special case of Theorem 2.9. Since completely separated sets in X have disjoint closures in βX, it is clear that (3) is equivalent to (4). The equivalence of (4) with (5) was pointed out in Corollary 14.22. Thus, the conditions (1) to (5) are mutually equivalent.

Next, we establish the cycle (4) → (6) → (7) → (8) → (9) → (4), from which it will follow that (1) to (9) are mutually equivalent.

(4) *implies* (6). To show that $X - \boldsymbol{Z}(h)$ is C^*-embedded in X, we apply the Urysohn extension theorem (1.17): we consider any two completely separated sets A and B in $X - \boldsymbol{Z}(h)$, and prove that they are completely separated in X. There exists $k \in C^*(X - \boldsymbol{Z}(h))$ such that k is positive on A and negative on B. Define f as follows:

$$f(x) = 0 \qquad (x \in \boldsymbol{Z}(h)),$$
$$f(x) = k(x)|h(x)| \qquad (x \in X - \boldsymbol{Z}(h)).$$

Since k is bounded, f is continuous on all of X. Obviously, $A \subset \operatorname{pos} f$ and $B \subset \operatorname{neg} f$. The hypothesis now implies that A and B are completely separated in X.

(6) *implies* (7). Given $\mathbf{0} \leqq f \leqq g$ in $C(X)$, we are to prove that f is a multiple of g. Define s as follows:

$$s(x) = \frac{f(x)}{g(x)} \qquad (x \in X - \boldsymbol{Z}(g));$$

then $s \in C^*(X - \boldsymbol{Z}(g))$. By hypothesis, s has an extension to a function $f_1 \in C(X)$. Since $f(x) = g(x) = 0$ when $x \in \boldsymbol{Z}(g)$, we have $f(x) = f_1(x)g(x)$ for every $x \in X$, i.e., $f = f_1 g$.

(7) *implies* (8). Since

$$-|f| - |g| \leqq f \leqq |f| + |g|,$$

convexity implies that f is a multiple of $|f| + |g|$. Similarly, g is a multiple of $|f| + |g|$. So $(f, g) \subset (|f| + |g|)$. Likewise, convexity implies that f is a multiple of $|f|$, whence $|f|$ is a multiple of f; and similarly, $|g|$ is a multiple of g. Hence $(|f| + |g|) \subset (f, g)$.

(8) *implies* (9). Trivial.

(9) *implies* (4). This follows from Corollary 14.22.

Thus, the conditions (1) to (9) are mutually equivalent. Finally, it is clear that (5), for example, will be valid when and only when the corresponding condition holds in C^*. Consequently, (9) is equivalent to (10).

14.26. The condition that all finitely generated ideals be principal is preserved under homomorphism. (For, if t is a homomorphism of a ring A, and $(a, b) = (d)$ in A, then, clearly, $(ta, tb) = (td)$.) Therefore, if $C(S)$ or $C^*(S)$ are homomorphic images of $C(X)$ or $C^*(X)$, and if X is an F-space, then S is an F-space. (See also 14A.)

If a subspace S is C^*-embedded in X, then $C^*(S)$ is a homomorphic image of $C^*(X)$. Consequently, every C^*-embedded subspace of an F-space is an F-space. In particular, every cozero-set in an F-space is an F-space.

If X is an F-space and I is a z-ideal containing an ideal $\boldsymbol{O}^p$, then $C(X)/I$ will satisfy (4) of Theorem 14.24, and hence the other conditions as well.

The combination of (5) with (7) in Theorem 14.25 shows that X is an F-space if and only if every ideal in C is *absolutely* convex. Also, we see from Corollary 14.22 that X is an F-space if and only if $(f, |f|)$ is principal for every $f \in C$.

Evidently, every discrete space D, and hence also βD, is an F-space. The next theorem provides us with another extensive class of F-spaces.

14.27. THEOREM. *If X is locally compact and σ-compact, then $\beta X - X$ is a compact F-space.*

PROOF. Since X is locally compact, it is open in βX (6.9(d)); hence $\beta X - X$ is compact, and so it is C-embedded in βX.

Given $\varphi \in C(\beta X - X)$, we are to prove that the compact sets cl pos φ and cl neg φ are disjoint (Theorem 14.25(4)). The function φ has a continuous extension to all of βX, and the extension is of the form f^β, for suitable $f \in C^*(X)$. Now,

$$\mathrm{cl}_{\beta X} \operatorname{pos} \varphi \subset \mathrm{cl}_{\beta X} \operatorname{pos} f \subset \mathrm{cl}_{\beta X} \boldsymbol{Z}(f \wedge \mathbf{0}),$$

and, by the Gelfand-Kolmogoroff theorem,

$$\text{(a)} \qquad \mathrm{cl}_{\beta X} \boldsymbol{Z}(f \wedge \mathbf{0}) = \{p : \boldsymbol{M}^p(f) \geqq 0\}.$$

Thus if $p \in \mathrm{cl} \operatorname{pos} \varphi$, then $\boldsymbol{M}^p(f) \geqq 0$. Similarly, if $p \in \mathrm{cl} \operatorname{neg} \varphi$, then $\boldsymbol{M}^p(f) \leqq 0$. Accordingly, we have only to consider points p of $\beta X - X$ for which $\boldsymbol{M}^p(f) = 0$. We shall show that all such points belong to the interior of $\boldsymbol{Z}(\varphi)$, and hence neither to cl pos φ nor to cl neg φ.

Since X is σ-compact, $\beta X - X$ is a compact G_δ, and hence a zero-set, in βX (3.11(b)). Let $g \in C^*(X)$ satisfy $\beta X - X = \boldsymbol{Z}(g^\beta)$. Then

g is a unit of $C(X)$, so that $\boldsymbol{M}^q(|g|) > 0$ for all $q \in \beta X$. When $\boldsymbol{M}^p(f) = 0$, the set

$$V = \{q \in \beta X : \boldsymbol{M}^q(|g| - |f|) > 0\}$$

contains p. But V is an open set (compare (a)). And for all $q \in V - X$, we have, using Theorem 7.6,

$$0 = |g^\beta(q)| \geqq |f^\beta(q)|,$$

i.e., $f^\beta(q) = 0$. Hence p is an interior point of $\boldsymbol{Z}(\varphi)$.

As an example, $\beta R^+ - R^+$, where R^+ denotes the space of nonnegative reals, is a compact, connected F-space (6.10), of cardinal $2^{\mathfrak{c}}$ (9.3).

14.28. The results on F-spaces lead to alternative proofs that $C^*(\mathbf{Q})$ is not a homomorphic image of $C^*(\mathbf{N})$ or $C^*(\mathbf{R})$. The first thing to notice is that $\mathbf{Q}$ is not an F-space. (For example, the identity function from $\mathbf{Q}$ into $\mathbf{R}$ does not satisfy condition (5) of Theorem 14.25.) But $\mathbf{N}$ *is* an F-space. Therefore $C^*(\mathbf{Q})$ is not a homomorphic image of $C^*(\mathbf{N})$. Next, we showed in 10.10 that if $C^*(\mathbf{Q})$ were a homomorphic image of $C^*(\mathbf{R})$, then $\beta\mathbf{R} - \mathbf{R}$ would have to contain a C^*-embedded copy of $\mathbf{Q}$. And this is impossible because $\beta\mathbf{R} - \mathbf{R}$ is an F-space.

14.29. The condition that every ideal $\boldsymbol{O}^p$ be prime characterizes X as an F-space. In the extreme case, every such ideal is *maximal.* This is evidently equivalent to the condition that every prime ideal be maximal, i.e., that X be a *P-space* as defined in 4J. Corresponding to Theorem 14.25 on F-spaces, we have the following characterizations of P-spaces.

THEOREM. *For any X, the following are equivalent.*

(1) *Every ideal $\boldsymbol{O}^p$ ($p \in \beta X$) is maximal.*

(2) *Every prime ideal is maximal (i.e., X is a P-space).*

(3) *Given $p \in \upsilon X$, and $f \in C(X)$, there is a zero-set of $\boldsymbol{O}^p$ on which f is constant.*

(4) *For each $f \in C$, $\boldsymbol{Z}(f)$ is open.*

(5) *Given $f \in C$, there exists $k \in C$ such that $f = kf^2$ (i.e., C is a regular ring).*

(6) *Every cozero-set in X is C-embedded.*

(7) *Every ideal in C is a z-ideal.*

(8) *For all $f, g \in C$, $(f, g) = (f^2 + g^2)$.*

(9) *Every finitely generated ideal in C is generated by an idempotent.*

(10) *υX is a P-space.*

By 7.15(a), (1) is equivalent to (2). Since $C(X)$ is isomorphic with $C(\upsilon X)$, (2) is equivalent with (10). Proofs of the remaining implications are omitted, as they were outlined in sufficient detail in 4J.

Additional equivalences will be found in 4J, 7L, 7Q.*4*, and 10K.*1*. Examples of nondiscrete P-spaces appear in 4N and 13P. The space Σ of 4M is an F-space that is not a P-space. Among totally ordered spaces, the classes are coextensive (5O.*2*). If X is an infinite P-space, then βX is an F-space (Theorem 14.25); but it is not a P-space, as every compact P-space is finite (4K.*1*).

PROBLEMS

14A. CONVEX IDEALS.

1. Let $\mathfrak{t}$ be a homomorphism from $C(Y)$ onto $C(X)$, and I an ideal in $C(X)$. If $\mathfrak{t}^{\leftarrow}[I]$ is convex, then I is convex. [5C.]

2. Use *1* to prove that if $C(X)$ is a homomorphic image of $C(Y)$, and Y is an F-space, then X is an F-space.

14B. PRIME IDEALS.

1. *If P and Q are prime ideals in C, then (P, Q) is prime (or all of C).*

2. *Every z-ideal is an intersection of prime z-ideals.*

3. If the prime ideals containing a z-ideal I form a chain, then I is prime.

4. X is a P-space if and only if every prime ideal in $C(X)$ is a z-ideal. [Apply 4I.*5*. Alternatively, if a maximal ideal contains another prime ideal, then it is a union of upper ideals, and these are not z-ideals.]

14C. GENERATORS OF IDEALS IN C/P.

Let P be a prime ideal in $C(X)$. In the ring C/P:

1. If $0 \leqq a \leqq b^r$, for some rational $r > 1$, then a is a multiple of b. [Lemma 13.5 and 1D.*3*.]

2. Let I be the finitely generated ideal $(a_1, \cdots, a_n)$, where $0 < a_1 < \cdots < a_n$. Then $a_n^{1/2} \notin I$. [If $b \in I$, then $b \leqq sa_n$ for some s. Every element $\geqq 1$ is a unit.]

3. Every upper ideal is countably generated; in fact, $P^a = (a, a^{1/2}, a^{1/3}, \cdots)$. But no nonzero prime ideal is finitely generated.

4. No nonzero lower ideal is countably generated.

14D. LOWER IDEALS IN C.

Let P be a prime ideal in $C(X)$.

1. P is a lower ideal if and only if there exists $f \geqq \mathbf{0}$ such that $\{P(f^n)\}_{n \in \mathbf{N}}$ is coinitial in the set of all positive elements of C/P.

2. P is a lower ideal if and only if P is maximal with respect to disjointness from a set of the form $\{f^n\}_{n \in \mathbf{N}}$. [*Sufficiency*. f belongs to all prime ideals containing P properly.]

3. If P is a lower ideal, and Q is its immediate successor, then $Q = (P, f, f^{1/3}, f^{1/5}, \cdots)$, for any $f \in Q - P$. [14C.*3*.]

4. A z-ideal is never a lower ideal. [With $m \in C(\mathbf{R})$ as in 4I.*6*, and with f as in *1*, define $g = m \circ f$, so that $\mathbf{Z}(g) = \mathbf{Z}(f)$. Apply 7.15(b).]

14E. $C(\mathbf{N})$ AND $C(\Sigma)$.

Let $\Sigma = \mathbf{N} \cup \{\sigma\}$, where $\sigma \in \beta\mathbf{N} - \mathbf{N}$.

1. The natural isomorphism of $C(\Sigma)$ into $C(\mathbf{N})$ carries $\boldsymbol{O}_\sigma$ (in $C(\Sigma)$) onto $\boldsymbol{M}^\sigma$ (in $C(\mathbf{N})$).

2. $C(\Sigma)/\boldsymbol{O}_\sigma$ is isomorphic with the subring of all elements in $C(\mathbf{N})/\boldsymbol{M}^\sigma$ whose absolute value is not infinitely large.

3. $C(\mathbf{N})/\boldsymbol{M}^\sigma$ is isomorphic with the field of quotients of $C(\Sigma)/\boldsymbol{O}_\sigma$.

14F. PRIME z-FILTERS.

1. A z-filter $\mathscr{Z}$ on X is prime if and only if there exists an ultrafilter $\mathscr{U}$ such that $\mathscr{Z}$ is the family of all zero-sets belonging to $\mathscr{U}$—i.e., in the notation of 10.17, $\mathscr{Z} = \delta^{\#}\mathscr{U}$. [Adjoin the complement of every zero-set not in $\mathscr{Z}$.]

2. The prime z-ideals contained in $\boldsymbol{M}^p$ are precisely the z-ideals P such that $\mathbf{Z}[P] = \mathscr{U} \cap \mathbf{Z}(X)$ for some ultrafilter $\mathscr{U}$ on X that converges to p.

3. On the space Σ of 14E, there are just two ultrafilters that converge to σ. Hence $\boldsymbol{O}_\sigma$ and $\boldsymbol{M}_\sigma$ are the only z-ideals containing $\boldsymbol{O}_\sigma$. (4M.*9*.)

14G. PRIME IDEALS IN $C(\mathbf{N}^*)$.

In the ring $C(\mathbf{N}^*)$, where $\mathbf{N}^* = \mathbf{N} \cup \{\omega\}$ is the one-point compactification of $\mathbf{N}$:

1. $\boldsymbol{O}_n = \boldsymbol{M}_n$ for all $n \in \mathbf{N}$.

2. $\boldsymbol{O}_\omega$ is not prime.

3. The mapping

$$\mathscr{V} \to P = \{f \in C(\mathbf{N}^*) \colon \mathbf{Z}(f) - \{\omega\} \in \mathscr{V}\}$$

is one-one from the family of all free ultrafilters on $\mathbf{N}$ onto the family of all nonmaximal, prime z-ideals contained in $\boldsymbol{M}_\omega$. (Cf. 14F.)

4. None of the ideals P of *3* contains another. Hence these are the minimal prime ideals contained in $\boldsymbol{M}_\omega$. They are also the maximal z-ideals contained properly in $\boldsymbol{M}_\omega$.

5. There are $2^{\mathfrak{c}}$ maximal chains of prime ideals contained in $\boldsymbol{M}_\omega$. Any two have only $\boldsymbol{M}_\omega$ in common. Compare 14I.*8*.

14H. PRIME IDEALS IN $C(\mathbf{T})$.

Let $\mathscr{F}$ be a z-filter on $\mathbf{T}$ (8.20) converging to t. As before, W and N will denote the top and right edges of $\mathbf{T}$, respectively.

1. $\mathscr{F} = \boldsymbol{A}^t$ if and only if $W \in \mathscr{F}$. Compare 10L.*4*.

2. If $\mathscr{F} \neq \boldsymbol{A}^t$, then every member of $\mathscr{F}$ meets N.

3. If $\mathscr{F}$ is prime but not maximal, then the trace of $\mathscr{F}$ on N is a free ultrafilter on N. Moreover, distinct prime z-filters have distinct traces. [If $Z_1 \cap N = Z_2 \cap N$, then there exists a zero-set-neighborhood Z of N such that $Z_1 \cap Z = Z_2 \cap Z$.]

4. The mapping

$$\mathscr{V} \to P = \{f \in C(\mathbf{T}) \colon Z(f) \cap N \in \mathscr{V}\}$$

is one-one from the family of all free ultrafilters on N onto the family of all nonmaximal, prime z-ideals in $C(\mathbf{T})$ contained in $\boldsymbol{M}^t$. Compare 14G.*3*.

5. The ideals P of *4* are the minimal prime ideals contained in $\boldsymbol{M}^t$, and the maximal z-ideals contained properly in $\boldsymbol{M}^t$.

14I. PRIME IDEALS IN $C(\mathbf{R})$.

1. The identity function $\mathbf{i}$ in $C(\mathbf{R})$ belongs to at least $2^{\aleph_1}$ prime ideals, all contained in $\boldsymbol{M}_0$.

2. If $\mathscr{F}$ is a prime z-filter on $\mathbf{R} - \{0\}$ that converges to 0, then $\{Z \cup \{0\}\colon Z \in \mathscr{F}\}$ is a prime z-filter on $\mathbf{R}$, contained properly in $\boldsymbol{Z}[\boldsymbol{M}_0]$.

3. Let $\mathscr{V}$ be a free ultrafilter on the set $\{1/n\colon n \in \mathbf{N}\}$. The set P of all $f \in C(\mathbf{R})$ for which $Z(f)$ contains a member of $\mathscr{V}$ is a prime z-ideal. It is maximal in the set of all z-ideals contained properly in $\boldsymbol{M}_0$.

4. The correspondence $\mathscr{V} \to P$ of *3* is one-one. Hence there exist $2^{\mathfrak{c}}$ such prime z-ideals P. The function $\mathbf{i}$ belongs to none of them.

5. Given $\mathscr{V}$, let I denote the set of all $f \in C(\mathbf{R})$ such that int $Z(f)$ contains a member of $\mathscr{V}$. Then I is a z-ideal, but is not prime. [If $s \in C(\mathbf{R})$ is as in 2G.*2*, then neither $s \vee \mathbf{0}$ nor $s \wedge \mathbf{0}$ belongs to I.]

6. Every nonmaximal z-ideal containing I is contained in P.

7. The prime ideals contained in P do not form a chain. [14B.*2*.]

8. Distinct maximal chains of prime ideals in $C(\mathbf{R})$ can have infinitely many members in common. Compare 14G.*5*.

14J. C/P AS AN η_1-SET.

1. Let P be a nonmaximal, prime ideal in C. The positive elements in C/P have a countable, coinitial subset if and only if P is a countable intersection of lower ideals. [If A is coinitial, then $(0) = \bigcap_{a \in A} P_a$.]

2. If P is a lower ideal, then neither C/P, nor the set of infinitely small elements of C/P, is an η_1-set.

3. Let P be an upper ideal. The set of infinitely small elements of C/P is always an η_1-set. C/P itself is an η_1-set if and only if the maximal ideal containing P is hyper-real.

14K. FINITELY GENERATED IDEALS CONTAINED IN $\boldsymbol{M}^p$.

Fix $p \in \beta X$. If $(g, |g|)$ is principal for every $g \in \boldsymbol{M}^p$, then $\boldsymbol{O}^p$ is prime. [Lemma 14.23. For $f \notin \boldsymbol{M}^p$, $\boldsymbol{O}^p(f)$ is a unit.]

14L. FINITELY GENERATED IDEALS CONTAINED IN A GIVEN IDEAL.

Let I be an ideal in a commutative ring A with unity. If every finitely generated ideal contained in I is principal, then the prime ideals contained in I form a chain. [For $p \in P - Q$, and $q \in Q - P$, there exist a, b, s, and t

such that $p = ba$, $q \in (a)$, and $a = sp + tq$. Then $b \in P$ and $sb - 1 \in Q$, whence $1 \in I$.]

14M. THE SPACE $\beta\mathbf{Q} - \mathbf{Q}$.

1. Let $f(x) = \sin \pi x$ ($x \in \mathbf{Q}$), and let $p \in \mathrm{cl}_{\beta\mathbf{Q}} \mathbf{N} - \mathbf{N}$. If V is an open set in $\beta\mathbf{Q}$ containing p, then f^β assumes both positive and negative values on V. [V is a neighborhood of some point of $\mathbf{N}$.]

2. $\beta\mathbf{Q} - \mathbf{Q}$ is not an F-space. [$V - \mathbf{Q}$ is dense in V.]

14N. SEQUENCES OF POINTS IN AN F-SPACE.

1. No point of an F-space is the limit of a sequence of distinct points. [Given $(x_n)_{n \in \mathbf{N}}$, there exists a continuous f such that $f(x_n) = (-1)^n/n$.] Cf. *5.*

2. In an F-space, any point with a countable base of neighborhoods is isolated.

3. Every metrizable subspace of an F-space is discrete.

4. X is an F-space if and only if any two disjoint cozero-sets are completely separated. Every basically disconnected space (1H) is an F-space; the converse fails.

5. Every countable set in an F-space is C^*-embedded; hence every infinite compact F-space contains a copy of $\beta\mathbf{N}$. [Argue as in 9H.]

14O. SOME SPECIAL F-SPACES.

1. For any space X, every zero-set in βX that does not meet X is an F-space.

2. If E is a cozero-set in a compact space, then $\beta E - E$ is an F-space.

3. If X is a locally compact F-space, then $\beta X - X$ is an F-space.

14P. P-SPACES.

Let X be a P-space.

1. Every subspace of υX that contains X is a P-space.

2. Every subspace of βX that contains X, but is not contained in υX, is an F-space, but not a P-space. [If $p \in \beta X - \upsilon X$, then there exists $f \in C(\beta X)$ such that $f(p) = 0$, while $0 \notin f[X]$.]

14Q. F-SPACES AND PRODUCT SPACES.

1. If $X \times Y$ is an F-space, then X or Y is a P-space. [Assuming the contrary, define $h(x, y) = |f(x)| - |g(y)|$ for suitably chosen $f \in C(X)$ and $g \in C(Y)$.]

2. If X and Y are infinite pseudocompact spaces, then $X \times Y$ is not an F-space. [4K.*2*.]

3. *There exists no homeomorphic mapping of* $\beta\mathbf{R} \times \beta\mathbf{R}$ *into* $\beta(\mathbf{R} \times \mathbf{R})$. [A homeomorphism would carry some compact set that is not an F-space into $\beta(\mathbf{R} \times \mathbf{R}) - (\mathbf{R} \times \mathbf{R})$.]

Chapter 15

UNIFORM SPACES

15.1. In this chapter, we present some of the interactions between the theories of uniform spaces and rings of continuous functions. We shall find that every realcompact space admits a complete structure. One of the outstanding successes of the theory of rings of continuous functions is Shirota's result that, barring measurable cardinals, the converse is also true, so that the spaces admitting complete structures are precisely the realcompact spaces.

From our point of view, the most efficient approach to uniform spaces is by way of pseudometrics, as they provide us with a large supply of continuous functions. Accordingly, we *define* a uniform structure to be a family of pseudometrics (satisfying appropriate closure conditions). This enables us to give complete proofs relatively quickly of all the facts about uniform spaces that are needed here.

15.2. *Pseudometrics.* A *pseudometric* on a set X is a function d on $X \times X$ into $\mathbf{R}$, satisfying, for all $x, y, z \in X$:

(i) $d(x, y) \geqq 0$;
(ii) $d(x, x) = 0$;
(iii) $d(x, y) = d(y, x)$; and
(iv) $d(x, z) \leqq d(x, y) + d(y, z)$ (triangle inequality).

Thus, a pseudometric differs from a *metric* only in that $d(x, y) = 0$ need not imply $x = y$. For example, the function $\mathbf{0}$ on $X \times X$ is a pseudometric on X. We write

$$d[A, y) = d(y, A] = \inf_{x \in A} d(x, y),$$

and

$$d[A, B] = \inf_{x \in A,\, y \in B} d(x, y)$$

(where $A, B \neq \emptyset$). Evidently, $d[A, B] = \inf_{y \in B} d[A, y)$. The *d-diameter* of A, denoted by $d\{A\}$, is defined by:

$$d\{A\} = \sup_{x, y \in A} d(x, y)$$

(with $d\{A\} = \infty$ in case $d(x, y)$ is unbounded for $x, y \in A$).

If d_1 and d_2 are pseudometrics, then so is $d = d_1 \vee d_2$: properties (i), (ii), and (iii) are trivial, while the triangle inequality is obvious from the relations

$$d_k(x, z) \leqq d_k(x, y) + d_k(y, z) \leqq d(x, y) + d(y, z) \qquad (k = 1, 2).$$

15.3. *Uniform spaces.* By a *uniform structure* on X, we shall mean a nonempty family $\mathscr{D}$ of pseudometrics on X, with the properties:

(i) if $d_1, d_2 \in \mathscr{D}$, then $d_1 \vee d_2 \in \mathscr{D}$; and
(ii) if e is a pseudometric, and if for every $\epsilon > 0$, there exist $d \in \mathscr{D}$ and $\delta > 0$ such that

$$d(x, y) \leqq \delta \quad \text{implies} \quad e(x, y) \leqq \epsilon$$

for all $x, y \in X$, then $e \in \mathscr{D}$.

This last condition may be expressed as follows: $d\{A\} \leqq \delta$ implies $e\{A\} \leqq \epsilon$. By (ii), if $d \in \mathscr{D}$, then $rd \in \mathscr{D}$ for every $r > 0$; and, if $d \in \mathscr{D}$ and $e \leqq d$, then $e \in \mathscr{D}$.

A uniform structure $\mathscr{D}$ is called a *Hausdorff* uniform structure if

(iii) whenever $x \neq y$, there exists $d \in \mathscr{D}$ for which $d(x, y) \neq 0$.

The intersection of any collection of uniform structures on X is, obviously, a uniform structure. (Since $\mathbf{0}$ belongs to every structure, the intersection is never empty.) Hence, if $\mathscr{S}$ is any nonempty family of pseudometrics on X, then there exists a smallest structure $\mathscr{D}$ containing $\mathscr{S}$. We call $\mathscr{S}$ a *subbase* for $\mathscr{D}$, and we say that $\mathscr{D}$ is generated by $\mathscr{S}$.

A subbase $\mathscr{B}$ is called a *base* for $\mathscr{D}$ if for every $e \in \mathscr{D}$ and $\epsilon > 0$, there exist $d \in \mathscr{B}$ and $\delta > 0$ such that $d(x,y) \leqq \delta$ implies $e(x,y) \leqq \epsilon$.

(a) *Given a subbase $\mathscr{S}$ for $\mathscr{D}$, let $\mathscr{B}$ be the family of all suprema of finite subsets of $\mathscr{S}$. Then $\mathscr{B}$ is a base for $\mathscr{D}$.*

This is so because the set of all pseudometrics e that satisfy condition (ii), with $d \in \mathscr{B}$, is a uniform structure, and, clearly, it is the smallest one containing $\mathscr{S}$.

A set X, equipped with a uniform structure $\mathscr{D}$, is called a *uniform space*, and is denoted by $[X; \mathscr{D}]$—for short, by X. If $\mathscr{D}$ is a Hausdorff structure, then $[X; \mathscr{D}]$ is called a Hausdorff uniform space.

A metric uniform structure is a uniform structure with a base consisting of a single metric. Reference to **R** as a uniform space will be (unless stated otherwise) to the standard structure, i.e., the metric uniform structure generated by the metric $(r, s) \to |r - s|$.

15.4. *Uniform topology.* A uniform structure $\mathscr{D}$ on X induces a topology on X, called the *uniform topology*, defined as follows: a subbase for the neighborhood system at x consists of the sets

(a) $$\{y: d(x, y) < \epsilon\}$$

($d \in \mathscr{D}$, $\epsilon > 0$). Indeed, these sets form a base—and it is enough for d to range over a base for $\mathscr{D}$. Clearly, then, if d ranges over a subbase for $\mathscr{D}$, then the sets (a) still constitute a subbase at x. Closures in the uniform topology are given by:

(b) $$\text{cl } A = \bigcap_{d \in \mathscr{D}} \{x: d(x, A] = 0\};$$

for, $x \in \text{cl } A$ if and only if every basic set (a) meets A. Evidently, the uniform topology is a Hausdorff topology if and only if $\mathscr{D}$ is a Hausdorff structure. The uniform topology induced on a metric space is the usual (metric) topology. References to the topology of a uniform space are always to the uniform topology.

(c) *The mapping $x \to d[A, x)$ is always continuous* ($d \in \mathscr{D}$).

For, every $z \in A$ satisfies

$$d[A, x) \leqq d(z, x) \leqq d(z, y) + d(x, y),$$

whence

$$d[A, x) \leqq d[A, y) + d(x, y).$$

Hence if y belongs to the neighborhood (a) of x, then

$$|d[A, x) - d[A, y)| \leqq d(x, y) < \epsilon.$$

It follows from (c) that the set

(d) $$d\text{-cl } A = \{x: d[A, x) = 0\}$$

(which may be called the *d-closure* of A) is a zero-set. Moreover,

$$d\{d\text{-cl } A\} = d\{A\};$$

for, if $x, y \in d\text{-cl } A$, then

$$d(x, y) \leqq d(x, x') + d\{A\} + d(y, y'),$$

for all $x', y' \in A$, whence $d(x, y) \leqq d\{A\}$. Also, since $A \subset \text{cl } A \subset d\text{-cl } A$, we have

$$d\{\text{cl } A\} = d\{A\}.$$

15.5. *Admissible uniform structures; $\mathscr{C}$ and $\mathscr{C}^*$.* Suppose X is given as an arbitrary *topological space* (not necessarily completely regular), and let $\mathscr{D}$ be a uniform structure on X. If the uniform topology coincides with the given topology, then X is said to *admit* the structure $\mathscr{D}$, and $\mathscr{D}$ is called an *admissible* uniform structure on X. There may exist one or more admissible structures, or none at all. Obviously, a Hausdorff space may admit only Hausdorff uniform structures.

When X is a topological space, we can use the functions in $C(X)$ for defining various uniform structures on X (not necessarily admissible). For each $f \in C(X)$, let ψ_f be defined as follows:

$$\psi_f(x, y) = |f(x) - f(y)|.$$

Clearly, ψ_f is a pseudometric (even without the continuity of f). The structure generated by all ψ_f, for $f \in C(X)$, will be denoted by $\mathscr{C} = \mathscr{C}(X)$; the structure generated by all ψ_f, for $f \in C^*(X)$, is denoted by $\mathscr{C}^* = \mathscr{C}^*(X)$. In each case, the pseudometrics ψ_f form, in general, only a subbase. A convenient base consists of all pseudometrics of the form

$$\psi_{f_1} \vee \cdots \vee \psi_{f_n}.$$

On the space $\mathbf{R}$, $\psi_{\mathbf{i}}$ is the usual metric, and generates the standard structure. For other admissible structures on $\mathbf{R}$, see 15A.

15.6. THEOREM. *A necessary and sufficient condition that a Hausdorff space X admit a uniform structure is that X be completely regular. Moreover, if X is completely regular, then both $\mathscr{C}$ and $\mathscr{C}^*$ are admissible uniform structures.*

PROOF. If X admits a structure $\mathscr{D}$, then

$$\operatorname{cl} A = \bigcap\nolimits_{d \in \mathscr{D}} d\text{-}\operatorname{cl} A,$$

so that every closed set is an intersection of zero-sets. Hence X is completely regular (Theorem 3.2).

On the other hand, the collection of all sets

$$\{y : \psi_f(x, y) < \epsilon\} = \{y : |f(x) - f(y)| < \epsilon\} \qquad (x \in X, \epsilon > 0),$$

for $f \in C(X)$ [resp. $f \in C^*(X)$], is a subbase in the uniform topology induced by $\mathscr{C}$ [resp. $\mathscr{C}^*$]; and if X is completely regular, these sets form a subbase (in fact, a base) for the topology of X (Theorem 3.6).

Because of this theorem, we may as well assume from now on that all given uniform spaces are *Hausdorff* spaces. Nevertheless, we shall usually include the adjective when it is really needed.

COMPLETE SPACES

15.7. *Cauchy z-filters.* Let $[X; \mathscr{D}]$ be a uniform space. A z-filter $\mathscr{F}$ is said to be a *Cauchy* z-filter if for every $d \in \mathscr{D}$ and $\epsilon > 0$, $\mathscr{F}$ contains a set of d-diameter $< \epsilon$—briefly, $\mathscr{F}$ contains *arbitrarily small* sets. Clearly, it suffices here to let d range over a base for $\mathscr{D}$.

The uniform space $[X; \mathscr{D}]$ is said to be *complete* if whenever $\mathscr{H}$ is a family of closed sets with the finite intersection property, and contains arbitrarily small sets, then $\bigcap \mathscr{H} \neq \emptyset$.

(a) *X is complete if and only if every Cauchy z-filter is fixed.*

Necessity is trivial. For the sufficiency, let $\mathscr{H}$ be as above. The family of zero-sets

$$\{d\text{-cl}\, H: d \in \mathscr{D}, H \in \mathscr{H}\}$$

is contained in a z-filter $\mathscr{F}$. Since $d\{d\text{-cl}\, H\} = d\{H\}$, $\mathscr{F}$ is a Cauchy z-filter. By hypothesis, $\bigcap \mathscr{F} \neq \emptyset$, and so we have

$$\bigcap \mathscr{H} = \bigcap_{H \in \mathscr{H}} \text{cl}\, H = \bigcap_{H \in \mathscr{H}} \bigcap_{d \in \mathscr{D}} d\text{-cl}\, H \supset \bigcap \mathscr{F} \neq \emptyset.$$

Since every z-filter is contained in a z-ultrafilter, we have, from (a):

(b) *X is complete if and only if every Cauchy z-ultrafilter is fixed.*

When $[X; \mathscr{D}]$ is a complete uniform space, we refer to $\mathscr{D}$ as a complete structure.

Note that, in any case, every *fixed* z-ultrafilter is a Cauchy z-filter—although an arbitrary z-filter whose intersection is just one point need not be a Cauchy z-filter. On the other hand, a Cauchy z-filter converges to each of its cluster points. In a *Hausdorff* space, a Cauchy z-filter converges to at most one point. Hence in a complete, Hausdorff uniform space, the intersection of a Cauchy z-filter contains exactly one point.

Assuredly, a *compact* space X is always complete. (By Theorem 15.6, $\mathscr{C}(X)$ is an admissible structure. It is also the *unique* admissible structure; see 15H.) The noncompact discrete space **N**, like any discrete space, is complete in its *discrete* structure, generated by the metric u for which $u(x, y) = 1$ whenever $x \neq y$. (On **N** itself, this is the same as the structure generated by the usual metric $(m, n) \rightarrow |m - n|$.) The noncompact space **R** is complete, since every closed set of finite diameter is compact. More generally, every sequentially complete metric space is complete (15F.*2*). As we shall see, $[\mathbf{R}; \mathscr{C}(\mathbf{R})]$ and $[\mathbf{N}; \mathscr{C}(\mathbf{N})]$ are also complete, while, on the other hand, neither $[\mathbf{R}; \mathscr{C}^*(\mathbf{R})]$ nor $[\mathbf{N}; \mathscr{C}^*(\mathbf{N})]$ is complete.

15.8. *Subspaces; completion.* Let X be a subset of a uniform space $[T; \mathscr{U}]$. Trivially, if $u \in \mathscr{U}$, then the restriction of u to $X \times X$ is a pseudometric on X. The *relative* uniform structure $\mathscr{D}$ on X is the structure generated by all such restrictions, for $u \in \mathscr{U}$. The space $[X; \mathscr{D}]$ is called a subspace of $[T; \mathscr{U}]$.

A check of the definitions shows that it is enough to have u range over a subbase for $\mathscr{U}$; and in case u ranges over a base, the induced pseudometrics will constitute a base for $\mathscr{D}$. It follows from this last that the uniform topology on $[X; \mathscr{D}]$ is the same as that induced by the uniform topology of $[T; \mathscr{U}]$. Consequently, references to the topology on a subspace of a uniform space are unambiguous.

A *completion* of a uniform space X is a complete Hausdorff uniform space in which X is dense. One of the key results in the theory of uniform spaces is the theorem that every Hausdorff uniform space X has a completion, γX; moreover, γX is essentially unique. We shall include a proof of this result, although it is not needed for the proof of Shirota's theorem (15.20).

The obvious way to produce γX is by generalizing the familiar construction of $\mathbf{R}$ as the completion of $\mathbf{Q}$. This generalization is outlined in the *Notes*. It will be seen to resemble the development of βX given in Chapter 6. This is not an accident: once the general completion theorem has been established, Theorem 15.13(a*) below may be used as the *definition* of βX.

What we shall do here is reverse the procedure by making use of βX —which has already been developed and explored—to construct γX. This approach has the advantage of exposing the relationship between βX and γX. Since γX is a topological space in which X is dense, it must be a quotient space of a suitable subspace of βX (Theorem 10.13). Now, each Cauchy z-filter must have a unique limit in γX. Accordingly, the construction of γX will be accomplished in two stages. In the first, the pseudometrics on X are extended to the subspace

$$cX = \{p \in \beta X: A^p \text{ is a Cauchy } z\text{-ultrafilter}\}$$

of βX. (Recall that A^p denotes the z-ultrafilter on X that converges to the point p of βX.) Here, we use the fact that X is C^*-embedded in βX. However, $X \times X$ is typically *not* C^*-embedded in $\beta X \times \beta X$ (see *Notes*, Chapter 6, *Compactification of a product*). Accordingly, the pseudometric—which is a function on $X \times X$—is extended in two steps, one variable at a time. The second stage in the construction of γX consists in identifying all points in cX that are cluster points of the same Cauchy z-filter.

15.9. COMPLETION THEOREM. *Every Hausdorff uniform space* $[X; \mathscr{D}]$ *has a completion* $\gamma X = \gamma[X; \mathscr{D}]$.

PROOF. We wish to extend every pseudometric $d \in \mathscr{D}$ to a pseudometric d^c on cX, in such a way that

(a) *For each fixed* $p \in cX$, *the mapping* $q \rightarrow d^c(p, q)$ *is continuous on* cX *into* $\mathbf{R}$.

Fix $y \in X$. The function $x \rightarrow d(x, y)$ belongs to $C(X)$ (15.4(c)); hence, by Stone's theorem (6.5), it has a continuous extension $p \rightarrow \bar{d}(p, y)$ from cX into the one-point compactification $\mathbf{R}^*$ of $\mathbf{R}$. This extension actually maps into $\mathbf{R}$ itself. For, select $Z \in A^p$ such that $d\{Z\}$ is finite. The triangle inequality implies that

$$\{d(z, y) : z \in Z\}$$

is a bounded set of real numbers. By continuity, $\bar{d}(p, y)$ belongs to the closure of this set (since $p \in \operatorname{cl} Z$).

From the relation

$$d(x, y) \leqq d(x, z) + d(y, z)$$

on X, we obtain, by continuity,

$$\text{(b)} \qquad \bar{d}(p, y) \leqq \bar{d}(p, z) + d(y, z)$$

for $p \in cX$. This implies (as in 15.4(c)) that the function $y \rightarrow \bar{d}(p, y)$ from X into $\mathbf{R}$ is continuous (for each fixed $p \in cX$). In turn, then, this function has a continuous extension $q \rightarrow d^c(p, q)$ from cX into $\mathbf{R}^*$. Using (b), we see that this, too, maps into $\mathbf{R}$. We have now established (a).

We still have to prove that the mapping $(p, q) \rightarrow d^c(p, q)$ is a pseudometric. Obviously, $d^c(p, q) \geqq 0$. To see that $d^c(p, p) = 0$, consider any $\epsilon > 0$, and choose Z in the Cauchy z-ultrafilter A^p such that $d\{Z\} < \epsilon$. By continuity, there exist $y \in Z$ such that

$$|d^c(p, p) - \bar{d}(p, y)| < \epsilon,$$

and $x \in Z$ such that

$$|\bar{d}(p, y) - d(x, y)| < \epsilon.$$

Therefore,

$$d^c(p, p) \leqq |d^c(p, p) - d(x, y)| + d(x, y) < 3\epsilon.$$

Finally to establish symmetry and the triangle inequality, we start with (b); two applications of continuity yield, successively,

$$d^c(p, q) \leqq \bar{d}(p, z) + \bar{d}(q, z),$$

and

$$d^c(p, q) \leqq d^c(p, r) + d^c(q, r),$$

for p, q, $r \in cX$. Putting $r = p$, we get $d^c(p, q) \leqq d^c(q, p)$; interchanging p with q leads to equality. Thus, d^c is a pseudometric.

We come to the second stage in the construction of γX. Define $p \equiv p'$ in cX to mean that $d^c(p, p') = 0$ for all $d \in \mathscr{D}$. This is obviously an equivalence relation, and each pseudometric d^c depends only upon the equivalence classes of its arguments. Denote the equivalence class of p by p^γ. The classes p^γ will be the points of γX. The equation

$$d^\gamma(p^\gamma, q^\gamma) = d^c(p, q)$$

defines d^γ as a pseudometric on γX, and the collection $\{d^\gamma : d \in \mathscr{D}\}$ generates a Hausdorff uniform structure on γX.

Now we have to prove that γX is a completion of X.

Since the restriction of d^c to $X \times X$ is d, and since X is a Hausdorff space, the mapping $x \to x^\gamma$, for $x \in X$, is one-one. As a matter of fact, $x^\gamma = \{x\}$, as is easily seen, and we shall identify x^γ with x. Under this identification, the restriction of d^γ to $X \times X$ is d. Since the pseudometrics d^γ form a subbase for the structure on γX, their restrictions form a subbase for the relative structure on X. But these restrictions generate (in fact, form) $\mathscr{D}$. Therefore, $\mathscr{D}$ is the relative structure on X, and so the mapping $x \to x^\gamma$ is an embedding of $[X; \mathscr{D}]$ in γX.

(c) *The mapping $p \to p^\gamma$ from cX onto γX is continuous.*

For, the inverse image of the subbasic open set

$$\{q^\gamma : d^\gamma(p^\gamma, q^\gamma) < \epsilon\}$$

is the open set $\{r : d^c(p, r) < \epsilon\}$.

Since X is dense in cX, it follows that X is dense in γX.

Finally, to see that γX is complete, consider any Cauchy z-filter $\mathscr{F}$ on γX. The collection of all zero-sets in X of the form

$$Z = \{x \in X : d^\gamma[F, x) \leqq \epsilon\},$$

where $F \in \mathscr{F}$, $d \in \mathscr{D}$, and $\epsilon > 0$, is embeddable in a z-ultrafilter A^p. Since $d\{Z\} \leqq d^\gamma\{F\} + 2\epsilon$, A^p is a Cauchy z-filter; so $p \in cX$. Now,

$$d^\gamma[F, p^\gamma) \leqq d^\gamma[Z, p^\gamma) + \epsilon = d^c[Z, p) + \epsilon = \epsilon.$$

It follows that $p^\gamma \in \operatorname{cl} F = F$. Thus, $p^\gamma \in \bigcap \mathscr{F}$.

REMARK. In case X itself is complete, so that $cX = X$, then, of course, $\gamma X = X$. More generally, if, for distinct p and q in cX, there exists $d \in \mathscr{D}$ such that $d^c(p, q) \neq 0$, then each equivalence class p^γ consists of the single point p, whence $\gamma X = cX$ (under the identification of p^γ with p). Moreover, in this case, the uniform topology on γX agrees with the topology of cX (see 15S.*3*).

UNIFORM CONTINUITY; UNIQUENESS OF COMPLETION

15.10. In the general theory of uniform spaces, the concept of *uniform continuity* is fundamental. If φ is a mapping from $[X; \mathscr{D}]$ to $[Y; \mathscr{E}]$, then clearly, for any $e \in \mathscr{E}$, the function $(x, x') \to e(\varphi x, \varphi x')$ is a pseudometric on X. If for every $e \in \mathscr{E}$, this pseudometric belongs to $\mathscr{D}$, then φ is said to be *uniformly continuous.* In terms of bases $\mathscr{D}'$ and $\mathscr{E}'$, respectively, the definition takes the familiar form: φ is uniformly continuous if and only if, for each $e \in \mathscr{E}'$ and $\epsilon > 0$, there exist $d \in \mathscr{D}'$ and $\delta > 0$ such that $d(x, x') \leqq \delta$ implies $e(\varphi x, \varphi x') \leqq \epsilon$ (for all $x, x' \in X$)—or, equivalently, $d\{A\} \leqq \delta$ implies $e\{\varphi[A]\} \leqq \epsilon$.

It is evident that a uniformly continuous mapping is continuous (with respect to the uniform topologies).

A function f from $[X; \mathscr{D}]$ into $\mathbf{R}$ is uniformly continuous if and only if $\psi_f \in \mathscr{D}$ (see 15.5). The function $x \to d[A, x)$, for $d \in \mathscr{D}$, is always uniformly continuous, as is shown by the proof of 15.4(c). Every f in $C(X)$ is uniformly continuous in the structure $\mathscr{C}$, and, moreover, $\mathscr{C}$ is the smallest such structure. Likewise, $\mathscr{C}^*$ is the smallest structure in which every f in $C^*(X)$ is uniformly continuous.

A one-one, uniformly continuous mapping whose inverse is also uniformly continuous is called a *uniform isomorphism.* Thus, a uniform isomorphism of $[X; \mathscr{D}]$ onto $[Y; \mathscr{E}]$ is a one-one mapping of X onto Y that induces (as above) a one-one mapping of $\mathscr{E}$ onto $\mathscr{D}$.

15.11. THEOREM. *If X is dense in a uniform space $[T; \mathscr{U}]$, then every uniformly continuous mapping φ from X to a complete Hausdorff uniform space $[Y; \mathscr{E}]$ has an extension to a uniformly continuous mapping $\bar{\varphi}$ from T into Y.*

PROOF. Given $p \in T$, let $\mathscr{F}_p$ denote the z-filter on X generated by the family of zero-sets

$$F = \{x \in X: u(x, p) \leqq \delta\},$$

for all $u \in \mathscr{U}$ and $\delta > 0$. Clearly, $\mathscr{F}_p$ is a Cauchy z-filter.

Consider the z-filter

$$\varphi^{\#}\mathscr{F}_p = \{Z \in \mathbf{Z}(Y): \varphi^{\leftarrow}[Z] \in \mathscr{F}_p\}$$

on Y. The restrictions to $X \times X$ of the pseudometrics in $\mathscr{U}$ form a base for the uniform structure on X. So, given $e \in \mathscr{E}$, and $\epsilon > 0$, there exist $u \in \mathscr{U}$ and $\delta > 0$ such that, for $A \subset X$, $u\{A\} \leqq 3\delta$ implies

$e\{\varphi[A]\} \leqq \epsilon$. Now, we can choose $A \in \mathscr{F}_p$ with $u\{A\} \leqq 3\delta$; then the zero-set e-cl $\varphi[A]$ belongs to $\varphi^{\#}\mathscr{F}_p$, and we have

$$e\{e\text{-cl}\,\varphi[A]\} = e\{\varphi[A]\} \leqq \epsilon.$$

Therefore, $\varphi^{\#}\mathscr{F}_p$ is a Cauchy z-filter on Y.

Since Y is a complete Hausdorff space, there exists a unique point q in $\bigcap \varphi^{\#}\mathscr{F}_p$. We define $\hat{\varphi}p = q$. When p is in X, it belongs to $\bigcap \mathscr{F}_p$, so that $\varphi p \in \bigcap \varphi^{\#}\mathscr{F}_p$. Therefore $\hat{\varphi}$ agrees with φ on X.

Finally, consider any subset S of T, with $u\{S\} \leqq \delta$. Define

$$S' = \{x \in X\colon u(x, S] \leqq \delta\}.$$

Then $u\{S'\} \leqq 3\delta$, and so $e\{\varphi[S']\} \leqq \epsilon$. Since $p \in S$ implies $S' \in \mathscr{F}_p$, so that $\hat{\varphi}p \in e$-cl $\varphi[S']$, we have $\hat{\varphi}[S] \subset e$-cl $\varphi[S']$. Therefore $e\{\hat{\varphi}[S]\} \leqq \epsilon$. This shows that $\hat{\varphi}$ is uniformly continuous.

15.12. COROLLARY. *If T is a completion of a uniform space X, then there exists a uniform isomorphism of γX onto T that leaves X pointwise fixed.*

The proof is like the proof of uniqueness of βX (Theorem 6.5).

Thus, we may refer to γX as *the* completion of X.

It follows from the corollary that a complete Hausdorff uniform space is its own completion. Also, if X is dense in a Hausdorff uniform space Y, then $\gamma X = \gamma Y$.

In contrast to the uniqueness of the completion, a completely regular space can have several compactifications. The correct analogue of completion is not arbitrary compactification, but rather Stone-Čech compactification. Corresponding to the extendability to βX of any continuous mapping into a compact space, we have here the extendability to γX of any uniformly continuous mapping into a complete space. As a matter of fact, the former result is a special case of the latter (15P.*2*).

βX AND υX AS COMPLETIONS

15.13. If a z-ultrafilter A^p on X is not real, then there exists a function f in $C(X)$ that is unbounded on every member of A^p (see 5.7(a) or 8.4). Then A^p contains no set of finite ψ_f-diameter. Thus, every Cauchy z-ultrafilter in the uniform structure $\mathscr{C}(X)$ is real. Consequently, if X is *realcompact*, then it is complete in the structure $\mathscr{C}(X)$.

THEOREM. *Let X be a completely regular space.*

(a) *The completion of X in the uniform structure $\mathscr{C}(X)$ is $[\upsilon X; \mathscr{C}(\upsilon X)]$.*

(a*) *The completion of X in the uniform structure $\mathscr{C}^*(X)$ is $[\beta X; \mathscr{C}^*(\beta X)]$.*

PROOF. (a). As noted above, $[\upsilon X; \mathscr{C}(\upsilon X)]$ is complete. Now, X is dense in υX, of course, and clearly the relative structure on X is $\mathscr{C}(X)$. Since the completion of $[X; \mathscr{C}(X)]$ is unique, it must be $[\upsilon X; \mathscr{C}(\upsilon X)]$.

The proof of (a*) is similar, since a compact space is always complete.

It follows from (a) that the Cauchy z-ultrafilters in the structure $\mathscr{C}(X)$ are the same as the real z-ultrafilters; and, by (a*), *every z-ultrafilter is a Cauchy z-filter in the structure* $\mathscr{C}^*(X)$. These results will be generalized in Theorems 15.21 and 15.16, respectively.

15.14. COROLLARY.

(a) *X is realcompact if and only if it is complete in the structure $\mathscr{C}$.*

(a*) *X is compact if and only if it is complete in the structure $\mathscr{C}^*$.*

COMPACT AND REALCOMPACT COMPLETIONS

15.15. Let $[X; \mathscr{D}]$ be a uniform space, and let $d \in \mathscr{D}$ and $\delta > 0$. A family of sets E_α will be called a *d-discrete* family, of *gauge* δ, provided that $d[E_\alpha, E_\sigma] \geqq \delta$ whenever $\alpha \neq \sigma$. A d-discrete set of points is defined similarly. A set A in X is said to be *d-closed* if $d\text{-cl}\, A = A$. Thus, every d-closed set is a zero-set.

(a) *The union of a d-discrete family of closed sets is closed. If each set is d-closed, then their union is d-closed and hence is a zero-set.*

In fact, for any x, the neighborhood $\{y: d(x, y) < \delta/2\}$ meets at most one set E_α (notation as above). Therefore,

$$\text{cl} \bigcup_\alpha E_\alpha \subset \bigcup_\alpha \text{cl}\, E_\alpha,$$

and

$$d\text{-cl} \bigcup_\alpha E_\alpha \subset \bigcup_\alpha d\text{-cl}\, E_\alpha.$$

In particular, every d-discrete set in X is closed.

(b) *Any d-discrete set S in X is C-embedded in X.*

The function $g \in C(S)$ has, as a continuous extension to X, the function f defined by:

$$f(x) = g(s)(1 - 3d(x, s)/\delta) \quad \text{for} \quad d(x, s) \leqq \delta/3,\ s \in S,$$

and

$$f(x) = 0 \quad \text{otherwise,}$$

where δ is a gauge for S.

15.16. THEOREM. *The following are equivalent for any Hausdorff uniform space* $[X; \mathscr{D}]$.

(1) γX *is compact (in which case* X *and* $\mathscr{D}$ *are said to be precompact).*

(2) *For each* $d \in \mathscr{D}$, *every d-discrete set in* X *is finite.*

(3) *For every* $d \in \mathscr{D}$ *and* $\epsilon > 0$, X *is a finite union of zero-sets of d-diameter* $\leqq \epsilon$ *(in which case* $[X; \mathscr{D}]$ *is said to be totally bounded).*

(4) *Every z-ultrafilter on* X *is a Cauchy z-filter, i.e.,* $\beta X = cX$.

PROOF. (1) *implies* (2). A d-discrete set in X is d^γ-discrete, and hence is closed and discrete in the compact space γX.

(2) *implies* (3). Suppose that for some $d \in \mathscr{D}$ and $\epsilon > 0$, X is not the union of any finite number of zero-sets of d-diameter $\leqq \epsilon$. Inductively, choose x_n in the complement of

$$\bigcup_{k=1}^{n-1} \{x: d(x, x_k) \leqq \epsilon/2\}.$$

The set $\{x_n\}_{n \in \mathbf{N}}$ is obviously infinite and d-discrete.

(3) *implies* (4). Every z-ultrafilter is prime.

(4) *implies* (1). Under the hypothesis, γX is a continuous image of βX (15.9(c)).

It follows from the theorem that *every admissible structure contained in a precompact structure is precompact.*

The theorem provides several characterizations of a uniform space whose completion is compact. (See also 15I, J, K, and Q.) We shall obtain an analogue for the realcompact case; however, the techniques of the proof will be quite different from those just employed. The main tool is presented in the next theorem.

15.17. THEOREM. *Let* $[X; \mathscr{D}]$ *be a Hausdorff uniform space, and let* $d \in \mathscr{D}$, *and* $\epsilon > 0$ *be given. There exist sets* $Z_{n,x}$ $(n \in \mathbf{N}, x \in X)$ *with the following properties:*

(i) $\bigcup_{n,x} Z_{n,x} = X$.

(ii) *Each set* $Z_{n,x}$ *is d-closed, and of d-diameter* $< \epsilon$.

(iii) *For each* $n \in \mathbf{N}$, *the family of sets* $Z_{n,x}$ $(x \in X)$ *is d-discrete.*

PROOF. We make tacit use of the conventions $d\{\emptyset\} = 0$, and $d[\emptyset, A] = +\infty > r$ (for $r \in \mathbf{R}$). Let $\prec$ be a well-ordering of X. For each fixed n, we proceed by induction on x. Let $\delta = \epsilon/2$. We put z in $Z_{n,x}$ if and only if

$$d[Z_{n,y}, z) \geqq \delta/n \quad \text{for all} \quad y \prec x,$$

and

$$d(x, z) \leqq \delta - \delta/n.$$

It is clear that $Z_{n,x}$ is d-closed, that $d\{Z_{n,x}\} < 2\delta = \epsilon$, and, for $y \prec x$, that $d[Z_{n,x}, Z_{n,y}] \geqq \delta/n$. This establishes (ii) and (iii).

To prove (i), consider any $z \in X$. There exists a least element x (in the well-ordering) for which $d(x, z) < \delta$. Select n so that $d(x, z) < \delta - \delta/n$. For each $y \prec x$, we have $d(y, z) \geqq \delta$ (by definition of x); hence, for every $w \in Z_{n, y}$,

$$d(w, z) \geqq d(y, z) - d(y, w) \geqq \delta - (\delta - \delta/n) = \delta/n,$$

which shows that $d[Z_{n, y}, z] \geqq \delta/n$. Therefore $z \in Z_{n, x}$.

15.18. For each n, every union of sets $Z_{n, x}$ is a zero-set (15.15(a)). In particular, for each n, the set $\bigcup_{x \in X} Z_{n, x}$ is a zero-set. Now, the union of this countable family of zero-sets is X. Hence, given a real z-ultrafilter $\mathscr{A}$, there exists $k \in \mathbf{N}$ such that $\bigcup_{x \in X} Z_{k, x} \in \mathscr{A}$ (5.15(a)). We summarize these facts as follows (relabeling the nonempty $Z_{k, x}$).

Given $d \in \mathscr{D}$, $\epsilon > 0$, and a real z-ultrafilter $\mathscr{A}$, there exists a family of nonempty sets Z_α with the following properties:

(i) $\bigcup_\alpha Z_\alpha \in \mathscr{A}$.
(ii) *Each Z_α is of d-diameter $< \epsilon$.*
(iii) *The family is d-discrete.*
(iv) *The union of every subfamily is a zero-set.*

Now we construct a set S by selecting one point s_α from each Z_α. Then S is d-discrete. Define $\mathscr{A}_S$ as follows: for $E \subset S$,

$$E \in \mathscr{A}_S \quad \text{if and only if} \quad \bigcup_{s_\alpha \in E} Z_\alpha \in \mathscr{A}.$$

We prove that $\mathscr{A}_S$ is a real z-ultrafilter on S. Since the sets Z_α are disjoint, the correspondence

$$E \rightarrow \bigcup_{s_\alpha \in E} Z_\alpha$$

preserves intersection as well as union. Consequently, $\mathscr{A}_S$ is a filter with the countable intersection property (since $\mathscr{A}$ has this property; see 5.15). Moreover, we see from (i) and (iv) that for any $E \subset S$, either $E \in \mathscr{A}_S$ or $S - E \in \mathscr{A}_S$; therefore $\mathscr{A}_S$ is an ultrafilter.

15.19. LEMMA. *Let $[X; \mathscr{D}]$ be a Hausdorff uniform space. If for all $d \in \mathscr{D}$, every d-discrete subspace is realcompact, then every real z-ultrafilter $\mathscr{A}$ on X is a Cauchy z-filter.*

PROOF. Consider any $d \in \mathscr{D}$ and $\epsilon > 0$. The set S defined as above is d-discrete, and therefore, by hypothesis, it is realcompact. Consequently, the real z-ultrafilter $\mathscr{A}_S$ is fixed, and so there exists α for which $\{s_\alpha\} \in \mathscr{A}_S$. Then $\mathscr{A}$ contains the set Z_α of d-diameter $< \epsilon$.

As the reader will recall from our discussion in Chapter 12, the requirement that a discrete space be realcompact is very mild: a discrete space fails to be realcompact if and only if its cardinal number is measurable. It is not known whether any measurable cardinals exist;

and any that may exist can be regarded as pathological, since they cannot be obtained from $\aleph_0$ and the other nonmeasurable cardinals by any of the standard processes of cardinal arithmetic. According to the preceding lemma, in order to conclude that every real z-ultrafilter is a Cauchy z-filter, in any structure, one has only to exclude the existence of d-discrete subspaces of measurable cardinal.

15.20. THEOREM (SHIROTA). *Let X be a completely regular space in which every closed discrete subspace has nonmeasurable cardinal. A necessary and sufficient condition that X admit a complete uniform structure is that it be realcompact.*

PROOF. Suppose that X admits a complete structure $\mathscr{D}$. By hypothesis, every closed discrete subspace, hence every d-discrete subspace (for all $d \in \mathscr{D}$), is realcompact. By the lemma, every real z-ultrafilter is a Cauchy z-filter, and hence fixed, since X is complete. Therefore X is realcompact. Conversely, if X is realcompact, then $\mathscr{C}(X)$ is an admissible complete structure (15.13).

The proof of sufficiency did not use the hypothesis about subspaces. But this is neither here nor there, because in a realcompact space, *every* closed set is realcompact (Theorem 8.10). Thus:

(a) *A complete space is realcompact if and only if every closed discrete subspace is realcompact.*

According to the theorem, if $|X|$ is nonmeasurable (in particular, if measurable cardinals do not exist), then X admits a complete structure if and only if it is realcompact. The condition on X is critical: should a measurable cardinal exist, then the discrete space of that cardinal is not realcompact, although, like any discrete space, it is complete in its discrete structure.

15.21. We now prove the analogue of Theorem 15.16. It includes a sharpened version of Shirota's theorem.

THEOREM. *The following are equivalent for any Hausdorff uniform space $[X; \mathscr{D}]$.*

(1) *γX is realcompact.*

(2) *For each $d \in \mathscr{D}$, the cardinal of every d-discrete set in X is nonmeasurable.*

(3) *For every $d \in \mathscr{D}$ and $\epsilon > 0$, X is a nonmeasurable union of zero-sets of d-diameter $\leqq \epsilon$.*

(4) *Every real z-ultrafilter on X is a Cauchy z-filter*, i.e., $\upsilon X \subset cX$.

PROOF. We shall prove that condition (2) is equivalent to each of the other three, in turn.

(2) *implies* (3). This follows as in Theorem 15.16, except that transfinite induction is now used in place of ordinary induction.

(3) *implies* (2). Let S be a d-discrete set. By hypothesis, X is a nonmeasurable union of sets of d-diameter smaller than a gauge for S; and each of these contains at most one point of S.

(2) *implies* (4). This is Lemma 15.19.

(4) *implies* (2). Let S be a d-discrete set in X. Then S is C-embedded in X (15.15(b)), so that $\text{cl}_{vX} S = vS$ (8.10(a)). Hence the hypothesis implies that $vS \subset cX$. Given $p \in vS$, let V be a neighborhood of p in cX whose d^c-diameter is less than a gauge for S. Then $V \cap S$ contains at most one point. But $p \in \text{cl}_{cX}(V \cap S)$, because $p \in \text{cl}\, S$. Therefore, $p \in V \cap S \subset S$. This shows that $vS \subset S$, i.e., the discrete space S is realcompact. Hence its cardinal is nonmeasurable.

(2) *implies* (1). We prove, first, that the condition corresponding to (2) holds for subsets of γX. Let T be any u-discrete subset of γX (where u belongs to the structure on γX). Construct $S \subset X$ by selecting, for each $t \in T$, a point $s \in X$ such that $u(t, s) < \delta/3$, where δ is a gauge for T. Then S is d-discrete (with gauge $\delta/3$), where d is the restriction of u. Hence the cardinal $|T| = |S|$ is nonmeasurable.

It follows that every real z-ultrafilter on γX is a Cauchy z-filter. Therefore the complete space γX is realcompact.

(1) *implies* (2). A d-discrete set in X is d^γ-discrete, and hence is closed in γX; and a closed set in a realcompact space is realcompact.

15.22. Let us examine what this theorem adds to our previous knowledge. We notice that only two of the proofs—(4) implies (2), and (2) implies (1)—required any effort; all the other implications listed followed quickly from what was already known. And, in two important special cases, everything had already been settled:

(i) *X is complete.* Then (1) and (4) are obviously equivalent; and the deep result that (2) implies (1) is Shirota's theorem.

(ii) $|X|$ *is nonmeasurable.* Then (2) is satisfied trivially; also, $|\gamma X|$ is nonmeasurable (see 9A). By Lemma 15.19, (4) holds, and by Shirota's theorem, (1) holds. Thus (1) and (4) are equivalent because they are both true.

The main new result in this theorem is that (4) implies (1) even if X contains sets of measurable cardinal. Although neither condition explicitly involves a restriction on cardinals, we do not know how to derive (1) from (4) without passing through (2). On the other hand, the converse implication is easy; see 15S.5.

15.23. Suppose that $\mathscr{D}$ and $\mathscr{E}$ are admissible structures on X, with $\mathscr{D} \subset \mathscr{E}$. Any Cauchy z-filter relative to $\mathscr{E}$ is certainly a Cauchy z-filter relative to $\mathscr{D}$. Therefore, if X is complete in $\mathscr{D}$, it is also complete in $\mathscr{E}$.

As an example, consider the structure $\mathscr{C}$ on $\mathbf{R}$. Since $\mathscr{C}(\mathbf{R})$ contains the metric $\psi_{\mathbf{i}}$, it contains the standard uniform structure. Since $\mathbf{R}$ is complete in this structure, it is also complete in $\mathscr{C}$. (This gives us yet another proof that $\mathbf{R}$ is realcompact.) In this example, $\mathscr{C}$ is far larger than the metric structure: *every* function in $C(\mathbf{R})$ is uniformly continuous in $\mathscr{C}$ (into $\mathbf{R}$ with the standard structure).

Every realcompact space X is complete in $\mathscr{C}(X)$. Hence, if X is realcompact, then it is complete in every admissible structure that contains $\mathscr{C}$. Shirota's theorem states a sort of converse: barring measurable cardinals, if there exists *any* admissible structure in which X is complete, then X is also complete in $\mathscr{C}$. As we have just seen, this tells us very little in the particular case $X = \mathbf{R}$. The power of the result resides in the fact that the structure $\mathscr{C}$, in more general spaces, is likely to be relatively *small*. In one sense, Shirota's theorem may be regarded as a generalization of the theorem that any $\{0, 1\}$-valued measure must be $\mathfrak{m}$-additive for every nonmeasurable $\mathfrak{m}$ (12.3(a)), as we shall now explain in some detail.

Let us apply Theorem 15.21 to the structure $\mathscr{C}$. Since condition (1) is satisfied, so is (3): for $d \in \mathscr{C}$, X is a nonmeasurable union of zero-sets of d-diameter $\leqq \epsilon$. But an overwhelmingly stronger statement can be made:

(a) *For every $d \in \mathscr{C}$ and $\epsilon > 0$, X is a countable union of zero-sets of d-diameter $\leqq \epsilon$.*

To prove this, we look first at the case in which d is a subbasic pseudo metric ψ_f, and consider the sets

$$Z_n = \{x \in X : n\epsilon \leqq f(x) \leqq (n + 1)\epsilon\},$$

n ranging over the set of all integers. Obviously, each Z_n is a zero-set of ψ_f-diameter $\leqq \epsilon$; and $X = \bigcup_n Z_n$.

For arbitrary $d \in \mathscr{C}$, there exist $f_1, \cdots, f_n \in C(X)$, and $\delta > 0$, such that $d\{A\} \leqq \epsilon$ whenever $\psi_{f_k}\{A\} \leqq \delta$ for all k. For each k, X is the union of a countable family of zero-sets of ψ_{f_k}-diameter $\leqq \delta$. Hence X is the union of the countable collection of zero-sets of the form

$$A_1 \cap \cdots \cap A_n,$$

where A_k belongs to the kth family; and these sets are of d-diameter $\leqq$

The force of this result shows up particularly clearly in the case of discrete spaces. By definition, the discrete structure $\mathscr{U}$ is that generated by the metric u for which $u(x, y) = 1$ whenever $x \neq y$. It is manifest that $\mathscr{U}$ contains every pseudometric on the space, and therefore is the largest structure; in particular, $\mathscr{U} \supset \mathscr{C}$. Since the sets of u-diameter < 1 are single points, (a) shows that if the space X is not countable, $\mathscr{C}$ must fall vastly short of $\mathscr{U}$:

(b) *For discrete* X, $\mathscr{C} = \mathscr{U}$ *(if and) only if* $|X| \leq \aleph_0$.

Yet, Shirota's theorem tells us that if the cardinal of X is arbitrary, so long as it is not measurable, then, because X is complete in $\mathscr{U}$, it is also complete in $\mathscr{C}$.

15.24. *Metric spaces.* Let $\mathscr{D}$ and $\mathscr{E}$ be uniform structures on X. The smallest structure containing both is denoted by $\mathscr{D} \vee \mathscr{E}$. One sees with little effort that $\mathscr{D} \vee \mathscr{E}$ is generated by all pseudometrics of the form $d \vee e$, for $d \in \mathscr{D}$ and $e \in \mathscr{E}$, and hence that $\mathscr{D} \vee \mathscr{E}$ is admissible if both $\mathscr{D}$ and $\mathscr{E}$ are. Furthermore, it is clear that a z-filter on X is a Cauchy z-filter in $\mathscr{D} \vee \mathscr{E}$ if and only if it is a Cauchy z-filter both in $\mathscr{D}$ and in $\mathscr{E}$. In symbols,

$$c[X; \mathscr{D} \vee \mathscr{E}] = c[X; \mathscr{D}] \cap c[X; \mathscr{E}].$$

This leads to the following result.

THEOREM. *Every metrizable space admits a complete uniform structure. Hence every metrizable space of nonmeasurable cardinal is realcompact.*

PROOF. It is clear, first of all, that the completion of a metric space $[X; \mathscr{D}]$ is metric: if d generates $\mathscr{D}$, then d^γ generates the uniform structure on γX. Now consider any Cauchy z-ultrafilter $\mathscr{A}$ on X, and suppose that its limit p^γ in γX belongs to $\gamma X - X$. The function g, defined by

$$g(q^\gamma) = d^\gamma(p^\gamma, q^\gamma),$$

vanishes only at p^γ. Set $f = (g|X)^{-1}$. Then for every $Z \in \mathscr{A}$, f is unbounded on Z, whence $\psi_f\{Z\} = \infty$. Thus, $\mathscr{A}$ is not a Cauchy z-filter in the structure $\mathscr{C}(X)$. This shows that if a z-ultrafilter is a Cauchy z-filter both in $\mathscr{D}$ and in $\mathscr{C}$, then it is fixed. Therefore X is complete in the structure $\mathscr{C} \vee \mathscr{D}$.

PROBLEMS

15A. ADMISSIBLE UNIFORM STRUCTURES ON **R**.

The following are distinct, admissible uniform structures on **R**. For each (except the first), describe the set of all uniformly continuous functions, and describe the completion. Order the structures by set inclusion.

(i) The standard structure.

(ii) The structure generated by ψ_f, where $f = \arctan$. [Note that f is a uniform isomorphism onto an open interval in the standard structure; also, ψ_f belongs to the standard structure.]

(iii) The structure generated by ψ_s and ψ_c, where $s(x) = 2x/(1 + x^2)$, and $c(x) = (1-x^2)/(1 + x^2)$. Compare with 15K.*4*,*7*.

(iv) $\mathscr{C}(\mathbf{R})$.

(v) $\mathscr{C}^*(\mathbf{R})$.

15B. *z*-FILTERS ON A SUBSPACE.

Let X be a subspace of a Hausdorff uniform space T.

1. Every *z*-filter on X with a limit in T is a Cauchy *z*-filter.

2. If X is dense in T, and every Cauchy *z*-filter on X has a limit in T, then T is complete.

3. For a compact space, the result corresponding to *2* was stated in 6F.*4*. The analogue is not valid for realcompact spaces however: if every real *z*-ultrafilter on X has a limit in T, T need not be realcompact. (See 8F.)

4. Use the above to prove that X is precompact if and only if every *z*-ultrafilter is Cauchy.

15C. CLOSED SUBSPACE; COMPLETE SUBSPACE.

1. A closed subspace of a complete space is complete.

2. A complete subspace of a Hausdorff uniform space is closed. [15B.*1*.] Hence a complete space is its own unique completion.

3. If X is a subspace of a Hausdorff uniform space T, then $\gamma X = \mathrm{cl}_{\gamma T} X$. Hence again, if X is dense in T, then $\gamma X = \gamma T$.

4. A closed subspace X of a realcompact space T is realcompact. (This is Theorem 8.10.) [X is complete in the relative structure from $\mathscr{C}(T)$, which is contained in $\mathscr{C}(X)$.]

15D. BOUNDED PSEUDOMETRICS.

1. Let $e = d \wedge \mathbf{1}$, where d belongs to a uniform structure $\mathscr{D}$. Then e is a pseudometric, and, for every $\epsilon < 1$, $e(x, y) \leqq \epsilon$ if and only if $d(x, y) \leqq \epsilon$, so that $e \in \mathscr{D}$. Thus, the bounded pseudometrics in $\mathscr{D}$ constitute a base.

2. Every pseudometric belonging to a given uniform structure is bounded when and only when every uniformly continuous function into $\mathbf{R}$ is bounded. [If d is unbounded, then $x \to d(p, x)$ is unbounded for each p.]

3. Let E be the unit ball in real, separable, Hilbert space, i.e., E is the set of all sequences $x = (x_n)_{n \in \mathbf{N}}$ such that $\sum_n x_n^2 \leqq 1$, with the metric

$$d(x, y) = \left(\sum_n (x_n - y_n)^2\right)^{1/2}.$$

The uniform space E is not precompact. [Theorem 15.16.]

4. Every uniformly continuous function on E is bounded. [If $d(x, y) \leqq \delta$ implies $|f(x) - f(y)| \leqq 1$, then

$$|f(x)| \leqq |f((0))| + 1 + 1/\delta$$

for all x.] Hence, in spite of *3*, every pseudometric belonging to the structure on E is bounded.

15E. TOTAL BOUNDEDNESS OF $(X; \mathscr{C}^*)$.

Let X be a completely regular space.

1. In Theorem 15.16, (2) and (3) are equivalent for any one d.

2. Prove directly, as in 15.23(a), that $[X; \mathscr{C}^*]$ is totally bounded.

3. If $f \in C(X)$, and f is unbounded, then $d = \psi_f \wedge \mathbf{1}$, which is a bounded pseudometric, does not belong to $\mathscr{C}^*$. [X has an infinite d-discrete subset.]

15F. METRIC UNIFORM STRUCTURE.

1. A Hausdorff uniform structure $\mathscr{D}$ is metric if and only if $\mathscr{D}$ has a countable base. [Use 15D.*1*, and consider a convergent sum of pseudometrics.]

2. A sequence $(x_n)_{n \in \mathbf{N}}$ in a metric space X is a Cauchy sequence (in the familiar sense) if and only if the family of sets $\{x_n\}_{n \geqq k}$, for $k \in \mathbf{N}$, contains arbitrarily small sets. X is said to be *sequentially* complete if every Cauchy sequence converges to a point in X. A metric space X is complete if and only if it is sequentially complete. [*Sufficiency*. From a given Cauchy z-filter, select Z_n of diameter $\leqq 1/n$, and choose $x_n \in Z_n$. Alternatively, use the fact that every point in the metric space γX is a limit of a Cauchy sequence in X.]

15G. CONTINUITY OF PSEUDOMETRICS; THE UNIVERSAL STRUCTURE.

1. If d is a pseudometric, then

$$|d(a, b) - d(x, y)| \leqq d(a, x) + d(b, y).$$

2. If a pseudometric on a topological space X is continuous in one variable, then it is continuous jointly. Hence if $\mathscr{D}$ is admissible, then every $d \in \mathscr{D}$ is continuous on $X \times X$.

3. If $\mathscr{D}$ is admissible, and e is a continuous pseudometric, then the structure generated by $\mathscr{D} \cup \{e\}$ is admissible.

4. If X is completely regular, then the set of all continuous pseudometrics is an admissible structure, and evidently the largest; it is called the ***universal*** structure.

5. Every continuous mapping from X into a uniform space is uniformly continuous in the universal structure on X.

15H. A COMPACT SPACE ADMITS A UNIQUE UNIFORM STRUCTURE.

1. For a compact space, the universal structure (15G.*4*) is the only admissible one. [Let u be a continuous pseudometric on a compact space

$[X; \mathscr{D}]$, and let $\epsilon > 0$. For each x, there exists $d_x \in \mathscr{D}$ such that $d_x(x, y) \leqq 1$ implies $u(x, y) \leqq \epsilon$. Apply compactness of X, and (i) and (ii) of 15.3.]

2. Every continuous mapping from a compact space into a uniform space is uniformly continuous.

15I. GENERATORS FOR A PRECOMPACT STRUCTURE.

Let X be a completely regular space.

1. An admissible uniform structure on X that is generated by a family of pseudometrics ψ_f, with $f \in C^*(X)$, is precompact. [Argue as in 15E.*2*. Alternatively, apply 15E.*2*.]

2. Conversely, every admissible precompact structure on X is generated by such a family. [Consider $C(\gamma X)$, and recall 15H.]

3. $\mathscr{C}^*$ is the largest admissible precompact structure.

4. If an admissible structure $\mathscr{D}$ on X is generated by a family of pseudometrics ψ_f, with $f \in C(X)$, then $\gamma[X; \mathscr{D}]$ is realcompact. [15.23(a) and Theorem 15.21. Alternatively, 15.23(a) may be applied to γX; it follows that every real z-ultrafilter on γX is Cauchy.] Here the converse is not true, however, as is shown by 15.23(b). See also 15L.*4*,*5*.

15J. ADMISSIBLE PRECOMPACT STRUCTURES.

Let γX and $\gamma' X$ be the completions of X in the admissible structures $\mathscr{D}$ and $\mathscr{D}'$, respectively.

1. $\mathscr{D} = \mathscr{D}'$ if and only if there exists a uniform isomorphism of γX onto $\gamma' X$ that leaves X pointwise fixed.

2. γX can be homeomorphic with $\gamma' X$, under a mapping that leaves X pointwise fixed, even though $\mathscr{D} \neq \mathscr{D}'$.

3. For precompact structures, $\mathscr{D} \supset \mathscr{D}'$ if and only if $\gamma' X$ is a continuous image of γX, under a mapping that leaves X pointwise fixed. [15H.*2*.] Note that $\gamma X - X$ maps onto $\gamma' X - X$. [Lemma 6.11.]

4. Distinct precompact structures have topologically distinct completions. Hence there is a one-one correspondence between the admissible precompact structures on X and the compactifications of X.

5. Use *3* and Theorem 6.12 to show that $\mathscr{C}^*$ is the largest admissible precompact structure.

6. $\mathscr{C}^*$ is the unique precompact structure in which any two disjoint zero-sets can be separated by a uniformly continuous function. [Theorem 6.5(III).]

15K. MINIMAL ADMISSIBLE STRUCTURES.

1. Every admissible structure on a completely regular space contains an admissible precompact structure. [Consider all ψ_f, where $f(x) = d(a, x)$, $a \in X$, and d is bounded; see 15D.*1* and 15I.*1*.]

2. The completion of a discrete space X in the structure generated by all ψ_{f_x}, where f_x is the characteristic function of $\{x\}$, is the one-point compactification of X.

3. Generalize *2* to arbitrary locally compact spaces.

4. In a locally compact space, the structure obtained from the one-point compactification is the smallest admissible precompact structure [15J.*3*]; in fact, it is the smallest admissible structure.

5. If a completely regular space X has a smallest admissible structure, then X is locally compact. [10C.]

6. If a function on a space is uniformly continuous in a given structure, it is uniformly continuous in every larger structure.

7. A function f in $C(\mathbf{R})$ is uniformly continuous in every admissible structure if and only if $\lim_{x\to+\infty} f(x)$ and $\lim_{x\to-\infty} f(x)$ exist and are equal.

15L. THE SPACE Y.

Let E be the unit ball in real, separable Hilbert space (see 15D.*3*), and let Y be the subspace of E consisting of all points x having at most one coordinate x_n different from 0.

1. Y is complete. [If some member of a Cauchy z-filter $\mathscr{F}$ does not contain (0), then it is bounded away from (0). It follows that $\mathscr{F}$ contains a compact set. Alternatively, use 15F.*2*.]

2. In the metric topology, Y is realcompact but not compact.

3. Every uniformly continuous function on Y is bounded. [See 15D.*4*.] Hence every pseudometric belonging to the uniform structure on Y is bounded.

4. The uniform structure on Y is contained in $\mathscr{C}(Y)$. [Consider the pseudometric $\psi_f \vee \psi_{g_\epsilon}$, where $f(x) = \sum_n x_n$, and $g_\epsilon(x) = 2^n(|x_n| - \epsilon)$ for $|x_n| > \epsilon$.]

5. In spite of *4*, the uniform structure on Y is not generated by any family of pseudometrics of the form ψ_h, for $h \in C(Y)$. [15I.*1*.] Contrast with 15I.*2*.

15M. THE PRODUCT OF UNIFORM SPACES.

The product uniform structure $\mathscr{D}$ on a cartesian product

$$X = \times_\alpha [X_\alpha; \mathscr{D}_\alpha]$$

is defined to be the smallest structure in which every projection π_α is uniformly continuous.

1. $\mathscr{D}$ is generated by the family of all pseudometrics of the form $(x, y) \to d(x_\sigma, y_\sigma)$, where $x = (x_\alpha)$, $y = (y_\alpha)$, and $d \in \mathscr{D}_\sigma$.

2. The uniform topology of the product structure is the product of the uniform topologies.

3. *An arbitrary product of complete spaces is complete.* [Adapt either the proof in 8.11, or those in 4.14 and 8.12.]

4. $\times_\alpha \gamma X_\alpha = \gamma(\times_\alpha X_\alpha)$. Contrast with 6N.*2* and 9I.

5. A product of precompact spaces is precompact. Contrast with 9.15 and 15Q.*1*.

15N. UNIFORM CONTINUITY OF PSEUDOMETRICS.

1. If e is a pseudometric on a space $[X; \mathscr{D}]$, then $e \in \mathscr{D}$ if and only if e is a uniformly continuous function from $X \times X$ (15M) into $\mathbf{R}$. [*Necessity.* 15G.*1*. *Sufficiency.* Given $\epsilon > 0$, there exist $d \in \mathscr{D}$ and $\delta > 0$ such that $d(x, y) \leqq \delta$ implies $|e(x, y) - e(y, y)| \leqq \epsilon$.]

2. If X is a dense subspace of $[T; \mathscr{U}]$, then each $u \in \mathscr{U}$ is determined by its restriction to $X \times X$.

3. Every pseudometric in the structure on $\gamma[X; \mathscr{D}]$ is of the form d^γ for some $d \in \mathscr{D}$.

4. If $[X; \mathscr{D}]$ is dense in a Hausdorff uniform space $[T; \mathscr{U}]$, then every pseudometric in $\mathscr{D}$ has an extension to one in $\mathscr{U}$. [Use *1*. Alternatively, use *3* and the fact that $\gamma T = \gamma X$.]

15O. UNIFORM STRUCTURES THAT AGREE ON A DENSE SUBSPACE.

Let $\mathscr{U}$ and $\mathscr{U}'$ be Hausdorff uniform structures on T, and let X be a subset of T, dense in both uniform topologies, and having the same relative structure in $\mathscr{U}$ as in $\mathscr{U}'$.

1. The spaces $[T; \mathscr{U}]$ and $[T; \mathscr{U}']$ need not be uniformly isomorphic. [Take for X an open interval of $\mathbf{R}$ with one point deleted.]

2. If the uniform topologies on T are the same, then $\mathscr{U} = \mathscr{U}'$. [15N.*4*.]

3. A uniform isomorphism of $[T; \mathscr{U}]$ onto $[T; \mathscr{U}']$ that leaves X pointwise fixed is necessarily unique; but it need not be the identity, even when $\mathscr{U}$ and $\mathscr{U}'$ are complete.

15P. THE UNIFORM STRUCTURES $\mathscr{C}$ AND $\mathscr{C}^*$.

Let X, Y, and T be completely regular spaces, with $X \subset T$.

1. Every continuous mapping from X to Y is uniformly continuous from $[X; \mathscr{C}(X)]$ to $[Y; \mathscr{C}(Y)]$, and from $[X; \mathscr{C}^*(X)]$ to $[Y; \mathscr{C}^*(Y)]$.

2. Use *1* to derive Stone's theorem, and (I) of Theorem 8.7.

3. If $\mathscr{C}^*(X)$ is the relative structure obtained from $\mathscr{C}^*(T)$, then X is C^*-embedded in T. [15C.*3*.]

4. If $\mathscr{C}(X)$ is the relative structure obtained from $\mathscr{C}(T)$, and if X is dense, or if T or υT is normal, then X is C-embedded in T.

5. In Theorem 11.3, X was embedded as a subspace of $P = \mathbf{R}^{C(X)}$, and as a subspace of $P_* = \mathbf{R}^{C^*(X)}$. If P and P_* are given the product uniform structures (15M), then the relative structures on the copies of X are $\mathscr{C}$ and $\mathscr{C}^*$, respectively.

15Q. PSEUDOCOMPACT SPACES.

Let X be a completely regular space.

1. *X is pseudocompact if and only if every admissible structure is precompact.* [15.15(b).] Hence every complete pseudocompact space is compact. Thus, no pseudocompact, noncompact space (e.g., $\mathbf{W}$) admits a complete structure.

2. X is pseudocompact if and only if $\mathscr{C}(X) = \mathscr{C}^*(X)$.

15R. SPACES WITH UNIQUE UNIFORM STRUCTURE.

The following conditions are equivalent for any completely regular space X. [Successive implications are suggested in the hints.]

By (7) and 6J, another equivalent condition is $|\beta X - X| \leqq 1$.

In particular, **W**, **T**, and Ω (8L) are noncompact spaces having unique structures. Recall that **T** is not even countably compact.

(1) X admits a unique structure.

(2) Every continuous mapping from X into any uniform space is uniformly continuous in every admissible structure on X. [15G.*5*.]

(3) Every function in $C(X)$ is uniformly continuous in every admissible structure on X.

(4) Every function in $C^*(X)$ is uniformly continuous in every admissible structure on X.

(5) Every function in $C^*(X)$ is uniformly continuous in every admissible precompact structure on X.

(6) X admits only one precompact structure. [$\mathscr{C}^*$ is the smallest precompact structure. By 15I.*3*, it is also the largest.]

(7) X has a unique compactification. [(7) *implies* (1). By 6J, X is pseudocompact. Apply 15J.*4* and 15Q.*1*.]

15S. TOPOLOGICAL PROPERTIES OF γX.

Let $[X; \mathscr{D}]$ be a Hausdorff uniform space.

1. cX is the largest subspace of βX to which the identity mapping on X into γX has a continuous extension. [15B.*1*.]

2. The mapping $p \to p^\gamma$ of cX onto γX is a closed mapping. [Theorem 10.13.]

3. If, for any distinct p, q in cX, there exists d such that $d^c(p, q) \neq 0$, then $\gamma X = cX$, topologically as well as pointwise.

4. If γX is realcompact, then cX is realcompact. [8B.*4*.] Hence, if $|X|$ is nonmeasurable, then cX is realcompact.

5. In Theorem 15.21, prove directly that (1) implies (4). [Corollary 8.5(b).]

6. If every closed, discrete subspace of X is realcompact, then every closed, discrete subspace of γX is realcompact. Compare 12A.

15T. REAL z-ULTRAFILTERS.

1. If d is any pseudometric belonging to the unique admissible structure on **W** (15R), then there exists a tail in **W** of d-diameter 0. Hence, directly, the real z-ultrafilter A^{ω_1} is Cauchy.

2. More generally, let X be any completely regular space, and let $d \in \mathscr{C}(X)$. Every real z-ultrafilter on X contains a set of d-diameter 0. [Theorem 5.14.]

3. Still more generally, let $\mathscr{D}$ be any admissible structure on X, and let $d \in \mathscr{D}$. Every real, Cauchy z-ultrafilter contains a set of d-diameter 0. Note that "Cauchy" is redundant except in the eventuality of a measurable cardinal.

15U. COMPLETE STRUCTURES CONTAINING $\mathscr{C}$.

Let $\mathscr{D}$ be an admissible structure on a completely regular space X.

1. X is complete in $\mathscr{C} \vee \mathscr{D}$ if and only if $vX \cap c[X; \mathscr{D}] = X$, i.e., every real Cauchy z-ultrafilter in $\mathscr{D}$ is fixed.

2. If every point in $\gamma[X; \mathscr{D}] - X$ is contained in a zero-set disjoint from X, then X is complete in $\mathscr{C} \vee \mathscr{D}$.

3. If there exists a continuous metric on X, then X admits a complete structure. [By 15G.*3*, $\mathscr{C} \vee \mathscr{E}$ is admissible, where $\mathscr{E}$ is the (not necessarily admissible) metric structure. Apply *1* and 15T.*3*. Alternatively, modify the proof of Theorem 15.24.]

15V. THE PSEUDOMETRIC d^c.

1. $d\{A \cup B\} \leqq d\{A\} + d\{B\} + d[A, B]$.

2. For $p, q \in c[X; \mathscr{D}]$,

$$\begin{aligned} d^c(p, q) &= \inf \{d\{Z\}: Z \in A^p \cap A^q\} \\ &= \sup \{d[Z_p, Z_q]: Z_p \in A^p, Z_q \in A^q\}. \end{aligned}$$

3. $d^c(p, q) = 0$ for all $d \in \mathscr{D}$ if and only if $A^p \cap A^q$ is a Cauchy z-filter.

15W. C-EMBEDDING OF d-DISCRETE SETS.

1. Let $[X; \mathscr{D}]$ be a uniform space and let $d \in \mathscr{D}$. If a set D is d-discrete, then there exist a family of sets (V_α) as in 3L.*1* and points $x_\alpha \in \text{int } V_\alpha$ such that $D = \{x_\alpha\}_\alpha$.

2. Conversely, let X be a completely regular space, let (V_α) be as in 3L.*1*, and let $D = \{x_\alpha\}_\alpha$, where $x_\alpha \in \text{int } V_\alpha$. Then D is d-discrete for some pseudometric d belonging to the universal structure (15G) on X. [Define $d = \sup_\alpha \psi_{f_\alpha}$ for suitably chosen f_α.]

Chapter 16

DIMENSION

16.1. This chapter is devoted to an exposition of Katětov's characterization of the dimension of X in terms of $C^*(X)$. We include little more of dimension theory than is needed for a self-contained proof of Katětov's theorem. The few additional properties of dimension that are found here should help the reader to examine some illuminating examples.

How to define the dimension of a topological space has long been a serious problem. There seems now to be widespread agreement that the definition best suited to normal spaces is the one due, essentially, to Lebesgue. The definition we shall use is a slight modification of the usual one for Lebesgue dimension; it leads to the same dimension of a normal space as the usual one, but avoids the anomalies that may arise when the space is not normal. In particular, it always assigns to βX the same dimension as to X.

Since the ring $C^*(X)$ determines βX as a topological space, and since the dimension of a space is a topological invariant, the preceding remark implies that $C^*(X)$ determines the dimension of X. The problem is to express this dimension directly in terms of $C^*(X)$. A clue to the successful approach is provided by the classical Weierstrass approximation theorem: if X is a cube in the n-dimensional euclidean space $\mathbf{R}^n$, then every continuous function on X is a uniform limit of polynomials in n real variables. To express this result in a way that is susceptible of generalization, $C^*(X)$ is made into a metric space by introducing the metric

$$d(f, g) = \sup_{x \in X} |f(x) - g(x)|.$$

The theorem then asserts that the subring of $C^*(X)$ generated by the n projections of X into the factors of $\mathbf{R}^n$ together with the constant functions on X, is dense in the metric space $C^*(X)$. It turns out, moreover, that any subring of $C^*(X)$ generated by fewer than n functions (together with the constants) will not be dense. Thus, the least number of generators of a dense subring of $C^*(X)$ is precisely the dimension of the cube X.

The dimension of an arbitrary completely regular space X will be found, in a similar way, among the properties of the metric ring $C^*(X)$. However, the relation will not be so simple as in the case of the cube. For instance, $\mathbf{R}$ is one-dimensional; but, as is easily seen, no subring of $C^*(\mathbf{R})$ with only one generator can be dense.

It is worth noting that the metric on $C^*(X)$—or, if one prefers, the norm $\|f\| = d(f, \mathbf{0})$—is determined by the algebra of the ring, as follows:

$$\|f\| = \sup |M(f)|,$$

M running through the set of maximal ideals in $C^*(X)$ (5R or 1J.6). So the metric topology, which will be used in an essential way to characterize dimension, is actually an algebraic invariant of $C^*(X)$.

THE STONE-WEIERSTRASS THEOREM

16.2. A crucial role in the development will be played by Stone's well-known generalization of the Weierstrass approximation theorem. We include a proof of this result, stated in the form best suited to our applications.

LEMMA. *If A is a subring of $C^*(X)$ that contains the constant functions, then* cl A, *its closure in the metric topology, is a sublattice of $C^*(X)$.*

PROOF. Evidently, cl A is a subring, and so we may assume in the proof that A itself is closed. Since $f \vee g = \mathbf{2}^{-1}(f + g + |f - g|)$, it suffices to show that $|f| \in A$ whenever $f \in A$. We may also suppose that $|f| \leqq \mathbf{1}$. Our object is to show that given $\epsilon > 0$, there exists a function in A whose distance from $|f|$ is at most ϵ. The familiar fact that the binomial expansion of $(1 - t)^{1/2}$ converges uniformly for $|t| \leqq 1$ implies the existence of a polynomial $\mathfrak{p}$ satisfying

$$|(1 - t)^{1/2} - \mathfrak{p}(t)| < \epsilon \quad \text{for} \quad |t| \leqq 1.$$

Then

$$\left||f| - \mathfrak{p}(\mathbf{1} - f^2)\right| = \left|\left(\mathbf{1} - (\mathbf{1} - f^2)\right)^{1/2} - \mathfrak{p}(\mathbf{1} - f^2)\right|$$
$$\leqq \boldsymbol{\epsilon},$$

that is, $d(|f|, \mathfrak{p}(\mathbf{1} - f^2)) \leqq \epsilon$. Thus, $\mathfrak{p}(\mathbf{1} - f^2)$ is the required element of A.

16.3. Lemma. *Let X be compact, and let A be a sublattice of $C(X)$. Then* cl A *contains every function f in $C(X)$ that can be approximated at each pair of points in X by a function from A.*

PROOF. Fix $\epsilon > 0$. For each pair of points p, $q \in X$, let g_{pq} be a function in A such that

$$|f(p) - g_{pq}(p)| < \epsilon \quad \text{and} \quad |f(q) - g_{pq}(q)| < \epsilon.$$

For each fixed q, the union over p of the open sets

$$\{x: g_{pq}(x) - f(x) < \epsilon\}$$

is all of X, and hence a finite number of them, say for $p_1, \cdots, p_s$, cover X. Let

$$g_q = g_{p_1q} \wedge \cdots \wedge g_{p_sq}.$$

Then $g_q \in A$, $g_q \leqq f + \epsilon$, and $g_q(x) > f(x) - \epsilon$ for all $x \in V_q$, where

$$V_q = \bigcap_{k \leqq s} \{x: g_{p_kq}(x) > f(x) - \epsilon\}.$$

Each set V_q is open; and $\bigcup_{q \in X} V_q = X$, since $q \in V_q$. Hence, a finite number of these sets cover X; let g denote the supremum of the corresponding functions g_q. Then $g \in A$, and $|f - g| \leqq \epsilon$, i.e., $d(f, g) \leqq \epsilon$.

16.4. Let A be a subfamily of $C(X)$; a subset of X on which every function in A is constant will be called a *stationary* set of A. A one-element set is stationary for any family of functions. For the family C^*, the one-element sets are the only stationary sets.

Theorem (stone-weierstrass). *Let X be a compact space, and let A be a subring of $C(X)$ that contains all the constant functions. Then* cl A *is the family of all functions in $C(X)$ that are constant on every stationary set of A.*

PROOF. Let A' denote the family in question. Now, evidently, the family of all functions in $C(X)$ that are constant on a given set is a closed subring. Since A' is an intersection of such families, it, too, is a closed subring. Hence cl $A \subset A'$. Consider any $f \in A'$, and any points p and q. If $f(p) = f(q)$, then there is a constant (in A) that agrees with f at p and q. If $f(p) \neq f(q)$, then $g(p) \neq g(q)$ for some $g \in A$; a suitable multiple of g, plus a constant, will agree with f at these points. Thus every function in A' can be approximated (in fact, duplicated) at every pair of points by a function in A. Since cl A is a closed sublattice of $C(X)$, we have, by the preceding lemma, cl $A \supset A'$.

DIMENSION OF A COMPLETELY REGULAR SPACE

16.5. Let X be a (nonempty) completely regular space. By a *cover* of X, we shall mean, throughout this chapter, a *finite open* cover, i.e., a finite collection of open sets whose union is X.

Let $\mathscr{U}$ and $\mathscr{V}$ be families of sets. We shall say that $\mathscr{V}$ *refines* $\mathscr{U}$, or is a *refinement* of $\mathscr{U}$, if for every $V \in \mathscr{V}$, there exists $U \in \mathscr{U}$ such that $V \subset U$. The definition applies to arbitrary families $\mathscr{U}$ and $\mathscr{V}$, not necessarily covers. A refinement of a *cover* $\mathscr{U}$ will usually be understood from context to mean a *cover* that refines $\mathscr{U}$.

The *order* of a cover $\mathscr{U}$ is the largest integer n for which there exist $n + 1$ members of $\mathscr{U}$ with nonempty intersection; evidently, $n \geqq 0$ (unless X is empty).

The (Lebesgue) dimension of a nonempty space is usually defined as the least cardinal $\mathfrak{m}$ such that every cover has a refinement of order at most $\mathfrak{m}$. We shall use a modified definition. While it is equivalent to the usual one for normal spaces (Corollary 16.9), it eliminates the pathology that may arise from the absence of normality (16.18).

Let a *basic* cover be a cover by cozero-sets (complements of zero-sets).

> The *dimension* of X—denoted by $\dim X$—is defined to be the least cardinal $\mathfrak{m}$ such that every *basic* cover of X has a *basic* refinement of order at most $\mathfrak{m}$.

Clearly, $0 \leqq \dim X \leqq \aleph_0$; and $\dim X = \aleph_0$ provided that for every $n \in \mathbf{N}$, there exists a basic cover having no basic refinement of order $\leqq n$.

16.6. THEOREM. *Let $\mathscr{V} = \{V_k\}_{k \leqq s}$ be a cover of X. If $\mathscr{V}$ is basic, or if X is normal, then there exist a basic refinement $\mathscr{W} = \{W_k\}_{k \leqq s}$ of $\mathscr{V}$, and zero-sets $Z_1, \cdots, Z_s$, such that*

$$W_k \subset Z_k \subset V_k \qquad (k \leqq s).$$

PROOF. The sets W_k and Z_k will be found inductively. We shall let

$$A_k = X - (W_1 \cup \cdots \cup W_{k-1} \cup V_{k+1} \cup \cdots \cup V_s),$$

for each $k \leqq s$. Suppose that the zero-sets $X - W_1, \cdots X - W_{k-1}$, $Z_1, \cdots, Z_{k-1}$ have been found, satisfying

$$W_i \subset Z_i \subset V_i \qquad (i < k),$$

and so that

$$\{W_1, \cdots, W_{k-1}, V_k, V_{k+1}, \cdots, V_s\}$$

is a cover of X. The closed sets A_k and $X - V_k$ are disjoint, and therefore, under either hypothesis of the theorem, they are completely separated (Theorem 1.15, and Urysohn's lemma (3.13)). In consequence, there exist zero-sets $X - W_k$ and Z_k such that

$$A_k \subset W_k \subset Z_k \subset V_k$$

(see 1.15(a)). It follows from the definition of A_k that

$$\{W_1, \cdots, W_k, V_{k+1}, \cdots, V_s\}$$

is still a cover of X.

16.7. The hypotheses in this theorem are essential: if F and F' are disjoint closed sets that are not completely separated, then the (nonbasic) cover

$$\mathscr{U} = \{X - F, X - F'\}$$

has no basic refinement. For, suppose that $\mathscr{V} = \{X - Z_k\}_{k \leqq s}$ refines $\mathscr{U}$, where each Z_k is a zero-set. For each k, Z_k contains either F or F'. Define

$$Z = \bigcap \{Z_k : Z_k \supset F\}, \quad \text{and} \quad Z' = \bigcap \{Z_k : Z_k \supset F'\}.$$

Then $Z \supset F$, and $Z' \supset F'$. Since F and F' are not contained in disjoint zero-sets, we have

$$\bigcap_{k \leqq s} Z_k = Z \cap Z' \neq \emptyset.$$

Therefore $\mathscr{V}$ is not a cover.

By definition, every nonnormal space contains disjoint closed sets that are not completely separated. Thus, the foregoing argument shows that:

(a) *Every cover of X has a basic refinement when and only when X is normal.*

Since every open-and-closed set is a zero-set, every cover of order 0 is basic. Hence, we have, as a particular case of (a):

(b) *If every cover of X has a refinement of order* 0, *then X is normal.*

16.8. Corollary. *Let $\mathscr{V}$ be a cover of X, where* $\dim X \leqq n$. *If $\mathscr{V}$ is basic, or if X is normal, then X has a basic cover of order $\leqq n$ whose closures refine $\mathscr{V}$.*

Proof. Evidently, in the theorem, the closures of $\mathscr{W}$ refine $\mathscr{V}$; and if $\dim X \leqq n$, then the basic refinement $\mathscr{W}$ has a basic refinement of order $\leqq n$.

16.9. Corollary. *If X is normal, then* $\dim X \leqq n$ *if and only if every cover has a refinement of order $\leqq n$.*

PROOF. Necessity is contained in Corollary 16.8. For the sufficiency, we observe that the order of the basic refinement $\mathscr{W}$ in the theorem is certainly not greater than the order of $\mathscr{V}$.

Thus, for normal spaces, the modified definition of dimension is equivalent to the usual definition. In the sequel, this result will be used continually, without further comment. As a matter of fact, our applications will be almost exclusively to *compact* spaces. The justification for this is the theorem, to which we now turn, that $\dim X = \dim \beta X$ for all completely regular X.

16.10. For $U \subset X$, define

$$U^\beta = \beta X - \mathrm{cl}_{\beta X}(X - U).$$

LEMMA. *If $\mathscr{U} = \{U_k\}_{k \leqq s}$ is a basic cover of X, then*

$$\mathscr{U}^\beta = \{U_k^\beta\}_{k \leqq s}$$

is a cover of βX, of the same order as $\mathscr{U}$.

REMARK. If U is open in X, then $U^\beta \cap X = U$, whence

$$U^\beta \subset \mathrm{cl}_{\beta X} U^\beta = \mathrm{cl}_{\beta X} U.$$

PROOF. By hypothesis, $X - U_k$ is a zero-set Z_k. Since $\mathscr{U}$ is a cover, $\bigcap_k Z_k = \emptyset$. Therefore $\bigcap_k \mathrm{cl}\, Z_k = \emptyset$, by (IV) of Theorem 6.5, whence $\mathscr{U}^\beta$ covers βX. Since

$$\bigcup_k \mathrm{cl}\, Z_k = \mathrm{cl} \bigcup_k Z_k,$$

over any common (finite) index set, the orders of $\mathscr{U}$ and $\mathscr{U}^\beta$ are the same.

As we know, $\mathscr{U}^\beta$ is not, in general, a *basic* cover of βX (6E.5). However, since βX is compact, we do not have to consider basic covers in discussing its dimension.

16.11. THEOREM. $\dim X = \dim \beta X$.

PROOF. [$\dim X \leqq \dim \beta X$.] Denote $\dim \beta X$ by n. Let $\mathscr{U}$ be any basic cover of X. By the lemma, $\mathscr{U}^\beta$ covers βX. By Corollary 16.8, $\mathscr{U}^\beta$ has a basic refinement $\mathscr{W}$ of order $\leqq n$. The trace of $\mathscr{W}$ on X is then a basic refinement, of order $\leqq n$, of the trace $\mathscr{U}$ of $\mathscr{U}^\beta$. Hence $\dim X \leqq n$.

[$\dim \beta X \leqq \dim X$.] Let $\mathscr{V}$ be any cover of βX. By Corollary 16.8, βX has a basic cover $\mathscr{W}$ whose closures refine $\mathscr{V}$. If $\dim X = n$, then the trace of $\mathscr{W}$ on X has a basic refinement $\mathscr{U}$ of order $\leqq n$. According to 16.10, the order of $\mathscr{U}^\beta$ is also $\leqq n$, and $\mathscr{U}^\beta$ refines the closures (in βX) of $\mathscr{U}$—whence $\mathscr{U}^\beta$ refines $\mathscr{V}$. Therefore $\dim \beta X \leqq n$.

ZERO-DIMENSIONAL SPACES

16.12. By a *partition* of a subset S of X, we shall mean a finite collection of disjoint, relatively open-and-closed subsets of S whose union is S. Thus, a partition of S is a cover of S of order 0. No connected subset can meet two members of a partition; and S is connected if and only if it has no partition by proper subsets.

Two sets S_1 and S_2 are said to be *separated by a partition* if there exists a partition $\{E_1, E_2\}$ of X such that $S_1 \subset E_1$ and $S_2 \subset E_2$.

We recall that a space is totally disconnected provided that no connected set has more than one point. (A space of fewer than two points is both connected and totally disconnected.) The components of a space are its maximal nonempty connected sets; their existence follows from the fact that an arbitrary union of connected sets with a point in common is connected. Thus, the component containing a given point x is simply the union of all the connected sets that contain x. Since the closure of a connected set is connected, every component is a closed set.

The results in the next few sections will help the reader to recognize zero-dimensional spaces. By definition, a space is zero-dimensional if every basic cover has a refinement by a partition of the space. It is useful to notice that if a cover $\{U_k\}_{k \leqq s}$ has a refinement by some partition $\mathscr{W}$, then it has a refinement by a partition $\{V_k\}_{k \leqq s}$ such that $V_k \subset U_k$ for each k: V_k is the union of all members of $\mathscr{W}$ that are contained in U_k, but not in U_i for any $i < k$.

16.13. LEMMA. *Let $\mathscr{K}$ be a family of compact sets, and let $\{K_1, K_2\}$ be a partition of $\bigcap \mathscr{K}$. Then there exist a finite subfamily $\mathscr{H}$ of $\mathscr{K}$, and a partition $\{H_1, H_2\}$ of $\bigcap \mathscr{H}$, such that $H_1 \supset K_1$ and $H_2 \supset K_2$.*

PROOF. The disjoint compact sets K_1 and K_2 are contained in disjoint open sets U_1 and U_2, respectively. Since $\bigcap \mathscr{K}$ is contained in the open set $U_1 \cup U_2$, there exists a finite subfamily $\mathscr{H}$ of $\mathscr{K}$ such that $\bigcap \mathscr{H} \subset U_1 \cup U_2$. Then the sets

$$H_k = (\bigcap \mathscr{H}) \cap U_k$$

($k = 1, 2$) have the required properties.

16.14. COROLLARY. *If the intersection of every finite subfamily of a family $\mathscr{K}$ of compact sets is connected, then $\bigcap \mathscr{K}$ is connected.*

16.15. THEOREM. *The component of a point in a compact space is the intersection of all the open-and-closed sets containing it.*

PROOF. Let $\mathscr{K}$ denote the family of open-and-closed sets containing a point x. We show first that $S = \bigcap \mathscr{K}$ is connected. Let $\{K_1, K_2\}$ be a partition of S, with $x \in K_1$, and let H_1 and H_2 be as in the lemma. Since $H_1 \cup H_2$ is the intersection of a *finite* family of open-and-closed sets, H_1 (as well as H_2) is open-and-closed in X. Now, $x \in K_1 \subset H_1$, and so $S \subset H_1$. Therefore $S = S \cap H_1 = K_1$; thus, S is connected.

If A is any set containing S, and $y \in A - S$, then there exists an open-and-closed set containing S but not y. Consequently, A is not connected, and, therefore, S is a maximal connected set.

16.16. THEOREM. *If a Lindelöf space X has a base of open-and-closed sets, then any two disjoint closed sets in X are separated by a partition.*

PROOF. Let H and K be disjoint closed sets. For each point x, let $U(x)$ be an open-and-closed neighborhood of x that fails to meet at least one of these sets. Since X is a Lindelöf space, the family of all sets $U(x)$ has a countable subfamily $\{U_k\}_{k \in \mathbf{N}}$ whose union is X. Define

$$V_k = U_k - \bigcup_{i<k} U_i;$$

then $\{V_k\}_{k \in \mathbf{N}}$ is a family of disjoint, open-and-closed sets whose union is X. Moreover, for each k, either $V_k \cap H = \emptyset$ or $V_k \cap K = \emptyset$. Let

$$W = \bigcup \{V_k : V_k \cap H = \emptyset\};$$

then

$$X - W = \bigcup \{V_k : V_k \cap H \neq \emptyset\},$$

so that $\{W, X - W\}$ is a partition of X. Obviously, $W \cap H = \emptyset$; and since $V_k \cap H \neq \emptyset$ implies $V_k \cap K = \emptyset$, we have $(X - W) \cap K = \emptyset$. Thus, H and K are separated by the partition.

16.17. THEOREM. *Each of the following conditions implies the next one.*

(a) $\dim X = 0$.

(b) *Any two disjoint zero-sets in X are separated by a partition.*

(c) *X has a base of open-and-closed sets.*

(d) *X is totally disconnected.*

Furthermore, (a) *and* (b) *are always equivalent; and if X is a Lindelöf space, they are also equivalent with* (c). *Finally, if X is compact, then all four conditions are equivalent.*

PROOF. (a) *implies* (b). Given disjoint zero-sets Z and Z', let $\mathscr{U}$ be a refinement of order zero of the basic cover $\{X - Z, X - Z'\}$. If U is the union of all the members of $\mathscr{U}$ contained in $X - Z'$, and U' the union of the others, then U and U' contain Z and Z', respectively, and $\{U, U'\}$ is a partition of X as required.

(b) *implies* (a). We prove, by induction on the number of sets in the cover, that every basic cover $\mathscr{U} = \{U_k\}_{k \leqq s}$ has a (basic) refinement of order 0. For $s = 1$, the result is trivial. For $s > 1$, the induction hypothesis implies that the basic cover

$$\{U_1, \cdots, U_{s-2}, U_{s-1} \cup U_s\}$$

has a refinement consisting of $s - 1$ disjoint sets $V_1, \cdots, V_{s-1}$, where $V_k \subset U_k$ for $k \leqq s - 2$, and

$$V_{s-1} \subset U_{s-1} \cup U_s.$$

By hypothesis (b), the disjoint zero-sets $V_{s-1} - U_{s-1}$ and $V_{s-1} - U_s$ are separated by a partition $\{W, X - W\}$. The cover

$$\{V_1, \cdots, V_{s-2}, V_{s-1} \cap W, V_{s-1} - W\}$$

is then a refinement of $\mathscr{U}$ of order 0.

(b) *implies* (c). If U is any neighborhood of a point x, then, by complete regularity, $\{x\}$ and $X - U$ are completely separated, and so the hypothesis implies that they are separated by a partition. The member of the partition containing x is an open-and-closed neighborhood of x contained in U.

That (c) implies (d) is trivial; and that (c) implies (b) for a Lindelöf space is the content of the preceding theorem.

To complete the proof, we have only to show that for a compact space, (d) implies (c). Let U be any neighborhood of a point x in a totally disconnected, compact space. It follows from Theorem 16.15 that the intersection of some finite number of open-and-closed neighborhoods of x is contained in U. Hence the open-and-closed sets form a base for the topology.

None of the other implications in the theorem can be reversed; see 16L, M.

16.18. EXAMPLE. An informative example is provided by the Tychonoff plank $\mathbf{T}$ (8.20). Clearly, the space $\beta\mathbf{T} = \mathbf{W}^* \times \mathbf{N}^*$ is totally disconnected. Therefore $\dim \beta\mathbf{T} = 0$, and hence $\dim \mathbf{T} = 0$ (Theorem 16.11). Since $\mathbf{T}$ is not normal, the unmodified definition of Lebesgue dimension would yield a dimension number greater than 0 (see 16.7(b)), i.e., greater than that of $\beta\mathbf{T}$.

THE EUCLIDEAN SPACES $\mathbf{R}^n$

16.19. The space $\mathbf{R}^n$—the topological product of n copies of $\mathbf{R}$ (where $n \in \mathbf{N}$)—is important not only as a space to be studied in its own right, but also, like $\mathbf{R}$, as a tool in the study of other spaces. It is well known that dim $\mathbf{R}^n = n$; and it is also well known that the proof of this fact is not easy. For the sake of completeness, we shall include here a proof of the one special dimension-theoretic result regarding $\mathbf{R}^n$ that will be needed in the development of the general theory—namely, the theorem that the dimension of any compact subset of $\mathbf{R}^n$ is at most n (Theorem 16.22).

It will be convenient to regard $\mathbf{R}^n$ as a vector space over $\mathbf{R}$, under the usual operations: if $a = (a_1, \cdots, a_n)$ and $b = (b_1, \cdots, b_n)$ belong to $\mathbf{R}^n$, and $r \in \mathbf{R}$, then

$$a + b = (a_1 + b_1, \cdots, a_n + b_n), \quad \text{and} \quad ra = (ra_1, \cdots, ra_n).$$

We also define a norm on $\mathbf{R}^n$:

$$\|a\| = \max \{|a_1|, \cdots, |a_n|\}$$

(for $n = 1$, $\|a\| = |a|$). The norm induces a metric in the usual way: $(a, b) \to \|a - b\|$. With these definitions, $\mathbf{R}^n$ is a complete metric vector space.

16.20. In a metric space, a set of the form

$$\{y : d(x, y) < \epsilon\}$$

(where d denotes the metric) will be called a *cell* of radius ϵ. In $\mathbf{R}^n$, this is the set

$$\{y : \|x - y\| < \epsilon\}.$$

LEMMA. *In* $\mathbf{R}^n$, *consider the cells of radius* $1/2 + 1/2^{n+1}$ *whose centers* $a = (a_1, \cdots, a_n)$ *satisfy*

(a) $$a_k = m_k + \tfrac{1}{2} \cdot a_{k-1}$$

($k = 1, \cdots, n$), *where each* m_k *is an integer, and* $a_0 = 0$. *Each point of* $\mathbf{R}^n$ *lies in at least one and in at most* $n + 1$ *of these cells.*

PROOF. For convenience, we denote the number $1/2 + 1/2^{n+1}$ by ρ.

Fix $x = (x_1, \cdots, x_n)$ in $\mathbf{R}^n$. The center of every cell of the stated type that contains x is found by solving the inequalities

(b) $$|m_j - (x_j - \tfrac{1}{2} \cdot a_{j-1})| < \rho$$

successively for integers m_j, and defining a_j by (a). For each j, the inequality clearly has an integral solution m_j, and so x belongs to at least one such cell.

We now show that x belongs to no more than $n + 1$ of these cells. For each $j \leqq n$, consider the set A_j of all j-tuples $(a_1, \cdots, a_j)$ that satisfy (a), along with the inequality

$$\text{(c)} \qquad |a_k - x_k| < \rho,$$

for all $k \leqq j$. Thus, A_n is the set of all centers of the cells in question that contain x.

We begin by deriving some relations among the coordinates of the members of A_j. Let $(a_1, \cdots, a_j)$ and $(a'_1, \cdots, a'_j)$ belong to A_j, and let m_k and m'_k $(k \leqq j)$ be the corresponding integers defined in (a). From (c), we have

$$\text{(d)} \qquad |a_k - a'_k| < 2\rho \qquad (k = 1, \cdots, j).$$

Next, for each k:

$$\text{(e)} \qquad \text{If } a_k = a'_k, \quad \text{then} \quad a_i = a'_i \quad \text{for every} \quad i \leqq k.$$

For, from (a), $a_{k-1} - a'_{k-1}$ is equal to $2(m'_k - m_k)$, an even integer. By (d), it can be only 0. The remaining equations follow inductively. Finally:

$$\text{(f)} \qquad \text{Either } m_k = m'_k, \quad \text{or} \quad m_k \text{ and } m'_k \text{ differ by 1.}$$

For, from (a) and (d),

$$|m_k - m'_k| \leqq |a_k - a'_k| + \tfrac{1}{2} \cdot |a_{k-1} - a'_{k-1}| < 2.$$

The proof that A_n has at most $n + 1$ members will proceed by induction. Clearly, A_1 has at most two members. Assume that A_{j-1} has at most j members. We shall show that there is only one value of a_{j-1} for which the inequality (b) can admit two solutions for m_j. By (f), two such solutions must be consecutive integers, say t and $t + 1$. We have, then,

$$-\rho < t - (x_j - \tfrac{1}{2} \cdot a_{j-1}), \quad \text{and} \quad t + 1 - (x_j - \tfrac{1}{2} \cdot a_{j-1}) < \rho.$$

Hence

$$2(x_j - t) - 2\rho < a_{j-1} < 2(x_j - t) - 2(1 - \rho),$$

so that a_{j-1} must lie in a given interval of length $2(2\rho - 1) = 1/2^{n-1}$. Since a_{j-1} is an integral multiple of $1/2^{j-2}$ (as follows inductively from (a)), and $j \leqq n$, there can be at most one such number in the interval.

It follows that A_j can have at most $j + 1$ members. This completes the induction.

16.21. LEMMA. *Let X be a compact metric space. For each cover $\mathcal{U}$, there exists $\epsilon > 0$ such that the cells in X of radius ϵ refine $\mathcal{U}$.*

PROOF. Let $\mathcal{U} = \{U_k\}_{k \leqq s}$. For each $k \leqq s$, write

$$f_k(x) = d(x, X - U_k] \qquad (x \in X),$$

d denoting the metric; this defines f_k as a continuous function on X, and we have $f_k(x) > 0$ for all $x \in U_k$. Since $\mathcal{U}$ is a cover, the continuous function $f_1 \vee \cdots \vee f_s$ is everywhere positive on X; since X is compact, this function has a positive lower bound ϵ. Hence, for each point x, there exists k for which $f_k(x) \geqq \epsilon$. It follows that the cell

$$\{y: d(x, y) < \epsilon\}$$

is contained in U_k.

16.22. THEOREM. *If X is a compact subset of* $\mathbf{R}^n$, *then* $\dim X \leqq n$.

PROOF. Let $\mathcal{U}$ be a cover of X. By Lemma 16.21, there exists $\epsilon > 0$ such that the cells (in X) of radius ϵ refine $\mathcal{U}$. Therefore, the system of cells of Lemma 16.20, scaled down by a factor of ϵ, and restricted to X, is a refinement of $\mathcal{U}$ of order $\leqq n$. Thus, $\dim X \leqq n$.

16.23. We shall need some other miscellaneous results about $\mathbf{R}^n$.

LEMMA. *Given nonempty open sets $T_1, \cdots, T_s$ in $\mathbf{R}^n$, there exist vectors $y^k \in T_k$ such that for any $n + 1$ of these vectors y^{k_i}, and any vector $y \in \mathbf{R}^n$, the equations*

$$\sum_{i=1}^{n+1} a_i y^{k_i} = y, \qquad \sum_{i=1}^{n+1} a_i = 1$$

have at most one solution $(a_1, \cdots, a_{n+1})$.

PROOF. Each set T_k contains a cell

$$\{y: \|y - x^k\| < \epsilon_k\}.$$

It is required to find numbers $y^k{}_j$ ($j \leqq n$, $k \leqq s$) that satisfy

(a) $$|y^k{}_j - x^k{}_j| < \epsilon_k,$$

and such that the determinant of none of the $n + 1$ by $n + 1$ submatrices of the matrix

$$\begin{bmatrix} y^1{}_1 & \cdots & y^s{}_1 \\ \vdots & & \vdots \\ y^1{}_n & \cdots & y^s{}_n \\ 1 & \cdots & 1 \end{bmatrix}$$

is zero. This can be achieved by choosing the $y^k{}_j$ successively, so that the determinant of no square submatrix, constructible from the entries already selected, is zero. Since there are infinitely many numbers satisfying (a), while the restriction on the determinants eliminates only finitely many possibilities, a suitable choice of $y^k{}_j$ can always be made. (The vectors y^k are said to be in *general position.*)

16.24. We shall denote the set of all continuous mappings from X into $\mathbf{R}^n$ by $C_n(X)$. (Thus, $C_1(X) = C(X)$.) We do not consider separately the set of bounded functions in $C_n(X)$, as, in all our applications, X will be compact.

LEMMA. *Let X be a compact space. Given $f_1, \cdots, f_m \in C_n(X)$, and given $\epsilon > 0$, there exists a cover $\mathscr{V}$ of X such that the diameter of $f_k[V]$ is $\leqq \epsilon$ for each $k \leqq m$ and each $V \in \mathscr{V}$.*

PROOF. For each $k \leqq m$, a finite collection of cells $T_{k,1}, \cdots, T_{k,s_k}$ in $\mathbf{R}^n$ of diameter ϵ cover the compact set $f_k[X]$. Their preimages under f_k cover X. For every choice of indices $j_1 \leqq s_1, \cdots, j_m \leqq s_m$, define

$$V_{j_1, \cdots, j_m} = \bigcap_{k=1}^m f_k^{\leftarrow}[T_{k,j_k}],$$

and take for $\mathscr{V}$ the collection of all such sets. Then $\mathscr{V}$ is a cover of X. The bound on diameters follows from the relations

$$f_k[V_{j_1, \cdots, j_m}] \subset f_k[f_k^{\leftarrow}[T_{k,j_k}]] = T_{k,j_k}.$$

16.25. For a compact space X, we consider the metric d on $C_n(X)$ defined by

$$d(f, g) = \sup_{x \in X} \|f(x) - g(x)\|.$$

The resulting metric space $C_n(X)$ is complete. This fact will allow us to apply the Baire category theorem:

THEOREM. *In a nonempty, complete metric space, every countable intersection of open, dense sets is nonempty.*

PROOF. Given such sets G_m ($m \in \mathbf{N}$), choose a cell of radius < 1 whose closure F_1 is contained in G_1. Inductively, choose a cell of radius $< 1/(m + 1)$ whose closure F_{m+1} is contained in $F_m \cap G_{m+1}$. Then the family $(F_m)_{m \in \mathbf{N}}$ contains arbitrarily small sets. Since the space is complete, we have $\emptyset \neq \bigcap_m F_m \subset \bigcap_m G_m$.

As a matter of fact, $\bigcap_m G_m$ is dense. To prove this, we simply add the stipulation (as we may) that F_1 be contained in a specified nonempty open set.

ENVELOPMENT

16.26. The algebraic characterization of dimension will require a prior analysis of the relation between covers of the space and certain connected subsets. The next three sections are devoted to this preliminary study. We shall say that a cover $\mathscr{U}$ *envelops* a set F if every *component* of F is contained in a member of $\mathscr{U}$ (i.e., the components of F refine $\mathscr{U}$). Clearly, if $\mathscr{U}$ envelops F, and $F \supset E$, then $\mathscr{U}$ also envelops E.

LEMMA. *If a cover $\mathscr{U}$ of X envelops a compact set F, then some partition of F refines $\mathscr{U}$.*

REMARK. The trace of this partition on a subset E of F is, evidently, a partition of E that refines $\mathscr{U}$.

PROOF. By hypothesis, each component F_α of F is contained in some member U_α of $\mathscr{U}$. Now, each F_α is an intersection of open-and-closed sets in the compact set F (Theorem 16.15); hence some finite intersection is contained in the open set U_α. Thus, there exists an open-and-closed set H_α in F, with $F_\alpha \subset H_\alpha \subset U_\alpha$. A finite collection of the sets H_α covers the compact set F; the required partition of F is obtained from this collection by taking intersections and differences where necessary.

We remark that the converse of the lemma is obvious: if a partition of a set F refines $\mathscr{U}$, then $\mathscr{U}$ envelops F.

A standard example shows that the conclusion of the lemma need not hold if F is not compact, even when X is. (See 16K.)

16.27. LEMMA. *Let X be a compact space, let $g \in C_n(X)$, and let $\mathscr{U}$ be a cover of X that envelops each set $g^{\leftarrow}(y)$, for $y \in g[X]$.*

(a) *There exists a cover $\mathscr{W}$ of $g[X]$ such that $\mathscr{U}$ envelops each set $g^{\leftarrow}[W]$, for $W \in \mathscr{W}$.*

(b) *$\mathscr{U}$ has a refinement of order $\leqq n$.*

PROOF. (a). Fix $y \in g[X]$. By the preceding lemma, there exist a partition $\{F_1, \cdots, F_m\}$ of the compact set $g^{\leftarrow}(y)$, and sets $U_1, \cdots, U_m \in \mathscr{U}$, such that $F_k \subset U_k$ $(k \leqq m)$. Since X is normal, there exist disjoint open sets $G_1, \cdots, G_m$, with $F_k \subset G_k$; certainly, we may assume that $G_k \subset U_k$ $(k \leqq m)$. The set $G = \bigcup_k G_k$ is a neighborhood of $g^{\leftarrow}(y)$. Now, $g^{\leftarrow}(y) = \bigcap_V g^{\leftarrow}[V]$, where V ranges over all closed neighborhoods of y; and since X is compact, a suitable finite intersection is contained in G. In consequence, there exists an open neighborhood V_y of y for which $g^{\leftarrow}[V_y] \subset G$. As the partition $\{G_1, \cdots, G_m\}$ of G refines $\mathscr{U}$, this implies that $\mathscr{U}$ envelops $g^{\leftarrow}[V_y]$.

If $\mathscr{W}$ is any finite collection of sets V_y, $y \in g[X]$, that cover the compact set $g[X]$, then $\mathscr{W}$ satisfies (a).

(b). The compact subset $g[X]$ of $\mathbf{R}^n$ is of dimension $\leqq n$ (Theorem 16.22). By Corollary 16.8, then, it has a cover $\mathscr{T}$ of order $\leqq n$ whose closures refine $\mathscr{W}$. Then $\mathscr{U}$ envelops the compact set $g^{\leftarrow}[\text{cl } T_i]$, for each $T_i \in \mathscr{T}$. By the preceding lemma (and remark), the subset $E_i = g^{\leftarrow}[T_i]$ has a partition $\{E_{i,j}\}_j$ that refines $\mathscr{U}$. Since all E_i, and hence all $E_{i,j}$, are open in X, the collection $\mathscr{E} = \{E_{i,j}\}_{i,j}$ is a cover—and it refines $\mathscr{U}$. Clearly, the order of $\mathscr{E}$ is that of $\{E_i\}_i$, which is that of $\mathscr{T}$; thus, $\mathscr{E}$ is a refinement of $\mathscr{U}$ of order $\leqq n$.

As the proof shows, the lemma holds, more generally, when g is a continuous mapping from X onto any space of dimension $\leqq n$.

16.28. THEOREM. *Let X be a compact space. For any cover $\mathscr{U}$ of X, the set*

$$G(\mathscr{U}) = \{g \in C_n(X) \colon \mathscr{U} \textit{ envelops } g^{\leftarrow}(y) \textit{ for all } y \in g[X]\}$$

is open in the metric space $C_n(X)$. If, further, $\dim X \leqq n$, *then $G(\mathscr{U})$ is also dense.*

PROOF. [*$G(\mathscr{U})$ is open.*] Given $g \in G = G(\mathscr{U})$, we are to find $\epsilon > 0$ such that G contains the neighborhood

$$\{f \colon d(g, f) < \epsilon\}$$

of g. By (a) of the preceding lemma, there is a cover $\mathscr{W}$ of $g[X]$ such that $\mathscr{U}$ envelops each set $g^{\leftarrow}[W]$, for $W \in \mathscr{W}$. By Lemma 16.21, there exists $\epsilon > 0$ such that the cells in $g[X]$ of radius 2ϵ refine $\mathscr{W}$. This ϵ will do the job. For, consider any f satisfying $d(g, f) < \epsilon$, and any $y \in f[X]$. Pick $a \in f^{\leftarrow}(y)$, and choose $W \in \mathscr{W}$ so that $\|z - g(a)\| < 2\epsilon$ implies $z \in W$. For all $x \in X$, we have

$$\|g(x) - f(x)\| < \epsilon;$$

hence, if $x \in f^{\leftarrow}(y)$, then $f(x) = y = f(a)$, and

$$\|g(x) - g(a)\| \leqq \|g(x) - f(x)\| + \|f(a) - g(a)\| < 2\epsilon,$$

so that $g(x) \in W$. Thus, $f^{\leftarrow}(y) \subset g^{\leftarrow}[W]$. It follows that $\mathscr{U}$ envelops $f^{\leftarrow}(y)$. Therefore $f \in G$.

[*If* $\dim X \leqq n$, *then $G(\mathscr{U})$ is dense.*] Given $f \in C_n(X)$, and $\epsilon > 0$, we are to find $g \in G$ for which $d(f, g) \leqq \epsilon$. By Lemma 16.24, there exists a cover $\mathscr{V}$ of X such that the diameter of $f[V]$ is at most $\epsilon/2$, for each $V \in \mathscr{V}$. The collection of all sets $V \cap U$, for $V \in \mathscr{V}$ and $U \in \mathscr{U}$, is a refinement of $\mathscr{U}$, and a refinement of $\mathscr{V}$; since $\dim X \leqq n$, it has a basic refinement $\mathscr{S} = \{S_k\}_{k \leqq s}$ of order $\leqq n$. The diameter of $f[S_k]$ is $\leqq \epsilon/2$, for each k. Since $\mathscr{S}$ is basic, there exist $h_k \in C(X)$, with $h_k \geqq \mathbf{0}$, such that $\mathbf{Z}(h_k) = X - S_k$ $(k \leqq s)$. Since $\mathscr{S}$ is a cover, $\sum_k h_k$ has no zeros on the compact space X, and so we may assume that $\sum_k h_k = \mathbf{1}$.

Next, choose any point x_k in S_k, and let T_k denote the cell in $\mathbf{R}^n$ with center $f(x_k)$ and radius $\epsilon/2$. For points $y^k \in T_k$, selected as in Lemma 16.23, the triangle inequality yields

$$\|f(x) - y^k\| < \epsilon \quad \text{for all} \quad x \in S_k.$$

Now define $g \in C_n(X)$ as follows:

$$g(x) = \sum_k h_k(x) y^k \qquad (x \in X)$$

(in which the symbols denote scalar multiplication and vector addition in $\mathbf{R}^n$). We shall show that $g \in G$, and that $d(f, g) \leqq \epsilon$. We begin with the second, which is simpler.

[$d(f, g) \leqq \epsilon$.] Consider any $x \in X$. If $x \in S_k$, then $\|f(x) - y^k\| < \epsilon$; and if $x \notin S_k$, then $h_k(x) = 0$. In either case, then,

$$h_k(x) \cdot \|f(x) - y^k\| \leqq \epsilon h_k(x).$$

Consequently, using the fact that $\sum_k h_k = \mathbf{1}$, we have:

$$\begin{aligned} \|f(x) - g(x)\| &= \|f(x) - \textstyle\sum_k h_k(x) y^k\| \\ &= \|\textstyle\sum_k h_k(x)(f(x) - y^k)\| \\ &\leqq \textstyle\sum_k h_k(x) \cdot \|f(x) - y^k\| \\ &\leqq \epsilon \textstyle\sum_k h_k(x) \\ &= \epsilon. \end{aligned}$$

Therefore $d(f, g) \leqq \epsilon$.

[$g \in G$.] Given $y \in g[X]$, we are to show that $\mathscr{U}$ envelops $g^{\leftarrow}(y)$; we shall show, in fact, that its refinement $\mathscr{S}$ has this property. Recall that the order of $\mathscr{S}$ is $\leqq n$, and that $S_k = X - \mathbf{Z}(h_k)$. These facts imply that for any $x \in X$, at most $n + 1$ of the numbers $h_k(x)$ are different from 0. If, now, $x \in g^{\leftarrow}(y)$, then

$$g(x) = \sum_k h_k(x) y^k = y.$$

But only one such linear combination of any $n + 1$ of the vectors y^k can yield y—by virtue of the choice of y^k in accordance with Lemma 16.23. It follows that the set of all vectors

$$h(x) = (h_1(x), \cdots, h_s(x)) \qquad (x \in X)$$

in $\mathbf{R}^s$, for which $g(x) = y$, must be finite. Expressed in terms of the mapping $h \in C_s(X)$ (defined by the above equation): the set $h[g^{\leftarrow}(y)]$ in $\mathbf{R}^s$ is finite. Call its distinct elements $b^1, \cdots, b^r$. Obviously,

$$\{h^{\leftarrow}(b^1), \cdots, h^{\leftarrow}(b^r)\}$$

is a partition of $g^{\leftarrow}(y)$; so to finish the proof, it suffices to show that this partition refines $\mathscr{S}$. For each $j \leqq r$, we have $\sum_k b^j{}_k = 1$, and hence $b^j{}_k$ is positive for some k. For such k, we have

$$h^{\leftarrow}(b^j) \subset \{x: h_k(x) = b^j{}_k > 0\} \subset X - \mathbf{Z}(h_k) = S_k.$$

This completes the proof of the theorem.

ANALYTIC DIMENSION OF $C^*(X)$

16.29. *Algebraic and analytic subrings.* A subring A of $C^*(X)$ will be called an *algebraic* subring provided that

(i) all constant functions belong to A; and
(ii) $f^2 \in A$ implies $f \in A$.

Trivially, C^* itself is an algebraic subring. An arbitrary intersection of algebraic subrings is algebraic. Next, if the range of a function $f \in C^*$ is finite, then f belongs to every algebraic subring. For, f is a linear combination (with constant coefficients) of continuous functions into $\{0, 2\}$; and each function g of the latter type belongs to every algebraic subring, since $(g - \mathbf{1})^2 = \mathbf{1}$. On the other hand, it is obvious that the set of all such functions f is an algebraic subring. So we have:

(a) *The set A_0 of all functions in $C^*(X)$ with finite range is the smallest algebraic subring of $C^*(X)$.*

According to 3L.*3*, A_0 is all of $C^*(X)$ only when X is finite.

An algebraic subring that is *closed* in the metric topology of $C^*(X)$ will be called an *analytic* subring.

It is evident that the algebraic subrings of $C^*(X)$ are in one-one correspondence with those of $C(\beta X)$, under the isomorphism $f \rightarrow f^{\beta}$. Since this mapping also preserves norm, analytic subrings of $C^*(X)$ correspond to analytic subrings of $C(\beta X)$.

16.30. THEOREM. *The set A of all functions in $C^*(X)$ that are constant on a given connected set S is an analytic subring.*

PROOF. Trivially, A is a subring. If a continuous function g^2 assumes only one value on S, then g can assume at most two; since S is connected, g must be constant. Therefore A is algebraic. Obviously, A is closed.

An analytic subring need not be of the above form, or even an intersection of subrings of this form. An example is the set of all functions f in $C^*(\mathbf{R})$ for which $\lim_{x \to +\infty} f(x)$ exists (16C.*1*). However, in case X is *compact*, then such intersections do comprise all analytic subrings, as we now proceed to demonstrate.

16.31. Let A be a subfamily of $C(X)$. We recall that a stationary set of A is a subset of X on which every member of A is constant. Any stationary set E is contained in a maximal such set, namely,

$$\{x : f(x) = f(p) \quad \text{for all} \quad f \in A\},$$

where p is any point of E.

LEMMA. *Let X be a compact space. If A is an algebraic subring of $C^*(X)$, then each maximal stationary set of A is connected.*

PROOF. If S is a maximal stationary set of A, then

$$S = \bigcap_{f \in A} \mathbf{Z}(f - \mathbf{r}),$$

where $\{r\} = f[S]$. Because A is a ring containing the constant functions, this reduces to $S = \bigcap_g \mathbf{Z}(g)$, where $g \in A$ and $g[S] = \{0\}$. Moreover, we need consider only nonnegative g, since $\mathbf{Z}(h^2) = \mathbf{Z}(h)$ for any h. Since

$$\mathbf{Z}(g) = \bigcap_{\epsilon > 0} \{x : g(x) \leqq \epsilon\}, \quad \text{for} \quad g \geqq \mathbf{0},$$

we get

$$S = \bigcap_{g, \epsilon} \{x : g(x) \leqq \epsilon\},$$

where $\epsilon > 0$ and $g \in A$, with $g \geqq \mathbf{0}$ and $g[S] = \{0\}$.

Now, evidently, any finite intersection of sets of the form $\{x : g(x) \leqq \epsilon\}$ contains a set of the same form. It follows from Lemma 16.13 that if $\{S_1, S_2\}$ is any partition of S, there exist g and ϵ (as above) for which

$$\{x : g(x) \leqq \epsilon\} = H_1 \cup H_2,$$

where H_1 and H_2 are disjoint closed sets containing S_1 and S_2, respectively. We now define f as follows:

$$f(x) = \epsilon - g(x) \quad \text{for } x \in H_1,$$
$$f(x) = g(x) - \epsilon \quad \text{for } x \in X - H_1.$$

Since $g(x) = \epsilon$ on the boundary of H_1, f is continuous on X. The relation $f^2 = (g - \boldsymbol{\epsilon})^2$ implies that f belongs to the algebraic subring A, and hence that f is constant on S. But $f(x) = \epsilon > 0$ for $x \in S_1$, while $f(x) = -\epsilon < 0$ for $x \in S_2$. Therefore either S_1 or S_2 must be empty. Thus, S is connected.

16.32. *Analytic base.* Let B be any subfamily of $C^*(X)$. Since intersections of analytic subrings are analytic, the intersection A of all those that contain B is the smallest analytic subring containing B. We refer to B as an (*analytic*) *base* for A. The smallest analytic subring of $C^*(X)$ is the one with base $\emptyset$; according to 16A.*1*, it is the closure of the algebraic subring A_0 of all functions with finite range.

For analytic subrings, the Stone-Weierstrass theorem takes the following form.

THEOREM. *Let X be a compact space, and B any subfamily of $C^*(X)$. The analytic subring with base B is precisely the family of all functions in $C^*(X)$ that are constant on every connected stationary set of B.*

PROOF. Let $\bar{B}$ denote the family in question. Evidently, it is an intersection of analytic subrings of the type considered in Theorem 16.30; therefore $\bar{B}$ is an analytic subring. Obviously, $\bar{B} \supset B$. Now consider any analytic subring A containing B. By Lemma 16.31, every stationary set of A is contained in a connected one. This is, of course, a stationary set of B, and therefore, by definition of $\bar{B}$, a stationary set of $\bar{B}$. Thus, every stationary set of A is a stationary set of $\bar{B}$. By the Stone-Weierstrass theorem, cl $A \supset \bar{B}$, i.e., $A \supset \bar{B}$.

REMARK. It follows from Theorem 16.30 that for arbitrary X, if $C^*(X)$ has no proper analytic subring, then X is totally disconnected. For compact X, the present theorem implies the converse: if X is totally disconnected, then $C^*(X)$ has no proper analytic subring. If, however, X is totally disconnected but not compact, then $C^*(X)$ can contain proper analytic subrings. For if X is a totally disconnected space of positive dimension (16L), then by Theorem 16.34 below, the smallest analytic subring of $C^*(X)$ (i.e., the one with base $\emptyset$) is a proper subring.

16.33. Each n-tuple of functions $g_1, \cdots, g_n \in C(X)$ determines an element $g \in C_n(X)$, defined by:

$$\text{(a)} \qquad g(x) = (g_1(x), \cdots, g_n(x)) \qquad (x \in X).$$

Conversely, given $g \in C_n(X)$, its coordinate functions g_i $(= \pi_i \circ g)$ belong to $C(X)$, and satisfy (a). Obviously, the maximal connected sets in X on which all g_i are constant are precisely the components of $g^{\leftarrow}(y)$, for $y \in g[X]$. Hence we have:

COROLLARY. *Let X be compact. For $g \in C_n(X)$, the analytic subring A_g with base $\{g_1, \cdots, g_n\}$ (as in (a)) is precisely the set of all functions that are constant on every component of $g^{\leftarrow}(y)$, for all $y \in g[X]$.*

16.34. *Analytic dimension.* Let X be an arbitrary completely regular space.

> The *analytic dimension* of $C^*(X)$—denoted by ad $C^*(X)$—is defined to be the least cardinal $\mathfrak{m}$ such that every *countable* family in $C^*(X)$ is contained in an analytic subring having a base of power $\leqq \mathfrak{m}$.

Some comments on this definition are desirable. Observe, first, that we do not require that every analytic subring with countable base have a base of power $\leqq \mathfrak{m}$, but only that it be *contained* in an analytic subring having such a base; see 16.36.

Secondly, we shall prove, in 16.35, that the definition is equivalent to the following: ad $C^*(X)$ is the least cardinal $\mathfrak{m}$ such that every *finite* family in $C^*(X)$ is contained in an analytic subring with a base of power $\leqq \mathfrak{m}$. In this form, the definition is more clearly analogous to the definition of dim X.

Evidently, $0 \leqq$ ad $C^*(X) \leqq \aleph_0$. The main result in this chapter is that dim $X =$ ad $C^*(X)$ for every completely regular X (Theorem 16.35). It is clear from the remarks made in 16.29 that ad $C^*(X) =$ ad $C^*(\beta X)$. As it is also true that dim $X =$ dim βX, we see that it is enough to handle the problem for compact X.

The result for dimension 0 can be established readily.

THEOREM. dim $X = 0$ *if and only if* ad $C^*(X) = 0$, *i.e.,* $C^*(X)$ *has no proper analytic subring.*

PROOF. We assume that X is compact. Then dim $X = 0$ if and only if X is totally disconnected. As remarked in 16.32, if X is totally disconnected, then $C^*(X)$ has no proper analytic subring.

Conversely, if X contains a connected subset S of more than one point, then the set of all functions constant on S is an analytic subring (Theorem 16.30), and it is not all of $C^*(X)$.

In the proof of the general case, all the accumulated machinery will be brought to bear.

16.35. THEOREM (KATĚTOV). *The following are equivalent for any completely regular space X.*

(1) dim $X \leqq n$.

(2) ad $C^*(X) \leqq n$—*i.e., every countable subfamily of $C^*(X)$ is contained in an analytic subring having a base of cardinal $\leqq n$.*

(3) *Every finite subfamily of $C^*(X)$ is contained in an analytic subring having a base of cardinal $\leqq n$.*

PROOF. As was pointed out in the preceding section, we may assume in the proof that X is compact.

The result for $n = 0$ has already been established; and for $\aleph_0$, it is trivial, because (1), (2), and (3) are always true. So we may assume that $n \in \mathbf{N}$.

(1) *implies* (2). Let $\{f_k\}_{k \in \mathbf{N}}$ be a countable set in $C(X)$. To each $m \in \mathbf{N}$, we apply Lemma 16.24; accordingly, there exists a cover $\mathscr{U}_m$ of X such that, for every $k \leqq m$ and $U \in \mathscr{U}_m$, the diameter of $f_k[U]$ is at most $1/m$. Since $\dim X \leqq n < \aleph_0$, there exists an open, dense subset $G(\mathscr{U}_m)$ of $C_n(X)$ as described in Theorem 16.28.

Since the metric space $C_n(X)$ is complete, we have

$$\bigcap_{m \in \mathbf{N}} G(\mathscr{U}_m) \neq \emptyset,$$

by the Baire category theorem (16.25). Choose g in this intersection. Then $\mathscr{U}_m$ envelops $g^{\leftarrow}(y)$, for all $m \in \mathbf{N}$ and all $y \in g[X]$.

We now show that the analytic subring A_g with base $\{g_1, \cdots, g_n\}$ (see 16.33) contains every f_k—as required in (2). Given $y \in g[X]$, let S be any component of $g^{\leftarrow}(y)$. Consider any m; since $\mathscr{U}_m$ envelops $g^{\leftarrow}(y)$, we have $S \subset U$ for some $U \in \mathscr{U}_m$. Then $f_k[S] \subset f_k[U]$, whence, for $k \leqq m$, the diameter of $f_k[S]$ is $\leqq 1/m$. As this holds for all m, $f_k[S]$ must have diameter 0. Thus, f_k is constant on S. By Corollary 16.33, $f_k \in A_g$.

(2) *implies* (3). This is trivial.

(3) *implies* (1). Let $\mathscr{U} = \{U_k\}_{k \leqq s}$ be any cover of X. By Theorem 16.6, there exist a refinement $\mathscr{W} = \{W_k\}_{k \leqq s}$ of $\mathscr{U}$, and functions $f_1, \cdots, f_s \in C(X)$, such that

$$W_k \subset \mathbf{Z}(f_k) \subset U_k \qquad (k = 1, \cdots, s).$$

By hypothesis, the s functions f_k are contained in an analytic subring A_g, for suitable $g \in C_n(X)$. Given $y \in g[X]$, let S be any component of $g^{\leftarrow}(y)$. Now, S meets some member W_k of the cover $\mathscr{W}$, whence S meets $\mathbf{Z}(f_k)$. But, since $f_k \in A_g$, f_k is constant on S (Corollary 16.33); so $\mathbf{Z}(f_k) \supset S$. Hence $U_k \supset S$. This shows that $\mathscr{U}$ envelops $g^{\leftarrow}(y)$. By Lemma 16.27(b), $\mathscr{U}$ has a refinement of order $\leqq n$. Therefore $\dim X \leqq n$.

16.36. REMARKS. One may be tempted to define the analytic dimension as the smallest cardinal of an analytic base for all of $C^*(X)$. In case X is a compact metric space, this will, indeed, yield ad $C^*(X)$ (see 16G). However, there exist one-dimensional spaces whose function rings have no countable bases; in fact, the real line is an example of such a space (16F).

In another direction, it may seem reasonable to require that every analytic subring with countable base have a base of cardinal ad C^*, rather than that the subring be contained in one with such a base. But this requirement may also lead to too large a dimension number. Suppose that τ is a continuous mapping from a compact space X onto a space Y, and that $\tau^{\leftarrow}(y)$ is connected, for each $y \in Y$ (i.e., τ is a *monotone* mapping). It is known that there exist such mappings with the additional property that dim $Y >$ dim X. But the subring

$$\tau'[C(Y)] = \{g \circ \tau \colon g \in C(Y)\}$$

of $C(X)$ is isomorphic with $C(Y)$ (Theorem 10.3(a)). It follows that ad $\tau'[C(Y)] =$ dim Y, which is greater than dim X. Moreover, $\tau'[C(Y)]$ is an analytic subring of $C(X)$; in fact, it consists of all functions in $C(X)$ that are constant on each connected set $\tau^{\leftarrow}(y)$ (Theorems 10.11 and 16.30). Thus, $\tau'[C(Y)]$ is an analytic subring of $C(X)$; some analytic subring of $\tau'[C(Y)]$ with countable base has no base of cardinal dim X; but every such subring is contained in a subring of $C(X)$ with a base of cardinal dim X.

PROBLEMS

16A. CLOSURE OF AN ALGEBRAIC SUBRING.

1. If A is an algebraic subring of $C^*(X)$, then cl A is analytic [X may be assumed to be compact.]

2. dim $X = 0$ if and only if cl $A_0 = C^*(X)$ (16.29(a)).

16B. POLYNOMIALS OVER AN ANALYTIC SUBRING.

Let A be an analytic subring of C^*. If $g_1, \cdots, g_n \in A$, and if $f \in C^*$ satisfies the equation

$$f^n + g_1 f^{n-1} + \cdots + g_n = \mathbf{0},$$

then $f \in A$. [A polynomial with real coefficients has only a finite number of zeros.]

16C. SUBRINGS OF $C^*(\mathbf{R})$.

1. The set of all $f \in C^*(\mathbf{R})$ for which $\lim_{x \to +\infty} f(x)$ exists is an analytic subring. [Argue directly. Alternatively, apply Theorem 16.30 to $\beta\mathbf{R}$.]

2. The set of all functions in $C^*(\mathbf{R})$ that have continuous extensions to $\mathbf{R}^*$ is a closed subring, but is not algebraic.

16D. ZERO-DIMENSIONAL COMPACTIFICATION.

1. Let X be a compact space, and let Y be the quotient space whose points are the components of X. Then Y is compact and totally disconnected. [Theorem 16.15.]

2. A space S has a zero-dimensional compactification if and only if it has a base of open-and-closed sets. [Apply 6L.*2* to S. *Caution*: It is not enough that any two points of S can be separated by a partition; see 16L.]

16E. CLOSED SUBRINGS OF C^*.

1. The image of $C^*(Y)$, under a homomorphism into $C^*(X)$, is a closed subring of $C^*(X)$. [Apply Theorem 10.8; then, either argue directly, or invoke Theorem 10.11. Alternatively, use 10D.*4*, 1J.*6*, and 15C.*2*.]

2. For X and Y compact, Y is a continuous image of X if and only if $C(Y)$ is isomorphic with a closed subring of $C(X)$ that includes the constants. [10.9(b).]

3. A closed subring A of $C^*(X)$ that contains the constants is isomorphic to $C(Y)$, where Y is a suitable continuous image of βX. [3I.*1*.]

4. Use the Stone-Weierstrass theorem to prove that any compact subset is C^*-embedded.

16F. DIMENSION OF **R**.

1. A subspace S ($\neq \emptyset$) of $\mathbf{R}$ is zero-dimensional if and only if its complement is dense. [*Sufficiency.* S has a base of open-and-closed sets.] In particular, $\dim \mathbf{Q} = \dim (\mathbf{R} - \mathbf{Q}) = 0$.

2. $\dim [0, 1] = 1$.

3. $\dim \mathbf{R} = 1$. [For each integer t, the trace of a given cover $\{U_k\}_{k \leqq s}$ on $[t, t + 1]$ has a refinement of order $\leqq 1$ by relatively open intervals. From these, obtain a countable collection of open intervals to refine $\{U_k\}_{k \leqq s}$, and let the kth member of the required refinement be the union of all those contained in U_k but not in $\bigcup_{i<k} U_i$.]

4. Any analytic base for $C^*(\mathbf{R})$ is uncountable. [10N.*5*.]

16G. AN ANALYTIC BASE FOR ALL OF C^*.

1. If X is a compact metric space, then $C(X)$ has a countable, dense subfamily. [X has a countable base of open sets. Construct a countable subfamily B of $C(X)$ that distinguishes points of X, and then a countable subfamily that is dense in the ring generated by B and the constants.]

2. If X is a compact metric space, $C(X)$ has an analytic base of cardinal ad $C(X)$.

3. A compact metric space X has dimension $\leqq n$ if and only if there exists a continuous mapping $g: X \to \mathbf{R}^n$ such that for each $y \in \mathbf{R}^n$, $g^{\leftarrow}(y)$ is totally disconnected (i.e., g is a *light* mapping). [Corollary 16.33.]

4. Let X be the quotient space of $\mathbf{W}^* \times [0, 1]$ obtained by identifying all points of the form $(\alpha, 0)$. Then X is compact and connected, but not metrizable. $C(X)$ has a base of cardinal 1.

16H. THE LONG LINE.

Let L denote the totally ordered space (3O) obtained by assigning the lexicographic order to $\mathbf{W} \times R^+$.

1. L is Dedekind-complete, and contains no consecutive elements. Hence L is connected.

2. Every function in $C(L)$ is constant on a tail. [5.12(c).] In fact, any countable set of continuous functions are constant on a common tail.

3. Every analytic base for $C^*(L)$ is uncountable.

4. If $q \neq (0, 0)$, then $\{p \in L : p \leqq q\}$ is homeomorphic with $[0, 1]$. [13B.]

5. $\dim L = 1$. [*2* and *4*.]

6. βL is the one-point compactification of L.

16I. THE SUBSPACE THEOREM.

1. If X is C^*-embedded in Y, then $\dim X \leqq \dim Y$. [Apply Theorem 16.35. Alternatively, argue from the fact that βX is a closed set in the normal space βY]. (When X is not C^*-embedded, the conclusion can fail; see 16M.) If, in addition, X is dense, then $\dim X = \dim Y$.

2. More generally, if $C^*(X)$ is a homomorphic image of $C^*(Y)$, then $\dim X \leqq \dim Y$. [10.9(c*).]

16J. THE PRODUCT THEOREM.

Let X and Y be compact spaces.

1. The ring generated in $C(X \times Y)$ by the functions that depend on only one coordinate is dense in $C(X \times Y)$.

2. Any countable subfamily of $C(X \times Y)$ is contained in a closed subring generated by a countable family of functions that depend on only one coordinate. [Given $h \in C(X \times Y)$ and $n \in \mathbf{N}$, there exists a finite set in $C(X)$ such that for each $y \in Y$, some member f of the set satisfies $|h(x,y) - f(x)| < 1/n$ for all $x \in X$. There also exists a corresponding set in $C(Y)$; apply the Stone-Weierstrass theorem. Alternatively, make use of a "partition of unity."]

3. $\dim (X \times Y) \leqq \dim X + \dim Y$.

16K. COMPONENTS NOT SEPARATED BY A PARTITION.

In the unit square $X = [0, 1] \times [0, 1]$, let $u = (0, 0)$, $v = (0, 1)$, and

$$F = \big(\{1/n : n \in \mathbf{N}\} \times [0, 1]\big) \cup \{u\} \cup \{v\}.$$

1. Every open-and-closed set in F that contains u also contains v.

2. $\{u\}$ is a component of F, but it is not an intersection of open-and-closed sets.

3. $\{X - \{u\}, X - \{v\}\}$ is a cover of X that envelops F, but which is refined by no partition of F.

16L. TOTALLY DISCONNECTED SPACE OF POSITIVE DIMENSION.

Let X be the metric space of all sequences $x = (x_n)_{n \in \mathbf{N}}$ of rational numbers for which $\sum_n {x_n}^2 < \infty$, with metric

$$d(x, y) = \big(\textstyle\sum_n (x_n - y_n)^2\big)^{1/2}$$

(i.e., X is the subspace of all points in real, separable Hilbert space with rational coordinates). Then X is totally disconnected; in fact, any two points are separated by a partition of X. However, X does not have a base of open-and-closed sets. [Let U be a bounded, open neighborhood of (0). Inductively, construct a sequence of rational numbers x_n such that each point

$$x^m = (x_1, \cdots, x_m, 0, 0, \cdots)$$

belongs to U, and $d(x^m, X - U] < 1/m$. Then $(x_n) \in \mathrm{cl}\, U - U$.]

16M. THE ZERO-DIMENSIONAL SPACE Δ_0 AND ITS ONE-DIMENSIONAL SUBSPACE Δ_1.

1. There exists an indexed set $\{r_\sigma\}_{\sigma<\omega_1}$ of distinct irrational numbers in $[0, 1]$ such that for each $\alpha < \omega_1$, the set $\{r_\sigma\}_{\sigma>\alpha}$ is dense in $[0, 1]$.

2. For each $\alpha < \omega_1$, the set $S_\alpha = [0, 1] - \{r_\sigma\}_{\sigma>\alpha}$ is dense in $[0, 1]$, and $\dim S_\alpha = 0$. [16F.*1*.]

3. The subspace

$$\Delta_1 = \{(\alpha, s): s \in S_\alpha, \alpha < \omega_1\}$$

of $\mathbf{W}^* \times [0, 1]$ has a base of open-and-closed sets.

4. For $g \in C(\Delta_1)$, there exists $\tau < \omega_1$ such that for every $\sigma > \tau$, $g(\alpha, s) = g(\sigma, s)$ for all $\alpha > \sigma$ and $s \in S_\sigma$. [Apply 5.12(c), first considering rational values of s.]

5. Δ_1 is dense and C-embedded in the subspace

$$\Delta = \{(\alpha, s): s \in S_\alpha, \alpha \leqq \omega_1\}, \quad \text{where} \quad S_{\omega_1} = [0, 1],$$

of $\mathbf{W}^* \times [0, 1]$. [*4* and 6H.]

6. $\dim \Delta_1 = \dim \Delta > 0$.

7. For each $\tau < \omega_1$, $\{(\alpha, s): s \in S_\alpha, \alpha \leqq \tau\}$ is an open-and-closed set in Δ, and is homeomorphic with a zero-dimensional subspace of $\mathbf{R}^2$. [13B.*2*.]

8. $\dim \Delta_1 = 1$. [In view of *4*, there exists $\tau < \omega_1$ such that a basic cover of Δ behaves on the complement of the set in 7 like a cover of [0, 1].]

9. Let Δ_0 be the quotient space of Δ obtained by identifying the points of $\Delta - \Delta_1$. Then Δ_1 is a subspace of Δ_0. But $\dim \Delta_0 = 0$. [*4*.] Thus, $\beta\Delta_0$ is a zero-dimensional compactification of Δ_1; cf. 16D.*2*.

16N. NORMALITY OF Δ_1 AND Δ_0.

1. The space Δ of 16M is normal. [If H and K are disjoint closed sets, then $H - \Delta_1$ and $K - \Delta_1$ have disjoint compact neighborhoods in $\Delta - \Delta_1$. Make use of 16M.7.]

2. Disjoint closed sets in Δ_1 have disjoint closures in Δ. [Argue as in 8J.*2*.]

3. Δ_1 and Δ_0 are normal.

16O. BASICALLY DISCONNECTED SPACES.

Every basically disconnected space (1H) is zero-dimensional. [Theorem 16.6. Alternatively, use 6M.*1*.]

16P. BASES OF OPEN-AND-CLOSED SETS.

1. If each point $p \in \beta X$ has a base of open-and-closed neighborhoods in $X \cup \{p\}$, then βX has a base of open-and-closed sets. [6L.*2*.]

2. If $\dim X > 0$, then there exists $p \in \beta X$ such that $X \cup \{p\}$ does not have a base of open-and-closed sets.

3. Specifically, let Δ_1 be the one-dimensional space of 16M, and let $p \in \Delta - \Delta_1 \subset \beta\Delta_1 - \Delta_1$. Then $\Delta_1 \cup \{p\}$ does not have a base of open-and-closed sets, although Δ_1 itself does have such a base.

NOTES

The notes that follow contain comments on the historical development of our subject. References are usually to the works in which (to the best of our knowledge) the results first appeared. Sometimes we refer instead to a standard text. As a general rule, references have been omitted when the result is well known—either from standard texts or as folklore—or when it is of only secondary importance for us; in the latter case, the result will often be found in a work cited in a related connection.

The groundwork for the theory of rings of continuous functions was laid in three papers. The first was Stone's [S_8], in which the basic theory of C^* was developed. The wealth of ideas that appeared for the first time in this paper will be evident from a reading of these notes. In the second paper, by Gelfand and Kolmogoroff [GK], it was shown that some of Stone's results could be obtained without considering, as Stone had, the metric structure (2M) of the ring C^*. This opened the way to a similar study of C, which they initiated. Finally, Hewitt, in [H_7], made the major contributions to our knowledge of the ring C, and set the direction for most of the subsequent research.

CHAPTER 1

Homomorphisms. The ring $C(X)$ may also be regarded as an algebra over the real field **R**, by setting $rf = \mathbf{r}f$. We do not make formal use of this fact, for, in the work presented here, the additional structure yields no additional information. This is because the field **R** has the very special property (not shared by the complex field, for example) that its only nonzero endomorphism is the identity; as a consequence, every (ring) homomorphism from $C(Y)$ into $C(X)$ is an algebra homomorphism (see 1I). For conceptual simplicity, we never multiply a function by a real number—only by another function. Some of the results on isomorphisms and homomorphisms appear in [GK] and [NB].

Zero-sets. Hewitt's paper [H_7] contains basic information about zero-sets (including Theorem 1.15). This is the first paper in which zero-sets were exploited in a systematic way in the study of $C(X)$.

C-embedding and C^*-embedding. Theorem 1.17 is an adaptation of Urysohn's theorem that any closed set in a normal space is C^*-embedded [U_2].

The space Λ, which was constructed by Katětov [K_7], makes emphatically clear the distinction between C-embedding and C^*-embedding of closed sets; this distinction was not always appreciated by earlier writers. The terms "C-embedding" and "C^*-embedding" were suggested by Kohls. The characterization given in Theorem 1.18 appears to be new.

Special classes of spaces. The definition of pseudocompact spaces and the results about these spaces (including several that appear in later chapters) are taken from [H_7]. Extremally disconnected spaces were defined and studied in [H_5], where it is pointed out that the concept arose earlier in a paper of Stone. The importance of these spaces lies in their connection with the completeness of $C(X)$ as a lattice (3N). Basically disconnected spaces (the term is introduced here) also arose in this connection ([S_{12}], [N_2]).

Direct sums. J. de Groot has communicated the following observation about direct sums of rings (see 1B). Examples are known of a pair of compact spaces X and Y, each homeomorphic to an open subset of the other, but with X not homeomorphic to Y [K_{15}]. By Theorem 4.9, $C(X)$ is not isomorphic to $C(Y)$. This, then, provides an example of a pair of rings, each isomorphic to a direct summand of the other, but yet not isomorphic to each other.

A relation between decompositions of $C(X)$ and components of X appears in [W_4, Theorem 9].

CHAPTER 2

z-filters. The theory of filters is presented in Bourbaki [B_5]. Filters and z-filters are dual ideals in the lattice of all subsets and the lattice of all zero-sets, respectively. Following Bourbaki, we shall refer below to a dual ideal in a lattice as a *prefilter.* The basic relations between ideals and z-filters are given in [H_7], where the importance of these relations, especially for the study of maximal ideals, is clearly demonstrated.

z-ideals and prime ideals. z-ideals, and also the term "z-filter," were introduced by Kohls in [K_{10}]. Results relating z-ideals to prime ideals, here and in later chapters, are taken from [K_{10}], [K_{11}], and [K_{12}]. The first example of a nonmaximal prime ideal in $C(X)$ was constructed by Hewitt (see [K_2, p. 176]). A special case of Theorem 2.11 appears in [K_2]; the general case is given in [GH_1] and in [S_3].

m-topology. The m-topology on $C(X)$ was defined and studied in [H_7, pp. 48–51, 73–74].

CHAPTER 3

Completely regular spaces. Tychonoff [T_5] demonstrated the importance of completely regular spaces by proving that they are precisely the subspaces of compact spaces. Earlier, Urysohn [U_2] had considered them briefly. The observation 3.2(b) will be found in [T_5]. The reduction to completely regular spaces (3.9) is due to Stone [S_8, p. 460] and Čech [C_1, p. 826]. An example of a regular space that is not completely regular was given in [T_5].

Later, Hewitt [H_6] produced a regular space on which every real-valued continuous function is constant. The result 3F appears in [S_7].

Urysohn's lemma; normal spaces. Urysohn's lemma was proved in [U_2]. From the point of view of the theory of rings of functions, it is natural to *define* a Hausdorff space to be normal whenever any two disjoint closed sets are completely separated (as is done in [B_5]). The definition given in the text seems to be more familiar. In any event, by Urysohn's lemma, the two are equivalent. The result 3D.*2* is in [H_7]; 3L.*5* is in [MP]. For the proof that every regular Lindelöf space is normal, consult [K_9, p. 113]. The space Γ of 3K is well known; see, e.g., [AH, p. 31]. The method of proof of nonnormality indicated in *4* (by counting functions) was used by Katětov in [K_4]. The same idea can be employed to prove that Sorgenfrey's product space (see [K_9, p. 134]) is not normal.

Completeness of the lattice $C(X)$. The results in 3N are taken from [S_9], [S_{12}], and [N_2].

CHAPTER 4

Fixed and free ideals. A systematic study of fixed and free ideals was initiated by Hewitt [H_7], who is also responsible for the terminology. The maximal ideals in $C^*(X)$, free as well as fixed, were characterized (4.6(a*) and (7.2)) by Stone [S_8, Theorems 79 and 80]. He also showed that in the compact case, all ideals are fixed. In addition, he proved that the closed ideals in C^* are precisely the intersections of maximal ideals (4O and 6A.*2*) [S_8, Theorem 85]. The characterization of all the maximal ideals in $C(X)$ (4.6(a) and 7.3) is due to Gelfand and Kolmogoroff [GK, Lemma 2]. Lemma 4.10 is [H_7, Theorem 37]. Example 4F is attributed to W. F. Eberlein in [EGH].

Isomorphism implies homeomorphism. The first result along the lines of Theorem 4.9 is due to Banach [B_1, p. 170], who proved that when X and Y are compact metric spaces, an isometry between $C(X)$ and $C(Y)$ (the metric being as defined in 2M) implies a homeomorphism between X and Y. Stone [S_8, Theorem 83] generalized this result to the case of arbitrary compact spaces ("Banach-Stone theorem"). The present theorem, whose hypothesis involves no metric on the rings, is due to Gelfand and Kolmogoroff [GK]; the proof in the text is theirs. Incidentally, the theorem follows from the Banach-Stone theorem and the fact that isomorphism implies isometry (1J.*6*).

Kaplansky proved that a compact space X is determined by the structure of $C(X)$ as a lattice (see [B_2, p. 175]); Milgram [M_9] proved the corresponding result for the multiplicative semigroup of $C(X)$. A survey of the literature on this subject appears in [DS, pp. 385–386]. See also the notes to Chapter 8.

Structure space. The Stone topology was defined first for Boolean rings [S_8, Theorem 1]. It was applied next to C and C^* [GK]. It has since been applied to general rings; see [J, Chapter 9].

Tychonoff's theorem. The first proofs of the product theorem appeared in [T_5] and [T_6]. For the proof by means of filters, consult [B_5, Chapter 1].

Problems. The original material on *P*-spaces and related topics is in [GH_1] and [GH_2, § 6 and § 8].

CHAPTER 5

Stone [S_8, Theorem 76] proved that C^*/M is always the real field. Hewitt [H_7, Theorem 41] proved that C/M is a totally ordered field containing $\mathbf{R}$, showed how the order in C/M is related to the zero-sets in the space X, and established the existence of nonarchimedean C/M whenever X is not pseudocompact. In [H_7, Theorem 50], he contributed the characterization of real ideals stated in 5.14. Theorem 5.5 about prime ideals is due to Kohls [K_{11}, Theorem 2.1].

Absolutely convex ideals are called *l-ideals* in [BP]. Problems about convex ideals, including 5G, were contributed by Kohls. The example 5I was communicated by Isbell; its novelty lies in choosing $\mathscr{E}$ to be maximal (see [K_6, p. 74]).

CHAPTER 6

Compactification. Tychonoff [T_5] proved that every completely regular space has a compactification. Later, Stone [S_8], and independently, Čech [C_1], produced the compactification βX. An expository account of the Stone-Čech compactification will be found in [S_{11}].

Čech's method consisted of embedding X in a product of intervals, as Tychonoff had done; see Chapter 11.

Stone's method depended upon a combination of the algebraic properties of $C^*(X)$ with properties of a certain Boolean ring of subsets of X. Each of these ingredients of Stone's method led, later, to a separate development of βX. First, Wallman [W_2] extracted the set-theoretic portion. His construction resembles the one given in the text, but is based upon maximal prefilters of arbitrary closed sets, rather than zero-sets. His space, known as the *Wallman compactification*, agrees with βX when X is normal, but not otherwise. (For nonnormal X, it fails to be a Hausdorff space.) Alexandroff [A_1] adopted a variant of Wallman's procedure to obtain βX for all completely regular X.

Gelfand and Kolmogoroff [GK], exploiting the algebraic aspects of Stone's proof, developed βX as the structure space of $C^*(X)$ (see 4.9 and 7.10). More unexpected was their development of βX as the structure space of $C(X)$ as well. Because of the simple connection between the maximal ideals in $C(X)$ and the z-ultrafilters on X, the construction via the structure space of $C(X)$ can be accomplished in purely set-theoretic terms. This is the method presented in 6.5 of the text.

One of the key devices for avoiding the hypothesis that X be normal is to work with zero-sets rather than with arbitrary closed sets (see 3.1). Applied to Wallman's prefilters, this again yields z-ultrafilters. Thus, the algebraic aspects and the set-theoretic aspects of Stone's construction of βX merge in the construction by z-ultrafilters.

Compactification theorem. Stone [S_8, Theorems 88 and 79] proved that βX satisfies (I) and (II) of Theorem 6.5 and showed that the two properties are equivalent. Čech [C_1, pp. 831–833] proved that βX satisfies (II), (III), and a special case of (I) (for τ a homeomorphism), and showed that these three properties are equivalent and determine βX uniquely; these results led him to 6.9(a) and 6.12. The analogues of (IV) and (V) for closed sets, relative to the Wallman compactification, will be found in [W_2].

It is worth noting that while (IV) generalizes immediately to any finite number of zero-sets, the corresponding generalization of (III) is by no means a trivial consequence of (III) (though it follows, of course, from the generalization of (IV)). Regarding the proof of Theorem 6.4, see [H_7, Theorem 20], [T_1], and [M_5, Corollary 2.2]. For 6G, see [M_5, Theorem 3.8].

Further information about the relation between Stone's theorem and Tychonoff's theorem (6.8) is contained in *Notes*, Chapter 11, *Axiom of choice.*

Compactification of a product. The result 6N was communicated by Henriksen. It is difficult, in any particular case, to determine whether $\beta(X \times Y)$ is homeomorphic with $\beta X \times \beta Y$. (See, however, 14Q.) A more manageable question is whether these spaces are identical, i.e., whether the Stone extension of the identity map on $X \times Y$, into $\beta X \times \beta Y$, is a homeomorphism. Glicksberg [G_3] has proved that this is the case, for infinite X and Y, when and only when $X \times Y$ is pseudocompact. The necessity was announced independently in [HI_2]. To establish the sufficiency, which is considerably more difficult, Glicksberg utilizes his earlier result [G_2, Theorem 2] that a space S is pseudocompact if and only if Ascoli's theorem holds in the metric space $C^*(S)$. Incidentally, homeomorphism of $\beta(X\times Y)$ with $\beta X\times\beta Y$ does not imply their equality; see [GJ].

The space $\beta\mathbf{N} - \mathbf{N}$. The discussion of $\beta\mathbf{N} - \mathbf{N}$ in 6S and 6V comes from Rudin's paper [R]. Regarding 6S.*7*,*8*, see also A. Gleason and E. E. Moise, quoted in [B_2, p. 39, Ex. 11]. The result that $\beta\mathbf{N} - \mathbf{N}$ is not basically disconnected (6W) is well known; the proof outlined in the text was communicated by Henriksen.

Miscellaneous problems. The results in 6J are due to Hewitt and Smirnov; see [H_8] and [M_{11}]. The spaces Λ (6P) and Π (6Q) were constructed by Katětov, in [K_7] and [K_4]. 6O.*6* is due to Čech; see [N_3].

CHAPTER 7

Maximal ideals. The maximal ideals in $C^*(X)$ were characterized (7.2) by Stone [S_8, Theorems 79 and 80]. The maximal ideals in $C(X)$ were characterized (7.3 and 7D) by Gelfand and Kolmogoroff [GK]. Consideration of the relation between M^p and M^{*p} was begun in [H_7, Theorems 45 and 48]; the extension f^* was introduced into this consideration in [GHJ].

Structure space. The structure spaces $\mathfrak{M}$ and $\mathfrak{M}^*$ were defined by Gelfand

and Kolmogoroff [GK] and used by them as models for βX. The approach to βX outlined in 7N is taken from [H_7, Theorem 46]. The content of 7N.*6* was employed by Gelfand [G_1] as an alternative way of defining a topology on $\mathfrak{M}^*$. Detailed discussion of the weak topology and the Stone topology on spaces of maximal ideals in Banach algebras may be found in [L_3].

The ideals O^p. Consideration of O^p stemmed from the result 7O.*2*, which is due to McKnight [M_6]. The content of 7.12–7.16 appeared in [GH_1], [GHJ], and [K_{11}].

Miscellaneous problems. 7C is in [P_2]. The bulk of 7E–7K comes from [K_{10}] and [K_{11}]. Hewitt's conjecture that closed ideals in the m-topology are always intersections of maximal ideals (7Q) was verified in [GHJ] and [S_3].

CHAPTER 8

Realcompact spaces. Realcompact spaces were defined and investigated by Hewitt [H_7]. (His original term was "Q-spaces." Other adjectives that have been employed by various writers are: e-complete, functionally closed, Hewitt, real-complete, saturated; and υX is sometimes called the "Nachbin completion" of X.) Hewitt demonstrated the importance of these spaces by proving the isomorphism theorem (8.3) and by establishing the existence, for any X, of a unique realcompactification in which X is C-embedded. He also derived many of the properties of realcompact spaces. Much of the theory of realcompact spaces was developed independently (but not published) by Nachbin within the framework of the theory of uniform spaces (15.13(a) and 15.14(a)); see Hewitt's review of [S_1] in Mathematical Reviews 14 (1953), p. 395.

Nachbin [N_1] and Shirota [S_4] characterized realcompact spaces X in terms of the topological vector space $C(X)$, where the topology is that of uniform convergence on compact sets. For a study of $C(X)$ from this point of view, see [W_3].

Realcompactness for totally ordered spaces is characterized in [GH_1].

Isomorphism implies homeomorphism. Shirota [S_1; S_2] established several analogues of Hewitt's isomorphism theorem (8.3), among which are the following: realcompact spaces X and Y are homeomorphic if $C(X)$ and $C(Y)$ are lattice-isomorphic or if there exists an isomorphism of the multiplicative semigroups of $C(X)$ and $C(Y)$. As a consequence of 8.8(a), Shirota's results imply that for any X and Y—not necessarily realcompact—$C(X)$ and $C(Y)$ are isomorphic as rings if and only if there is an isomorphism between them relative to either of these other structures. See [H_4].

Characterizations of $C(X)$. A related problem (not treated in the text) is to represent a given abstract object as a family of continuous functions. The method of attack is like that of 8.3, where one regards $C(X)$ as an abstract entity and constructs from it a topological space homeomorphic with X. There is an extensive literature devoted to the characterization of

Banach algebras, Banach spaces, et al., that are isomorphic with $C(X)$ for some *compact* X; summaries of these results, along with references to the literature, may be found in [D_1, pp. 87–104], [DS, pp. 396–398], and [K_1]. An algebraic characterization of $C(X)$ for arbitrary X is given in [AB].

Construction and properties of υX. Properties (II), (III), and (IV), and their mutual equivalence, are discussed in [K_{13}], [V], and [S_1]. The systematic application in the text of property (I) to the proofs of theorems about realcompact spaces is new. Many of these theorems were already known: 8.10 is in [K_6]; 8.10, 8.11 (previously stated in [H_7], but with an incorrect proof), 8.15, and most of 8.17 are in [S_1]; and 8.9 was announced in [W_6]. Analogies with compact spaces are emphasized in [EM]. Property (I) itself is a special case of a general result in the theory of uniform spaces (see Chapter 15); the particular proof given in the text that (6) implies (1) is taken from [M_5, Theorem 2.6].

Intersection of free maximal ideals. Kaplansky [K_2] proved that $C_K(X)$ is the intersection of all the free maximal ideals in $C(X)$ in the case of discrete X, and asked whether the equality holds in general. Theorem 8.19, which is new, shows that the equality holds for the class of realcompact spaces. On the other hand, the example of the plank shows that the equality may fail in pathological cases.

Nonnormal realcompact spaces. Hewitt [H_7] proved that the nonnormal space Γ is realcompact. Another example is afforded by any uncountable product of realcompact, noncompact spaces: by 8.11, such a product is realcompact; and by [S_6, Theorem 3 ff.], the product is not normal. The following theorem has been proved by Corson [C_2]. Let T be the product of uncountably many copies of $\mathbf{N}$, and let X be the subspace of T consisting of those points having all but a countable number of coordinates equal to 1; then X is normal, and $\upsilon X = T$.

Problems. The space $\upsilon_f X$ (8B) was introduced and applied in various problems by Henriksen. Example 8I is in [GI].

CHAPTER 9

Cardinal of βX *for discrete* X. The proof of Theorem 9.2 is taken from Hausdorff [H_2]. It does not seem to be widely known that the content of this paper is an evaluation of $|\beta X|$. (The problem treated there is necessarily couched in somewhat different terms: βX had not yet been invented.) We are indebted to P. Erdös for calling Hausdorff's paper to our attention. An explicit statement (with proof) of Theorem 9.2 appeared for the first time in [P_1]. The special argument in 9O is taken from [M_{12}].

Cardinals of closed sets contained in $\beta X - X$. The original results are due to Čech, to whom is due the idea of securing a lower bound for the cardinal of a set by finding in the set a copy of $\beta\mathbf{N} - \mathbf{N}$. In [$C_1$], he applied this idea to zero-sets, obtaining Theorem 9.5 (with $\mathfrak{c}$ as the stated lower bound) and its corollary, 9.6. Later (see [N_3]), he proved that every infinite closed set in $\beta\mathbf{N}$ contains a copy of $\beta\mathbf{N}$. Henriksen has observed that the proof can be adapted to show that every infinite closed set contained in $\beta X - X$ contains a copy of $\beta\mathbf{N}$ whenever X is a Lindelöf space (as well as whenever X is an arbitary discrete space). The results 9.9–9.12 are new.

Isomorphism implies homeomorphism. Cech [C_1] pointed out that a point having a countable base of neighborhoods in X also has a countable base in βX, so that, by his result 9.6, if X and Y both satisfy the first countability axiom, then a homeomorphism of βX onto βY will carry X onto Y. Gelfand and Kolmogoroff then had only to add their result 6A.*1* to obtain 9.7(a*). The analogous proof of 9.7(a) was noted in [H_3]; the result itself had been obtained earlier in [P_3] and [A_5]. The strengthened version of 9.7(a*) stated in 9N was communicated to us by Katětov.

In Chapter 4, Chapter 8, and again here, we have found that within certain classes of spaces, a space is determined topologically by the algebraic structure of its ring of continuous functions. We have also seen that this is not true for the class of *all* completely regular spaces. However, the following theorem has been proved by Pursell [P_4]: let X and Y be arbitrary completely regular spaces; if there exists an isomorphism of $C^*(X)$ onto $C^*(Y)$ that can be extended to an isomorphism of all of $\mathbf{R}^X$ onto $\mathbf{R}^Y$, then X is homeomorphic with Y.

Pseudocompact spaces. The product theorem (9.14) was proved by Glicksberg [G_3], its analogue (9J) by Katětov (see [N_3]). Example 9.15 is due to Terasaka [T_3]; see also Novák [N_3]. The result 9I was observed by Henriksen.

P-points and nonhomogeneity. The results in 9M were first established by Rudin [R] for locally compact, normal X. The general case is due to Isiwata [I_4].

CHAPTER 10

Duality. For a general theory of duality, see [M_2]. Its application to function rings is discussed in [I_3]. Theorem 10.8 has not been published before. Representation theorems of this type go back to Banach [B_1, p. 172]. The existence of such a representation indicates that whenever the isomorphism theorems 4.9 or 8.3 can be used to prove that two given spaces are homeomorphic, it is probable that a direct proof can be found more easily.

Stone's theorem. The proof given for Theorem 10.7 (for C^*) is Stone's original proof [S_8, Theorem 88].

CHAPTER 11

Embedding in a product. It was Tychonoff's idea to compactify a space X by embedding it in a product of intervals [T_5]. Every compactification of X can be obtained by this procedure (11C). Whereas Tychonoff was interested in keeping the number of factors in the product as small as possible, Čech [C_1] constructed βX by recognizing the particular advantages of taking what amounts to the largest possible product. The embedding of X in P to obtain υX is due to Hewitt [H_7, Theorem 59]; the sufficiency in 11.12 is due to Shirota [S_1].

Axiom of choice. Rubin and Scott [RS] announced the equivalence, in the absence of the axiom of choice, of a number of deep topological theorems; among them are Stone's theorem (6.5) and the Tychonoff product theorem. We saw in 6.8 that the former implies the latter. Conversely, by 11.5, the latter implies the former.

Kelley [K_8] proved that a strengthened form of Tychonoff's theorem—namely, for products of T_1-spaces, rather than only for Hausdorff spaces—implies the axiom of choice. Kelley's proof shows that the Tychonoff theorem for Hausdorff spaces implies a weakened form of the axiom of choice —namely, for arbitrary families of *finite* sets. Thus, any proof of existence of βX must depend upon some form of the axiom of choice. Whether it implies this axiom in full strength is not known. Nor is it known whether the axiom of choice is independent of the remaining axioms in standard systems of set theory, although results very close to this have been obtained; see [M_8].

CHAPTER 12

Realcompact discrete spaces. Theorem 12.2 was proved by Mackey [M_1]. Hewitt [H_9] obtained the following generalization: a completely regular space X is realcompact if and only if, for every $\{0, 1\}$-valued *Baire* measure μ on X, with $\mu(X) = 1$, there exists $p \in X$ such that the sets of measure 1 are precisely the Baire sets containing p. The result on extremally disconnected P-spaces (12H.*6*) is due to Isbell [I_2]; our proof was worked out in collaboration with Henriksen.

Measurable and inaccessible cardinals. Theorem 12.5 is due to Ulam [U_1] and Tarski [T_2, p. 153]; the proof in the text is taken from Ulam's paper. The quotation in 12.6 is from [H_1, p. 131]. Actually, the reference there is to the so-called *weakly* inaccessible numbers; however, for all we know, the number $\mathfrak{c}$, for instance, may be weakly inaccessible (see [ET, p. 326]). Regarding the nonexistence of inaccessible numbers, see [M_7] and the references contained therein. The argument given in the text apparently goes back to Kuratowski [K_{14}].

Residue class fields. Material about residue class fields of large cardinal is taken from [EGH]. The result 12B is due to Sierpiński [S_5, p. 448]. (In the bibliography of [S_5], the entry Sierpiński [26], which is the original source of the result, should read: *Sur les suites transfinies finalement disjointes*, Fund. Math. 28 (1937), p. 115–119.)

CHAPTER 13

The theory of field extensions is presented in [B_4, Chapter 5] and [W_1, Chapters 5 and 8]. Material on ordered fields appears in [B_4, Chapter 6, pp. 31–49], [W_1, Chapter 9], and the original paper of Artin and Schreier [AS]. The result 13G.*1* is [AS, Theorem 8]. Results about η_α-sets and cofinality are due to Hausdorff [H_1, pp. 129, 132, 180–185]. The development by dyadic sequences follows Sierpiński [S_5, Chapter 17, § 6].

Theorem 13.2 is due to Hewitt [H_7, Theorem 42]. The theorem that $C(X)/M$ is real-closed (13.4) was also stated by Hewitt, but was not correctly proved. The first complete proof was given for the case of a normal space, by Henriksen and Isbell [HI_1]. The proof for arbitrary X was obtained later by Isbell [I_1]. The isomorphism theorems were proved in [EGH]. Lemma 13.7 is stated in that paper for maximal ideals only; the general case appears in [K_{11}].

CHAPTER 14

The first general investigation of prime ideals in $C(X)$ was undertaken by Kohls [K_{11}; K_{12}]. The present chapter is devoted chiefly to an exposition of Kohls' results. Most of the material in the first 24 sections of the chapter, and many of the problems, come from the papers cited. Lemma 14.8, Theorems 14.9, 11, 16, 19, and 20, and 14F.*1* are new results. Material on F-spaces is taken from [GH_2]; characterizations of C for other classes of spaces are also included in this paper. 14B.*1* was contributed by Kohls, and 14Q.*1* by Henriksen and P. C. Curtis. 14Q.*3* is in [GJ].

CHAPTER 15

The theory of uniform spaces was founded by Weil [W_5]. Detailed expositions of the theory may also be found in [B_5, Chapters 2 and 9] and in [K_9, Chapter 6]. An entirely different approach to the subject was given by Tukey [T_4].

According to Weil's definition, a uniform structure on X is a certain kind of filter on $X \times X$. Weil proved that each such filter has a base of sets of the form

$$\{(x, y) : d(x, y) < 1\}$$

where d runs through a family of pseudometrics. The correspondence between these filters on $X \times X$, and those families of pseudometrics that satisfy (i) and (ii) of 15.3, is one-one; accordingly, the choice of which is to be regarded as the uniform structure is dictated only by convenience. As was remarked in the text, we choose to emphasize the pseudometrics because they supply us with continuous functions. This approach also has the technical advantage of bypassing Weil's theorem on the existence of pseudometrics. (The theorem is proved by the same kind of argument as in Urysohn's lemma (3.13).)

Our approach is not so well suited to the theory of topological groups. In order to apply it there, one would have to begin by proving the existence of sufficiently many pseudometrics that are invariant under translation. Cf. [K_9, p. 210].

Base for a uniform structure. Since $d \in \mathscr{D}$ implies $rd \in \mathscr{D}$ for all $r \geqq 0$, the collection of sets $\{y: d(x, y) < 1\}$ is the same as the collection $\{y: d(x, y) < \epsilon\}$. We retain the ϵ so as to allow d to range only over a base for $\mathscr{D}$, rather than over all of $\mathscr{D}$. As a consequence, the formulas look just like the classical ones for metric spaces. Indeed, in the classical case, the metric is a base for the uniform structure.

Because a d-discrete family (15.15) of gauge δ is $(\delta^{-1}d)$-discrete of gauge 1, we could deal exclusively with families of gauge 1. The term "gauge" would then be unnecessary. This, however, would complicate the statements of subsequent theorems.

Completion theorem. A completion of a uniform space X can be constructed as follows. For each pseudometric d on X, the d-distance between any two Cauchy filters is defined by the formulas in 15V.*2*, suitably modified. This yields a uniform structure on the family of all Cauchy filters. By reducing to equivalence classes, as in the text, one obtains a Hausdorff uniform space which can be identified as the completion of X.

Another method is to construct γX by embedding X in a product of complete metric spaces—corresponding to the development of βX given in Chapter 11. See [K_9] for details.

When the completion has been constructed by either of these methods, βX can be defined as the completion of $[X; \mathscr{C}^*(X)]$. Then, Stone's theorem can be derived as in 15P, and compactness of βX established by 15I.*1*. (See [B_5, Chapter 9, p. 14].) It is interesting to observe that in this construction of βX, the axiom of choice enters by way of Theorem 15.16, i.e., in the proof that every totally bounded space is precompact. (See the notes to Chapter 11, *Axiom of choice.*) The latter is equivalent to Tychonoff's theorem and to Stone's theorem, as was observed by Rubin and Scott [RS].

Shirota's theorem. Theorem 15.17 is due to A. H. Stone [S_6]. Shirota's theorem (15.20) was proved in [S_1]; see also [K_6, Theorem 3]. The proof that (4) implies (1) in 15.21 was supplied by Isbell.

Problems. Most of the results stated in the problems are taken from the literature. Example 15L was contributed by Isbell.

CHAPTER 16

Dimension theory. The standard reference for the theory of dimension of separable metric spaces is [HW]. An exposition based on the Lebesgue dimension will be found in [A_4, Chapters 5 and 6]. For a discussion of the dimension of general spaces, see [A_3]. The use of basic covers (16.5) goes back to Alexandroff [A_2]. The scheme for evaluating the dimension of a compact set in $\mathbf{R}^n$ (16.20) is essentially the one that Lebesgue suggested in [L_1] and carried out in detail in [L_2]. Within the framework of a general development of dimension theory, the result stated in Theorem 16.22 is obtained more efficiently than here, but via the introduction of additional concepts.

Stone-Weierstrass theorem. This theorem appeared in [S_8, p. 467]. An expository account of the theorem and its applications is presented in [S_{10}]. Generalizations of the theorem are discussed in [DS, pp. 383–385].

Dimension of βX. That $\dim \beta X = \dim X$ is due to Wallman [W_2], who proved the analogous statement with the usual definition of Lebesgue dimension and with the Wallman compactification in place of βX.

Katětov's theorem. The theorem appears in [K_5]. A good deal of the machinery of the proof is developed in the earlier paper [K_3]. (The example 16H and the general form of the subspace theorem (16I) are also in these papers.) Our algebraic subring is related to what Katětov calls an "algebraically closed" subring, but is somewhat simpler. The analytic subrings here are identical with his "analytically closed" subrings. (The term "analytic ring" was used by Stone in [S_8] in a broader sense.)

Dimension-raising mappings. Clearly, every compact metric space is a continuous image of $\beta\mathbf{N}$. Indeed, it is well known that every such space is a continuous image of the Cantor set [K_9, p. 166]. More significant is the fact that every compact metric space is a *monotone* image of some one-dimensional space; see [H_{10}] and [M_3].

Miscellaneous problems. The result in 16G.*3* is due to Hurewicz [H_{11}]. The counterexample 16G.*4* was supplied by R. F. Williams. We do not know whether 16J.*3* is valid for arbitrary spaces; known counterexamples for the analogous statement with the usual definition of Lebesgue dimension all involve nonnormal spaces (see [M_{10}]). The examples in 16M and 16N are taken from Dowker [D_2].

BIBLIOGRAPHY

This list contains only those works cited in the *Notes* or in Chapter 0.

[A_1] P. ALEXANDROFF (ALEKSANDROV), *Bikompakte Erweiterungen topologische Räume*, Mat. Sb. 5 (1939), 403–423. (Russian. German summary)

[A_2] ———, *Über die Dimension der bikompakten Räume*, Dokl. Akad. Nauk SSSR 26 (1940), 619–622.

[A_3] ———, *The present status of the theory of dimension*, Amer. Math. Soc. Transl. 1 (1955), 1–26.

[A_4] ———, *Combinatorial topology*, vol. 1, Rochester, 1956.

[A_5] F. W. ANDERSON, *A lattice characterization of completely regular G_δ-spaces*, Proc. Amer. Math. Soc. 6 (1955), 757–765.

[AB] F. W. ANDERSON AND R. L. BLAIR, *Characterizations of the algebra of all real-valued continuous functions on a completely regular space*, Illinois J. Math. 3 (1959), 121–133.

[AH] P. ALEXANDROFF AND H. HOPF, *Topologie*, Berlin, 1935.

[AS] E. ARTIN AND O. SCHREIER, *Algebraische Konstruktion reeller Körper*, Abh. Math. Sem. Univ. Hamburg 5 (1926), 85–99.

[B_1] S. BANACH, *Théorie des opérations linéaires*, Warsaw, 1932.

[B_2] G. BIRKHOFF, *Lattice theory*, New York, 1948.

[B_3] N. BOURBAKI, *Théorie des ensembles*, Actualités Sci. Ind. 1212 (1954); 1243 (1956); Paris.

[B_4] ———, *Algèbre*, Actualités Sci. Ind. 1144 (1951); 1102 (1950); 1179 (1952); Paris.

[B_5] ———, *Topologie générale*, Actualités Sci. Ind. 1142 (1951); 1045 (1958); Paris.

[BP] G. BIRKHOFF AND R. S. PIERCE, *Lattice-ordered rings*, An. Acad. Brasil. Ci. 28 (1956), 41–69.

[C_1] E. ČECH, *On bicompact spaces*, Ann. of Math. 38 (1937), 823–844.

[C_2] H. H. CORSON, *Normality in subsets of product spaces*, Amer. J. Math. 81 (1959), 785–796.

[D_1] M. M. DAY, *Normed linear spaces*, Ergebnisse der Mathematik und ihrer Grenzgebiete. Neue Folge. Heft 21. Reihe: Reelle Funktionen. Berlin-Göttingen-Heidelberg, 1958.

[D_2] C. H. DOWKER, *Local dimension of normal spaces*, Quart. J. Math. Oxford Ser. (2) 6 (1955), 101–120.

[DS] N. DUNFORD AND J. T. SCHWARTZ, *Linear operators, Part* I, New York, 1958.

[EGH] P. ERDÖS, L. GILLMAN, AND M. HENRIKSEN, *An isomorphism theorem for real-closed fields*, Ann. of Math. 61 (1955), 542–554.

[EM] R. ENGELKING AND S. MRÓWKA, *On E-compact spaces*, Bull. Acad. Polon. Sci. Cl. III 6 (1958), 429–436.

[ET] P. ERDÖS AND A. TARSKI, *On families of mutually exclusive sets*, Ann. of Math. 44 (1943), 315–329.

[G_1] I. GELFAND, *Normierte Ringe*, Mat. Sb. 9 (1941), 3–24.

[G_2] I. GLICKSBERG, *The representation of functionals by integrals*, Duke Math. J. 19 (1952), 253–261.

[G_3] ———, *Stone-Čech compactifications of products*, Trans. Amer. Math. Soc. 90 (1959), 369–382.

[GH_1] L. GILLMAN AND M. HENRIKSEN, *Concerning rings of continuous functions*, Trans. Amer. Math. Soc. 77 (1954), 340–362.

[GH_2] ———, *Rings of continuous functions in which every finitely generated ideal is principal*, Trans. Amer. Math. Soc. 82 (1956), 366–391.

[GHJ] ——— AND M. JERISON, *On a theorem of Gelfand and Kolmogoroff concerning maximal ideals in rings of continuous functions*, Proc. Amer. Math. Soc. 5 (1954), 447–455.

[GI] S. GINSBURG AND J. R. ISBELL, *Some operators on uniform spaces*, Trans. Amer. Math. Soc. 93 (1959), 145–168.

[GJ] L. GILLMAN AND M. JERISON, *Stone-Čech compactification of a product*, Arch. Math. 10 (1959), 443–446.

[GK] I. GELFAND AND A. KOLMOGOROFF, *On rings of continuous functions on topological spaces*, Dokl. Akad. Nauk SSSR 22 (1939), 11–15.

[H_1] F. HAUSDORFF, *Grundzüge der Mengenlehre*, Leipzig, 1914 (New York, 1949).

[H_2] ———, *Über zwei Sätze von G. Fichtenholz und L. Kantorovitch*, Studia Math. 6 (1936), 18–19.

[H_3] L. J. HEIDER, *A note concerning completely regular G_δ-spaces*, Proc. Amer. Math. Soc. 8 (1957), 1060–1065.

[H_4] M. HENRIKSEN, *On the equivalence of the ring, lattice, and semigroup of continuous functions*, Proc. Amer. Math. Soc. 7 (1956), 959–960.

[H_5] E. HEWITT, *A problem of set-theoretic topology*, Duke Math. J. 10 (1943), 309–333.

[H_6] ———, *On two problems of Urysohn*, Ann. of Math. 47 (1946), 503–509.

[H_7] ———, *Rings of real-valued continuous functions*, I, Trans. Amer. Math. Soc. 64 (1948), 54–99.

[H_8] ———, *A note on extensions of continuous functions*, An. Acad. Brasil. Ci. 21 (1949), 175–179.

[H_9] ———, *Linear functionals on spaces of continuous functions*, Fund. Math. 37 (1950), 161–189.

[H_{10}] W. HUREWICZ, *Über oberhalb-stetige Zerlegungen von Punktmengen in Kontinua*, Fund. Math. 15 (1930), 57–60.

[H_{11}] ———, *Über Abbildungen von endlichdimensionalen Räumen auf Teilmengen Cartesischer Räume*, Sitz. Akad. Berlin 24 (1933), 754–768.

[HI_1] M. HENRIKSEN AND J. R. ISBELL, *On the continuity of the real roots of an algebraic equation*, Proc. Amer. Math. Soc. 4 (1953), 431–434.

[HI_2] ———, *On the Stone-Čech compactification of a product of two spaces*, Bull. Amer. Math. Soc. 63 (1957), 145–146.

[HW] W. HUREWICZ AND H. WALLMAN, *Dimension theory*, Princeton, 1941.

[I_1] J. R. ISBELL, *More on the continuity of the real roots of an algebraic equation*, Proc. Amer. Math. Soc. 5 (1954), 439.

[I_2] ———, *Zero-dimensional spaces*, Tohoku Math. J. 7 (1955), 1–8.

[I_3] ———, *Algebras of uniformly continuous functions*, Ann. of Math. 68 (1958), 96–125.

[I_4] T. ISIWATA, *A generalization of Rudin's theorem for the homogeneity problem*, Sci. Rep. Tokyo Kyoiku Daigaku. Sect. A 5 (1957), 300–303.

[J] N. JACOBSON, *Structure of rings*, Providence, 1956.

[K_1] R. V. KADISON, *A representation theory for commutative topological algebra*, Mem. Amer. Math. Soc. No. 7, 1951.

[K_2] I. KAPLANSKY, *Topological rings*, Amer. J. Math. 69 (1947), 153–183.

[K_3] M. KATĚTOV, *On rings of continuous functions and the dimension of compact spaces*, Časopis Pěst. Mat. Fys. 75 (1950), 1–16. (Russian. English and Czech summaries)

[K_4] ———, *On nearly discrete spaces*, Časopis Pěst. Mat. Fys. 75 (1950), 69–78.

[K_5] ———, *A theorem on the Lebesgue dimension*, Časopis Pěst. Mat. Fys. 75 (1950), 79–87.

[K_6] ———, *Measures in fully normal spaces*, Fund. Math. 38 (1951), 73–84.

[K_7] ———, *On real-valued functions in topological spaces*, Fund. Math. 38 (1951), 85–91; 40 (1953), 203–205.

[K_8] J. L. KELLEY, *The Tychonoff product theorem implies the axiom of choice*, Fund. Math. 37 (1950), 75–76.

[K_9] ———, *General topology*, New York, 1955.

[K_{10}] C. W. KOHLS, *Ideals in rings of continuous functions*, Fund. Math. 45 (1957), 28–50.

[K_{11}] ———, *Prime ideals in rings of continuous functions*, Illinois J. Math. 2 (1958), 505–536.

[K_{12}] ———, *Prime ideals in rings of continuous functions*, II, Duke Math. J. 25 (1958), 447–458.

[K_{13}] A. A. KUBENSKIĬ, *On functionally-closed spaces*, Uspehi Mat. Nauk 6 (1951) (4), 202. (Russian)

[K_{14}] C. KURATOWSKI, *Sur l'état actuel de l'axiomatique de la théorie des ensembles*, Ann. Soc. Polon. Math. 3 (1924), 146–147.

[K_{15}] ———, *On a topological problem connected with the Cantor-Bernstein theorem*, Fund. Math. 37 (1950), 213–216.

[L_1] H. Lebesgue, *Sur la non-applicabilité de deux domaines appartenant respectivement à des espaces à n et n + p dimensions*, Math. Ann. 70 (1911), 166–168.

[L_2] ———, *Sur les correspondances entre les points de deux espaces*, Fund. Math. 2 (1921), 256–285.

[L_3] L. H. Loomis, *An introduction to abstract harmonic analysis*, New York, 1953.

[M_1] G. W. Mackey, *Equivalence of a problem in measure theory to a problem in the theory of vector lattices*, Bull. Amer. Math. Soc. 50 (1944), 719–722.

[M_2] S. MacLane, *Duality for groups*, Bull. Amer. Math. Soc. 56 (1950), 485–516.

[M_3] S. Mazurkiewicz, *Sur les images continues des continus*, Fund. Math. 17 (1931), 330.

[M_4] N. H. McCoy, *Rings and ideals*, Buffalo, 1948.

[M_5] R. H. McDowell, *Extension of functions from dense subspaces*, Duke Math. J. 25 (1958), 297–304.

[M_6] J. D. McKnight, Jr., *On the characterization of rings of functions*, Thesis, Purdue University, 1953.

[M_7] E. Mendelson, *Some proofs of independence in axiomatic set theory*, J. Symb. Logic 21 (1956), 291–303.

[M_8] ———, *The axiom of Fundierung and the axiom of choice*, Math. Logik und Grundlagenforschung 4 (1958), 66–70.

[M_9] A. N. Milgram, *Multiplicative semigroups of continuous functions*, Duke Math. J. 16 (1949), 377–383.

[M_{10}] K. Morita, *On the dimension of product spaces*, Amer. J. Math. 75 (1953), 205–223.

[M_{11}] S. Mruvka (Mrówka), *Remark on P. Aleksandrov's work "On two theorems of Yu. Smirnov,"* Fund. Math. 43 (1956), 399–400. (Russian)

[M_{12}] ———, *On the potency of subsets of βN*, Colloq. Math. 7 (1959), 23–25.

[MP] S. Mardešić and P. Papić, *Sur les espaces dont toute transformation réelle continue est bornée*, Glasnik Mat.-Fiz. Astr. (2) 10 (1955), 225–232.

[N_1] L. Nachbin, *Topological vector spaces of continuous functions*, Proc. Nat. Acad. Sci. U. S. A. 40 (1954), 471–474.

[N_2] H. Nakano, *Über das System aller stetigen Funktionen auf einem topologischen Raum*, Proc. Imp. Acad. Tokyo 17 (1941), 308–310.

[N_3] J. Novák, *On the cartesian product of two compact spaces*, Fund. Math. 40 (1953), 106–112.

[NB] G. Nöbeling and H. Bauer, *Über die Erweiterungen topologischer Räume*, Math. Ann. 130 (1955), 20–45; 132 (1957), 451.

[P_1] B. Pospíšil, *Remark on bicompact spaces*, Ann. of Math. 38 (1937), 845–846.

[P_2] V. PTÁK, *Weak compactness in convex topological linear spaces*, Čehoslovack Mat. Ž. 4 (79) (1954), 175–186.

[P_3] L. E. PURSELL, *An algebraic characterization of fixed ideals in certain function rings*, Pacific J. Math. 5 (1955), 963–969.

[P_4] ———, *The ring $C(X, R)$ considered as a subring of the ring of all real-valued functions*, Proc. Amer. Math. Soc. 8 (1957), 820–821.

[R] W. RUDIN, *Homogeneity problems in the theory of Čech compactifications*, Duke Math. J. 23 (1956), 409–419, 633.

[RS] H. RUBIN AND D. SCOTT, *Some topological theorems equivalent to the Boolean prime ideal theorem*, Bull. Amer. Math. Soc. 60 (1954), 389.

[S_1] T. SHIROTA, *A class of topological spaces*, Osaka Math. J. 4 (1952), 23–40.

[S_2] ———, *A generalization of a theorem of I. Kaplansky*, Osaka Math. J. 4 (1952), 121–132.

[S_3] ———, *On ideals in rings of continuous functions*, Proc. Japan Acad. 30 (1954), 85–89.

[S_4] ———, *On locally convex vector spaces of continuous functions*, Proc. Japan Acad. 30 (1954), 294–298.

[S_5] W. SIERPIŃSKI, *Cardinal and ordinal numbers*, Warsaw, 1958.

[S_6] A. H. STONE, *Paracompactness and product spaces*, Bull. Amer. Math. Soc. 54 (1948), 977–982.

[S_7] M. H. STONE, *Applications of the theory of Boolean rings to general topology*, Mat. Sb. 1 (1936), 765–772.

[S_8] ———, *Applications of the theory of Boolean rings to general topology*, Trans. Amer. Math. Soc. 41 (1937), 375–481.

[S_9] ———, *A general theory of spectra*, I, Proc. Nat. Acad. Sci. U. S. A. 26 (1940), 280–283.

[S_{10}] ———, *The generalized Weierstrass approximation theorem*, Math. Mag. 21 (1947–1948), 167–183, 237–254.

[S_{11}] ———, *On the compactification of topological spaces*, Ann. Soc. Polon. Math. 21 (1948), 153–160.

[S_{12}] ———, *Boundedness properties in function lattices*, Canad. J. Math. 1 (1949), 176–186.

[T_1] A. D. TAĬMANOV, *On extension of continuous mappings of topological spaces*, Mat. Sb. 31 (1952), 459–463. (Russian)

[T_2] A. TARSKI, *Drei Überdeckungssätze der allgemeinen Mengenlehre*, Fund. Math. 30 (1938), 132–155.

[T_3] H. TERASAKA, *On cartesian product of compact spaces*, Osaka Math. J. 4 (1952), 11–15.

[T_4] J. W. TUKEY, *Convergence and uniformity in topology*, Princeton, 1940.

[T_5] A. TYCHONOFF, *Über die topologische Erweiterung von Räumen*, Math. Ann. 102 (1929), 544–561.

[T_6] ———, *Ein Fixpunktsatz*, Math. Ann. 111 (1935), 767–776.

[U_1] S. ULAM, *Zur Masstheorie in der allgemeinen Mengenlehre*, Fund. Math. 16 (1930), 140–150.

[U_2] P. URYSOHN, *Über die Mächtigkeit der zusammenhängenden Mengen*, Math. Ann. 94 (1925), 262–295.

[V] B. Z. VULIH, *On the extension of continuous functions in topological spaces*, Mat. Sb. 30 (1952), 167–170. (Russian)

[W_1] B. L. VAN DER WAERDEN, *Algebra* (4th edition), Berlin, 1955.

[W_2] H. WALLMAN, *Lattices and topological spaces*, Ann. of Math. 39 (1938), 112–126.

[W_3] S. WARNER, *The topology of compact convergence on continuous function spaces*, Duke Math. J. 25 (1958), 265–282.

[W_4] ———, *Characters of cartesian products of algebras*, Canad. J. Math. 11 (1959), 70–79.

[W_5] A. WEIL, *Sur les espaces à structure uniforme et sur la topologie générale*, Actualités Sci. Ind. 551, Paris, 1937.

[W_6] C. WENJEN, *A characterization of Hewitt's Q-spaces*, Notices Amer. Math. Soc. 5 (1958), 300–301.

LIST OF SYMBOLS

(References are to sections)

Special symbols

$\leftarrow$ (inverse of mapping), 0.1–2

$'$ (induced mapping), 10.2, *see Index:* Homomorphism of C (induced)

(induced mapping), 4.12, *see Index:* Sharp mapping

0,1 (constant functions), 1.1

Other symbols

∞ (infinity), 7.5

$^{-1}$ (multiplicative inverse), 0.1

$|\ |$ (absolute value), 0.19, 1.2

$|\ |$ (cardinal), 0.2

$(\ ,\ )$ (generators), 0.14, 2.1

$[\ ]$ (image of set), 0.2

$\vee$, $\wedge$ (sup, inf), 0.5, 15.24

$\circ$ (composition), 0.2

$/$ (residue class ring), 0.14

$|$ (restriction), 0.2

$\cup$, $\cap$ (union, intersection), 0.2

INDEX

(References are to sections or problems)

A CATALOG OF SELECTED

DOVER BOOKS

IN SCIENCE AND MATHEMATICS

Mathematics–Bestsellers

HANDBOOK OF MATHEMATICAL FUNCTIONS: with Formulas, Graphs, and Mathematical Tables, Edited by Milton Abramowitz and Irene A. Stegun. A classic resource for working with special functions, standard trig, and exponential logarithmic definitions and extensions, it features 29 sets of tables, some to as high as 20 places. 1046pp. 8 x 10 1/2. 0-486-61272-4

ABSTRACT AND CONCRETE CATEGORIES: The Joy of Cats, Jiri Adamek, Horst Herrlich, and George E. Strecker. This up-to-date introductory treatment employs category theory to explore the theory of structures. Its unique approach stresses concrete categories and presents a systematic view of factorization structures. Numerous examples. 1990 edition, updated 2004. 528pp. 6 1/8 x 9 1/4. 0-486-46934-4

MATHEMATICS: Its Content, Methods and Meaning, A. D. Aleksandrov, A. N. Kolmogorov, and M. A. Lavrent'ev. Major survey offers comprehensive, coherent discussions of analytic geometry, algebra, differential equations, calculus of variations, functions of a complex variable, prime numbers, linear and non-Euclidean geometry, topology, functional analysis, more. 1963 edition. 1120pp. 5 3/8 x 8 1/2. 0-486-40916-3

INTRODUCTION TO VECTORS AND TENSORS: Second Edition--Two Volumes Bound as One, Ray M. Bowen and C.-C. Wang. Convenient single-volume compilation of two texts offers both introduction and in-depth survey. Geared toward engineering and science students rather than mathematicians, it focuses on physics and engineering applications. 1976 edition. 560pp. 6 1/2 x 9 1/4. 0-486-46914-X

AN INTRODUCTION TO ORTHOGONAL POLYNOMIALS, Theodore S. Chihara. Concise introduction covers general elementary theory, including the representation theorem and distribution functions, continued fractions and chain sequences, the recurrence formula, special functions, and some specific systems. 1978 edition. 272pp. 5 3/8 x 8 1/2. 0-486-47929-3

ADVANCED MATHEMATICS FOR ENGINEERS AND SCIENTISTS, Paul DuChateau. This primary text and supplemental reference focuses on linear algebra, calculus, and ordinary differential equations. Additional topics include partial differential equations and approximation methods. Includes solved problems. 1992 edition. 400pp. 7 1/2 x 9 1/4. 0-486-47930-7

PARTIAL DIFFERENTIAL EQUATIONS FOR SCIENTISTS AND ENGINEERS, Stanley J. Farlow. Practical text shows how to formulate and solve partial differential equations. Coverage of diffusion-type problems, hyperbolic-type problems, elliptic-type problems, numerical and approximate methods. Solution guide available upon request. 1982 edition. 414pp. 6 1/8 x 9 1/4. 0-486-67620-X

VARIATIONAL PRINCIPLES AND FREE-BOUNDARY PROBLEMS, Avner Friedman. Advanced graduate-level text examines variational methods in partial differential equations and illustrates their applications to free-boundary problems. Features detailed statements of standard theory of elliptic and parabolic operators. 1982 edition. 720pp. 6 1/8 x 9 1/4. 0-486-47853-X

LINEAR ANALYSIS AND REPRESENTATION THEORY, Steven A. Gaal. Unified treatment covers topics from the theory of operators and operator algebras on Hilbert spaces; integration and representation theory for topological groups; and the theory of Lie algebras, Lie groups, and transform groups. 1973 edition. 704pp. 6 1/8 x 9 1/4. 0-486-47851-3

A SURVEY OF INDUSTRIAL MATHEMATICS, Charles R. MacCluer. Students learn how to solve problems they'll encounter in their professional lives with this concise single-volume treatment. It employs MATLAB and other strategies to explore typical industrial problems. 2000 edition. 384pp. 5 3/8 x 8 1/2. 0-486-47702-9

NUMBER SYSTEMS AND THE FOUNDATIONS OF ANALYSIS, Elliott Mendelson. Geared toward undergraduate and beginning graduate students, this study explores natural numbers, integers, rational numbers, real numbers, and complex numbers. Numerous exercises and appendixes supplement the text. 1973 edition. 368pp. 5 3/8 x 8 1/2. 0-486-45792-3

A FIRST LOOK AT NUMERICAL FUNCTIONAL ANALYSIS, W. W. Sawyer. Text by renowned educator shows how problems in numerical analysis lead to concepts of functional analysis. Topics include Banach and Hilbert spaces, contraction mappings, convergence, differentiation and integration, and Euclidean space. 1978 edition. 208pp. 5 3/8 x 8 1/2. 0-486-47882-3

FRACTALS, CHAOS, POWER LAWS: Minutes from an Infinite Paradise, Manfred Schroeder. A fascinating exploration of the connections between chaos theory, physics, biology, and mathematics, this book abounds in award-winning computer graphics, optical illusions, and games that clarify memorable insights into self-similarity. 1992 edition. 448pp. 6 1/8 x 9 1/4. 0-486-47204-3

SET THEORY AND THE CONTINUUM PROBLEM, Raymond M. Smullyan and Melvin Fitting. A lucid, elegant, and complete survey of set theory, this three-part treatment explores axiomatic set theory, the consistency of the continuum hypothesis, and forcing and independence results. 1996 edition. 336pp. 6 x 9. 0-486-47484-4

DYNAMICAL SYSTEMS, Shlomo Sternberg. A pioneer in the field of dynamical systems discusses one-dimensional dynamics, differential equations, random walks, iterated function systems, symbolic dynamics, and Markov chains. Supplementary materials include PowerPoint slides and MATLAB exercises. 2010 edition. 272pp. 6 1/8 x 9 1/4. 0-486-47705-3

ORDINARY DIFFERENTIAL EQUATIONS, Morris Tenenbaum and Harry Pollard. Skillfully organized introductory text examines origin of differential equations, then defines basic terms and outlines general solution of a differential equation. Explores integrating factors; dilution and accretion problems; Laplace Transforms; Newton's Interpolation Formulas, more. 818pp. 5 3/8 x 8 1/2. 0-486-64940-7

MATROID THEORY, D. J. A. Welsh. Text by a noted expert describes standard examples and investigation results, using elementary proofs to develop basic matroid properties before advancing to a more sophisticated treatment. Includes numerous exercises. 1976 edition. 448pp. 5 3/8 x 8 1/2. 0-486-47439-9

THE CONCEPT OF A RIEMANN SURFACE, Hermann Weyl. This classic on the general history of functions combines function theory and geometry, forming the basis of the modern approach to analysis, geometry, and topology. 1955 edition. 208pp. 5 3/8 x 8 1/2. 0-486-47004-0

THE LAPLACE TRANSFORM, David Vernon Widder. This volume focuses on the Laplace and Stieltjes transforms, offering a highly theoretical treatment. Topics include fundamental formulas, the moment problem, monotonic functions, and Tauberian theorems. 1941 edition. 416pp. 5 3/8 x 8 1/2. 0-486-47755-X

Mathematics–Logic and Problem Solving

PERPLEXING PUZZLES AND TANTALIZING TEASERS, Martin Gardner. Ninety-three riddles, mazes, illusions, tricky questions, word and picture puzzles, and other challenges offer hours of entertainment for youngsters. Filled with rib-tickling drawings. Solutions. 224pp. 5 3/8 x 8 1/2. 0-486-25637-5

MY BEST MATHEMATICAL AND LOGIC PUZZLES, Martin Gardner. The noted expert selects 70 of his favorite "short" puzzles. Includes The Returning Explorer, The Mutilated Chessboard, Scrambled Box Tops, and dozens more. Complete solutions included. 96pp. 5 3/8 x 8 1/2. 0-486-28152-3

THE LADY OR THE TIGER?: and Other Logic Puzzles, Raymond M. Smullyan. Created by a renowned puzzle master, these whimsically themed challenges involve paradoxes about probability, time, and change; metapuzzles; and self-referentiality. Nineteen chapters advance in difficulty from relatively simple to highly complex. 1982 edition. 240pp. 5 3/8 x 8 1/2. 0-486-47027-X

SATAN, CANTOR AND INFINITY: Mind-Boggling Puzzles, Raymond M. Smullyan. A renowned mathematician tells stories of knights and knaves in an entertaining look at the logical precepts behind infinity, probability, time, and change. Requires a strong background in mathematics. Complete solutions. 288pp. 5 3/8 x 8 1/2. 0-486-47036-9

THE RED BOOK OF MATHEMATICAL PROBLEMS, Kenneth S. Williams and Kenneth Hardy. Handy compilation of 100 practice problems, hints and solutions indispensable for students preparing for the William Lowell Putnam and other mathematical competitions. Preface to the First Edition. Sources. 1988 edition. 192pp. 5 3/8 x 8 1/2. 0-486-69415-1

KING ARTHUR IN SEARCH OF HIS DOG AND OTHER CURIOUS PUZZLES, Raymond M. Smullyan. This fanciful, original collection for readers of all ages features arithmetic puzzles, logic problems related to crime detection, and logic and arithmetic puzzles involving King Arthur and his Dogs of the Round Table. 160pp. 5 3/8 x 8 1/2. 0-486-47435-6

UNDECIDABLE THEORIES: Studies in Logic and the Foundation of Mathematics, Alfred Tarski in collaboration with Andrzej Mostowski and Raphael M. Robinson. This well-known book by the famed logician consists of three treatises: "A General Method in Proofs of Undecidability," "Undecidability and Essential Undecidability in Mathematics," and "Undecidability of the Elementary Theory of Groups." 1953 edition. 112pp. 5 3/8 x 8 1/2. 0-486-47703-7

LOGIC FOR MATHEMATICIANS, J. Barkley Rosser. Examination of essential topics and theorems assumes no background in logic. "Undoubtedly a major addition to the literature of mathematical logic." – *Bulletin of the American Mathematical Society.* 1978 edition. 592pp. 6 1/8 x 9 1/4. 0-486-46898-4

INTRODUCTION TO PROOF IN ABSTRACT MATHEMATICS, Andrew Wohlgemuth. This undergraduate text teaches students what constitutes an acceptable proof, and it develops their ability to do proofs of routine problems as well as those requiring creative insights. 1990 edition. 384pp. 6 1/2 x 9 1/4. 0-486-47854-8

FIRST COURSE IN MATHEMATICAL LOGIC, Patrick Suppes and Shirley Hill. Rigorous introduction is simple enough in presentation and context for wide range of students. Symbolizing sentences; logical inference; truth and validity; truth tables; terms, predicates, universal quantifiers; universal specification and laws of identity; more. 288pp. 5 3/8 x 8 1/2. 0-486-42259-3

Mathematics–Algebra and Calculus

Mathematics-Probability and Statistics

BASIC PROBABILITY THEORY, Robert B. Ash. This text emphasizes the probabilistic way of thinking, rather than measure-theoretic concepts. Geared toward advanced undergraduates and graduate students, it features solutions to some of the problems. 1970 edition. 352pp. 5 3/8 x 8 1/2. 0-486-46628-0

PRINCIPLES OF STATISTICS, M. G. Bulmer. Concise description of classical statistics, from basic dice probabilities to modern regression analysis. Equal stress on theory and applications. Moderate difficulty; only basic calculus required. Includes problems with answers. 252pp. 5 5/8 x 8 1/4. 0-486-63760-3

OUTLINE OF BASIC STATISTICS: Dictionary and Formulas, John E. Freund and Frank J. Williams. Handy guide includes a 70-page outline of essential statistical formulas covering grouped and ungrouped data, finite populations, probability, and more, plus over 1,000 clear, concise definitions of statistical terms. 1966 edition. 208pp. 5 3/8 x 8 1/2. 0-486-47769-X

GOOD THINKING: The Foundations of Probability and Its Applications, Irving J. Good. This in-depth treatment of probability theory by a famous British statistician explores Keynesian principles and surveys such topics as Bayesian rationality, corroboration, hypothesis testing, and mathematical tools for induction and simplicity. 1983 edition. 352pp. 5 3/8 x 8 1/2. 0-486-47438-0

INTRODUCTION TO PROBABILITY THEORY WITH CONTEMPORARY APPLICATIONS, Lester L. Helms. Extensive discussions and clear examples, written in plain language, expose students to the rules and methods of probability. Exercises foster problem-solving skills, and all problems feature step-by-step solutions. 1997 edition. 368pp. 6 1/2 x 9 1/4. 0-486-47418-6

CHANCE, LUCK, AND STATISTICS, Horace C. Levinson. In simple, non-technical language, this volume explores the fundamentals governing chance and applies them to sports, government, and business. "Clear and lively ... remarkably accurate." – *Scientific Monthly*. 384pp. 5 3/8 x 8 1/2. 0-486-41997-5

FIFTY CHALLENGING PROBLEMS IN PROBABILITY WITH SOLUTIONS, Frederick Mosteller. Remarkable puzzlers, graded in difficulty, illustrate elementary and advanced aspects of probability. These problems were selected for originality, general interest, or because they demonstrate valuable techniques. Also includes detailed solutions. 88pp. 5 3/8 x 8 1/2. 0-486-65355-2

EXPERIMENTAL STATISTICS, Mary Gibbons Natrella. A handbook for those seeking engineering information and quantitative data for designing, developing, constructing, and testing equipment. Covers the planning of experiments, the analyzing of extreme-value data; and more. 1966 edition. Index. Includes 52 figures and 76 tables. 560pp. 8 3/8 x 11. 0-486-43937-2

STOCHASTIC MODELING: Analysis and Simulation, Barry L. Nelson. Coherent introduction to techniques also offers a guide to the mathematical, numerical, and simulation tools of systems analysis. Includes formulation of models, analysis, and interpretation of results. 1995 edition. 336pp. 6 1/8 x 9 1/4. 0-486-47770-3

INTRODUCTION TO BIOSTATISTICS: Second Edition, Robert R. Sokal and F. James Rohlf. Suitable for undergraduates with a minimal background in mathematics, this introduction ranges from descriptive statistics to fundamental distributions and the testing of hypotheses. Includes numerous worked-out problems and examples. 1987 edition. 384pp. 6 1/8 x 9 1/4. 0-486-46961-1

Mathematics-Geometry and Topology

PROBLEMS AND SOLUTIONS IN EUCLIDEAN GEOMETRY, M. N. Aref and William Wernick. Based on classical principles, this book is intended for a second course in Euclidean geometry and can be used as a refresher. More than 200 problems include hints and solutions. 1968 edition. 272pp. 5 3/8 x 8 1/2. 0-486-47720-7

TOPOLOGY OF 3-MANIFOLDS AND RELATED TOPICS, Edited by M. K. Fort, Jr. With a New Introduction by Daniel Silver. Summaries and full reports from a 1961 conference discuss decompositions and subsets of 3-space; n-manifolds; knot theory; the Poincaré conjecture; and periodic maps and isotopies. Familiarity with algebraic topology required. 1962 edition. 272pp. 6 1/8 x 9 1/4. 0-486-47753-3

POINT SET TOPOLOGY, Steven A. Gaal. Suitable for a complete course in topology, this text also functions as a self-contained treatment for independent study. Additional enrichment materials make it equally valuable as a reference. 1964 edition. 336pp. 5 3/8 x 8 1/2. 0-486-47222-1

INVITATION TO GEOMETRY, Z. A. Melzak. Intended for students of many different backgrounds with only a modest knowledge of mathematics, this text features self-contained chapters that can be adapted to several types of geometry courses. 1983 edition. 240pp. 5 3/8 x 8 1/2. 0-486-46626-4

TOPOLOGY AND GEOMETRY FOR PHYSICISTS, Charles Nash and Siddhartha Sen. Written by physicists for physics students, this text assumes no detailed background in topology or geometry. Topics include differential forms, homotopy, homology, cohomology, fiber bundles, connection and covariant derivatives, and Morse theory. 1983 edition. 320pp. 5 3/8 x 8 1/2. 0-486-47852-1

BEYOND GEOMETRY: Classic Papers from Riemann to Einstein, Edited with an Introduction and Notes by Peter Pesic. This is the only English-language collection of these 8 accessible essays. They trace seminal ideas about the foundations of geometry that led to Einstein's general theory of relativity. 224pp. 6 1/8 x 9 1/4. 0-486-45350-2

GEOMETRY FROM EUCLID TO KNOTS, Saul Stahl. This text provides a historical perspective on plane geometry and covers non-neutral Euclidean geometry, circles and regular polygons, projective geometry, symmetries, inversions, informal topology, and more. Includes 1,000 practice problems. Solutions available. 2003 edition. 480pp. 6 1/8 x 9 1/4. 0-486-47459-3

TOPOLOGICAL VECTOR SPACES, DISTRIBUTIONS AND KERNELS, François Trèves. Extending beyond the boundaries of Hilbert and Banach space theory, this text focuses on key aspects of functional analysis, particularly in regard to solving partial differential equations. 1967 edition. 592pp. 5 3/8 x 8 1/2. 0-486-45352-9

INTRODUCTION TO PROJECTIVE GEOMETRY, C. R. Wylie, Jr. This introductory volume offers strong reinforcement for its teachings, with detailed examples and numerous theorems, proofs, and exercises, plus complete answers to all odd-numbered end-of-chapter problems. 1970 edition. 576pp. 6 1/8 x 9 1/4. 0-486-46895-X

FOUNDATIONS OF GEOMETRY, C. R. Wylie, Jr. Geared toward students preparing to teach high school mathematics, this text explores the principles of Euclidean and non-Euclidean geometry and covers both generalities and specifics of the axiomatic method. 1964 edition. 352pp. 6 x 9. 0-486-47214-0

Mathematics–History

THE WORKS OF ARCHIMEDES, Archimedes. Translated by Sir Thomas Heath. Complete works of ancient geometer feature such topics as the famous problems of the ratio of the areas of a cylinder and an inscribed sphere; the properties of conoids, spheroids, and spirals; more. 326pp. 5 3/8 x 8 1/2. 0-486-42084-1

THE HISTORICAL ROOTS OF ELEMENTARY MATHEMATICS, Lucas N. H. Bunt, Phillip S. Jones, and Jack D. Bedient. Exciting, hands-on approach to understanding fundamental underpinnings of modern arithmetic, algebra, geometry and number systems examines their origins in early Egyptian, Babylonian, and Greek sources. 336pp. 5 3/8 x 8 1/2. 0-486-25563-8

THE THIRTEEN BOOKS OF EUCLID'S ELEMENTS, Euclid. Contains complete English text of all 13 books of the Elements plus critical apparatus analyzing each definition, postulate, and proposition in great detail. Covers textual and linguistic matters; mathematical analyses of Euclid's ideas; classical, medieval, Renaissance and modern commentators; refutations, supports, extrapolations, reinterpretations and historical notes. 995 figures. Total of 1,425pp. All books 5 3/8 x 8 1/2.
Vol. I: 443pp. 0-486-60088-2
Vol. II: 464pp. 0-486-60089-0
Vol. III: 546pp. 0-486-60090-4

A HISTORY OF GREEK MATHEMATICS, Sir Thomas Heath. This authoritative two-volume set that covers the essentials of mathematics and features every landmark innovation and every important figure, including Euclid, Apollonius, and others. 5 3/8 x 8 1/2.
Vol. I: 461pp. 0-486-24073-8
Vol. II: 597pp. 0-486-24074-6

A MANUAL OF GREEK MATHEMATICS, Sir Thomas L. Heath. This concise but thorough history encompasses the enduring contributions of the ancient Greek mathematicians whose works form the basis of most modern mathematics. Discusses Pythagorean arithmetic, Plato, Euclid, more. 1931 edition. 576pp. 5 3/8 x 8 1/2.
0-486-43231-9

CHINESE MATHEMATICS IN THE THIRTEENTH CENTURY, Ulrich Libbrecht. An exploration of the 13th-century mathematician Ch'in, this fascinating book combines what is known of the mathematician's life with a history of his only extant work, the Shu-shu chiu-chang. 1973 edition. 592pp. 5 3/8 x 8 1/2.
0-486-44619-0

PHILOSOPHY OF MATHEMATICS AND DEDUCTIVE STRUCTURE IN EUCLID'S ELEMENTS, Ian Mueller. This text provides an understanding of the classical Greek conception of mathematics as expressed in Euclid's Elements. It focuses on philosophical, foundational, and logical questions and features helpful appendixes.
400pp. 6 1/2 x 9 1/4. 0-486-45300-6

BEYOND GEOMETRY: Classic Papers from Riemann to Einstein, Edited with an Introduction and Notes by Peter Pesic. This is the only English-language collection of these 8 accessible essays. They trace seminal ideas about the foundations of geometry that led to Einstein's general theory of relativity. 224pp. 6 1/8 x 9 1/4. 0-486-45350-2

HISTORY OF MATHEMATICS, David E. Smith. Two-volume history – from Egyptian papyri and medieval maps to modern graphs and diagrams. Non-technical chronological survey with thousands of biographical notes, critical evaluations, and contemporary opinions on over 1,100 mathematicians. 5 3/8 x 8 1/2.
Vol. I: 618pp. 0-486-20429-4
Vol. II: 736pp. 0-486-20430-8

Physics

THEORETICAL NUCLEAR PHYSICS, John M. Blatt and Victor F. Weisskopf. An uncommonly clear and cogent investigation and correlation of key aspects of theoretical nuclear physics by leading experts: the nucleus, nuclear forces, nuclear spectroscopy, two-, three- and four-body problems, nuclear reactions, beta-decay and nuclear shell structure. 896pp. 5 3/8 x 8 1/2. 0-486-66827-4

QUANTUM THEORY, David Bohm. This advanced undergraduate-level text presents the quantum theory in terms of qualitative and imaginative concepts, followed by specific applications worked out in mathematical detail. 655pp. 5 3/8 x 8 1/2. 0-486-65969-0

ATOMIC PHYSICS AND HUMAN KNOWLEDGE, Niels Bohr. Articles and speeches by the Nobel Prize–winning physicist, dating from 1934 to 1958, offer philosophical explorations of the relevance of atomic physics to many areas of human endeavor. 1961 edition. 112pp. 5 3/8 x 8 1/2. 0-486-47928-5

COSMOLOGY, Hermann Bondi. A co-developer of the steady-state theory explores his conception of the expanding universe. This historic book was among the first to present cosmology as a separate branch of physics. 1961 edition. 192pp. 5 3/8 x 8 1/2. 0-486-47483-6

LECTURES ON QUANTUM MECHANICS, Paul A. M. Dirac. Four concise, brilliant lectures on mathematical methods in quantum mechanics from Nobel Prize-winning quantum pioneer build on idea of visualizing quantum theory through the use of classical mechanics. 96pp. 5 3/8 x 8 1/2. 0-486-41713-1

THE PRINCIPLE OF RELATIVITY, Albert Einstein and Frances A. Davis. Eleven papers that forged the general and special theories of relativity include seven papers by Einstein, two by Lorentz, and one each by Minkowski and Weyl. 1923 edition. 240pp. 5 3/8 x 8 1/2. 0-486-60081-5

PHYSICS OF WAVES, William C. Elmore and Mark A. Heald. Ideal as a classroom text or for individual study, this unique one-volume overview of classical wave theory covers wave phenomena of acoustics, optics, electromagnetic radiations, and more. 477pp. 5 3/8 x 8 1/2. 0-486-64926-1

THERMODYNAMICS, Enrico Fermi. In this classic of modern science, the Nobel Laureate presents a clear treatment of systems, the First and Second Laws of Thermodynamics, entropy, thermodynamic potentials, and much more. Calculus required. 160pp. 5 3/8 x 8 1/2. 0-486-60361-X

QUANTUM THEORY OF MANY-PARTICLE SYSTEMS, Alexander L. Fetter and John Dirk Walecka. Self-contained treatment of nonrelativistic many-particle systems discusses both formalism and applications in terms of ground-state (zero-temperature) formalism, finite-temperature formalism, canonical transformations, and applications to physical systems. 1971 edition. 640pp. 5 3/8 x 8 1/2. 0-486-42827-3

QUANTUM MECHANICS AND PATH INTEGRALS: Emended Edition, Richard P. Feynman and Albert R. Hibbs. Emended by Daniel F. Styer. The Nobel Prize–winning physicist presents unique insights into his theory and its applications. Feynman starts with fundamentals and advances to the perturbation method, quantum electrodynamics, and statistical mechanics. 1965 edition, emended in 2005. 384pp. 6 1/8 x 9 1/4. 0-486-47722-3